Advances in Biological Tissue Biomechanics

Advances in Biological Tissue Biomechanics

Editors

Chung-Hao Lee
Jun Liao

MDPI • Basel • Beijing • Wuhan • Barcelona • Belgrade • Manchester • Tokyo • Cluj • Tianjin

Editors

Chung-Hao Lee
The University of Oklahoma
USA

Jun Liao
University of Texas at Arlington
USA

Editorial Office
MDPI
St. Alban-Anlage 66
4052 Basel, Switzerland

This is a reprint of articles from the Special Issue published online in the open access journal *Bioengineering* (ISSN 2306-5354) (available at: https://www.mdpi.com/journal/bioengineering/special_issues/tissue_biomechanics).

For citation purposes, cite each article independently as indicated on the article page online and as indicated below:

LastName, A.A.; LastName, B.B.; LastName, C.C. Article Title. *Journal Name* **Year**, *Article Number*, Page Range.

ISBN 978-3-03943-150-2 (Hbk)
ISBN 978-3-03943-151-9 (PDF)

Contents

About the Editors

Chung-Hao Lee is an assistant professor of Aerospace and Mechanical Engineering at the University of Oklahoma (OU). His Biomechanics and Biomaterials Laboratory (BBDL, http://www.ou.edu/coe/ame/bbdl) focuses on heart valve experimental tissue biomechanics and multiscale modeling as currently and previously funded by the American Heart Association (AHA), the Presbyterian Health Foundation, Oklahoma Center for the Advancement of Science and Technology (OCAST), and Oklahoma Shared Clinical and Translational Resources (OSCTR). Prior to his faculty career at OU, Dr. Lee received BS/MS from the National Taiwan University (2003 and 2005, respectively) and his Ph.D. in Civil Engineering (Structural and Computational Mechanics) from UCLA in 2011, followed by a 4-year ICES/AHA postdoctoral fellowship at the University of Texas at Austin, affording him a solid foundation in the field of soft tissue biomechanics. During his tenure at OU, Dr. Lee has established a unique and strong undergraduate research program that regularly trains undergrads with cutting-edge cardiovascular biomechanics research and produces 7 MS and 2 Ph.D. students for fostering a sustainable lab.

Jun Liao is an Associate Professor in the Department of Bioengineering at the University of Texas at Arlington. His research advances the understanding of structural–functional mechanisms of cardiac tissues, heart valves, and many other soft tissues. Dr. Liao received his Ph.D. in biomedical engineering from the Cleveland Clinic Foundation/Cleveland State University. After his postdoctoral training at the University of Pittsburgh, Dr. Liao joined Mississippi State University as an Assistant Professor. His contributions are evidenced by 90 peer-reviewed journal publications, 8 book chapters, 162 conference presentations/posters, and a book entitled: "Advances in Heart Valve Biomechanics." His research has been supported by NIH, AHA, and DoD, covering projects in tissue biomechanics, tissue regeneration, injury biomechanics, and computational modeling. Owing to his scientific achievements, Dr. Liao was elected as a Fellow of the American Heart Association in 2014.

bioengineering

MDPI

Editorial

Introduction to the Special Issue on *Advances in Biological Tissue Biomechanics*

Chung-Hao Lee [1,2,*] and Jun Liao [3,*]

1 Biomechanics and Biomaterials Design Laboratory (BBDL), School of Aerospace and Mechanical Engineering, The University of Oklahoma, Norman, OK 73019, USA
2 Institute for Biomedical Engineering, Science and Technology (IBEST), The University of Oklahoma, Norman, OK 73019, USA
3 Tissue Biomechanics and Bioengineering Laboratory (TBBL), Department of Bioengineering, University of Texas at Arlington, Arlington, TX 76010, USA
* Correspondence: ch.lee@ou.edu (C.-H.L.); jun.liao@uta.edu (J.L.)

Received: 13 August 2020; Accepted: 14 August 2020 ; Published: 17 August 2020

Keywords: in vivo stress/strain quantification; constitutive models; soft tissues; growth and remodeling (G & R); multiscale biomechanics; patient-specific modeling; biomechanics

1. Introduction

Advanced experimental and computational biomechanics have become essential components to better understand the physiological and pathological conditions of biological tissues in the human body. Recent advances in medical imaging modalities, image segmentation, tissue characterization experiments, and predictive computer simulations have made major contributions to transforming current therapeutic paradigms towards the facilitation of patient-specific diagnostics and individualized surgery planning. This *Bioengineering* special issue on *Advances in Biological Tissue Biomechanics*, therefore, focuses on research dealing with cutting-edge experimental and computational methodologies for biomechanical investigations of tissues in the human body system across multiple spatial and temporal scales.

2. Individual Contributions

2.1. Experimental Characterizations and Computational Modeling of Cardiac Heart Valves

Valvular heart disease (VHD) encompasses a number of common cardiovascular conditions that account for 10 to 20% of all cardiac surgical procedures in the United States [1]. A better understanding of the natural history coupled with the major advances in diagnostic imaging, interventional cardiology, and surgical approaches have resulted in accurate diagnosis and appropriate patient–treatment matching for therapeutic interventions. A thorough understanding of the various valvular disorders is important to aid in the management of patients with VHD, especially as the U.S. population ages [2]. This special issue collects biomechanical studies of the tricuspid valve (TV), the aortic valve (AV), and mechanical heart valve replacement devices via groundbreaking experimental and computational technologies.

Lee et al. (2019) [3] provided a comprehensive review (the first of its kind) on the current diagnoses and treatments for functional tricuspid regurgitation from the clinical perspective, the state of the art for in-vivo and in-vitro biomechanical investigations of the subvalvular components of the tricuspid heart valve (e.g., TV annular dynamics, biaxial mechanical properties of the TV leaflets, pressure-induced changes in the collagen fiber organization, and TV chordae tendineae mechanics), and advanced in-silico biomechanical modeling techniques for the TV, including isogeometric analysis-based geometry construction, tissue-structure-informed constitutive model developments, and bio-solid and fluid-structure interaction simulations of the TV closing behaviors. Potential future research directions and developments of innovative technologies for improved management of functional tricuspid regurgitation were also discussed.

Novel multiscale characterizations for the three TV leaflets were presented by Hudson et al. (2020) [4]. In this pilot study, Hudson et al. used an integrated experimental approach, which combined a biaxial testing apparatus with a polarized spatial frequency domain imaging device, to observe the load-dependent changes in the collagen fiber architectures (CFAs) for the three TV leaflets from representative porcine hearts. Their results from this pilot study could provide key insight into the connections between the microstructure and tissue mechanics of the TV. The methods employed in their study also show promise for future works to investigate the load-dependent changes in diseased TV leaflets' CFAs, which would provide a better understanding of the effects of disease on the microstructure–mechanics relationships of the cardiac heart valve tissues and help inform TV computational models for guiding clinical therapeutics, for example, in terms of the feasibility or effectiveness of valve repair techniques.

In soft tissue biomechanics, characterizations of a tissue's mechanical properties have traditionally been facilitated through bench-top experiments, such as uniaxial or biaxial tensile extension tests, in which tissue samples are immersed in a buffer solution (e.g., phosphate buffered saline (PBS)). Salinas et al. (2019) [5] conducted an in-vitro experiment to examine whether maintaining the physiological osmolarity of the tissue may alter the mechanical response of the tricuspid valve (TV) anterior leaflet. Their study demonstrated the influence of a hypo-osmotic environment on the quantified mechanical behaviors of TV anterior leaflet tissue. The imbalance in ions leads to water absorption in the valvular tissue that can alter its mechanical responses. As such, it is essential to use isotonic buffer solution in ex-vivo experiments to characterize the mechanical properties of native valve tissues.Their work provided quantitative justification for the use of isotonic buffers and will provide a valuable citation in guiding future studies of biomechanical characterizations of tricuspid valve tissue.

Next, Ross et al. (2020) [6] reviewed the current status of tissue biomechanics for the chordae tendineae of the atrioventricular heart valves (AHVs), i.e., the mitral and tricuspid valves. Besides a comprehensive summary of the existing biomechanical characterizations and microstructural quantifications of the AHVs' chordae tendineae, Ross et al. also presented novel results from their studies [7] on the chordae-leaflet insertion for the chordae tissues. In this work, the coupled polarized spatial frequency domain imaging and uniaxial testing method (similar to the integrated approach in Hudson et al. [4]) was used to analyze the changes in the load-dependent collagen fiber orientations of the strut chordae insertions of porcine AHV anterior leaflets under varied uniaxial tensile loads. Ross et al. concluded this review article by summarizing perspectives for future studies that could help better understand the mechanics–microstructure linking of chordae tendineae tissues and improve the therapeutics for valvular heart diseases caused by chordae failure.

The progression of calcific aortic valve disease (CAVD) is the consequence of extracellular matrix (ECM) remodeling, leading to structural abnormalities and improper valve function. Tesfamariam et al. (2019) [8] provided insight to the effects of ECM remodeling on the aortic valve leaflet axial curvature changes by performing elastin degradation via enzymatic digestion. They found that the elastin-degraded leaflets had significant increases in the mean curvature as compared to the undegraded (control) specimens, while this significant change was not observed when the measurements were performed using either the minimum or maximum curvature. These results suggest that the mean axial curvature metric can be used to detect distinct spatial changes in the aortic valve ECM arising from the loss in the bulk content and/or structure of elastin. Therefore, the instance of maximum leaflet flexure during the cardiac cycle could be targeted for the mean curvature measurements, serving as a potential biomarker for elastin degradation in early CAVD remodeling.

Artificial heart valves may become dysfunctional, leading to thrombus and/or pannus formations. Computational fluid dynamics (CFD) is a promising tool for improved understanding of heart valve hemodynamics, allowing quantification of detailed flow velocities and turbulent stresses to complement in-vivo clinical Doppler measurements. Khalili et al. (2018) [9] investigated the flow characteristics and hemodynamic performance of a bileaflet mechanical heart valve via CFD simulations with a more realistic aortic sinus geometry than was used previously in similar works. The modeling results of this study suggested that higher levels of valve dysfunction are accompanied with flow separation at the leaflet surfaces and growing eddies, especially downstream of the valve in the aortic sinuses. Additionally, the principal turbulent stresses for immobile leaflets could exceed the threshold values for elevated risk of hemolysis and platelet activation, further leading to potential development of thrombosis, especially around the normal leaflet. This computational modeling approach may be adapted to serve as a patient-specific tool to identify adverse conditions that are associated with an increased risk of hemolysis and thrombus formation, capturing a more complete picture of the valve status in clinical management of patients with dysfunctional valves.

2.2. *Biomechanical Investigations of Cardiovascular/Vascular Tissues*

The etiology of abdominal aortic aneurysm (AAA) development is believed to be multi-factorial, in that the pathology is initiated at the molecular level (protease- and enzyme-related), builds up to the tissue level through extracellular matrix (ECM) and structural changes, and manifests as geometrical-, biomechanical-, and blood-flow-related alterations in the abdominal aorta. Of the numerous etiological theories of AAA pathology, the degraded ECM theory is most widely accepted, as human AAA specimens usually exhibit a reduction in elastin content and an increase in collagen cross-linking [10]. AAA porcine models based on elastase–collagenase combination treatments are uncommon, but their combined effects produce maximal damage to the ECM and pronounced in-vivo inflammatory infiltration. In this work, Patnaik et al. [11] quantified pentagalloyl glucose (PGG)-mediated biomechanical restoration of degenerated ECM, which serves as a "proof-of-concept" investigation of the cross-linking properties of PGG specificality with respect to the degenerated arterial ECM. In other words, PGG leads to the cross-linking between the ECM proteins, improving the biomechanical strength of enzymatically degraded tissues. The primary contributions of this work are the quantification of the biomechanical restoration potential of PGG and its relation to the PGG binding to arterial ECM.

Despite the advances in modern clinical management, heart failure (HF) maintains a high mortality and morbidity in the United States. More than 5 million Americans have HF, and around 550,000 new cases occur every year [12]. Liu and Wang [13] reviewed the current understanding of the biomechanics of ventricular tissues. They also presented the common methods for characterizations of the ventricles, the known ventricular mechanical properties including the viscoelasticity of the tissue, the existing computational models, and the clinical relevance of the ventricular mechanical properties. Furthermore, Liu and Wang suggested future research directions, including characterizing the viscoelastic properties of the ventricles at different physiological and pathological conditions and examining how the acellular and cellular components affect the ventricular tissue's viscoelastic properties. These future research efforts will elucidate the roles of ventricular biomechanics in ventricular dysfunction, enhancing current insights of the pathogenesis of RV failure or biventricular failure and to inspire new therapies for patients with heart failure.

2.3. *Hydration Effect on the Compressive Tissue Mechanics of Brain*

Traumatic brain injury (TBI), primarily induced by mechanical impact to the head, is a leading cause of death and life-long disability in the United States. Around 5.3 million Americans currently have long-term disabilities after sustaining a TBI [14]. Designing protective systems for the human head and the brain requires a better fundamental understanding of the brain's microstructural response to mechanical insults. Prabhu et al. [15] assessed the effects of hydration on the mechanical behavior of porcine brain over a range of strain rates (quasi-static and high dynamic rates) and complemented these results with an analysis of tissue hydration effects using finite element analyses (FEAs) at high strain rates. Their experimental results showed a strong strain rate dependence for the wet brain (~80% m/m), whereas the dry brain's tangent modulus, elastic–inelastic transition stress, and the strain at yield point were strain-rate insensitive. In addition, micromechanical FEA considering various proportions of water in the dry brain further demonstrated that water plays a major role in the initial hardening trend of the brain's biomechanical behaviors. These novel results highlight the importance of incorporating the hydration effect into simulations of the brain's mechanical responses under injury scenarios or virtual human-centric protective headgear design.

2.4. High Strain Rate Mechanical Responses of Liver Tissue

In automobile accidents, abdominal injuries are often life-threatening yet not apparent at the time of initial injury. The liver is the most commonly injured abdominal organ from this type of trauma owing to its fragile material properties. Efforts to determine the optimal safety measures for automobile-related accidents have largely relied on crash dummies, which exhibit significant limitations in recapitulating injury impact to humans. Since the 1970s, there have been no substantial changes in assessing injury. In this study, Chen et al. [16] utilized a polymeric split-Hopkinson pressure bar (PSHPB) testing system to explore the effects of strain rates on the material behaviors of porcine liver tissue. Three key findings from this study are: (i) the liver tissue response at high-rate compression was characterized by an initial hardening peak, followed by softening, and then by strain hardening to failure; (ii) the liver's mechanical stiffness increased as the applied strain rate increased; and (iii) an isotropic high-rate material behavior was observed along all three orthogonal directions and was confirmed by the liver histological microstructure. Experimental data of this work coupled with finite element modeling can be implemented in large-scale simulations of the human body considering extreme, high-strain-rate scenarios, such as automobile accidents, to guide automobile safety measures that will reduce the risk of abdominal injuries in high-impact situations.

2.5. Linking Tissue Mechanics and Extracellular Matrix for Venous Valve Tissue

Jugular venous valve incompetence has no long-term remedy and symptoms of transient global amnesia and/or intracranial hypertension continue to discomfort patients. Benson and Huang [17] examined the synergy of the collagen and elastin microstructure that compose the bi-layer extracellular matrix (ECM) of the jugular venous valve. In this study, Benson and Huang investigated the jugular venous valve and related the tissue-level mechanical properties, fibril orientation, and fibril composition. Light microscopy was used to deduce that the venous valve's isolated elastin microstructure was unaffected by their drying methods. In addition, light microscopy of the collagen microstructure allowed characterizations of the crimp effects on the venous valve's mechanical properties, indicating that collagen aligning circumferentially and elastin orienting radially attributing to a stiffer response in the circumferential direction. Force-controlled mechanical testing also provided the proof that the elastin's cross-linked mesh accounts for the circumferential direction's mechanical properties at low strains. In short, this work outlined the contribution of both the collagen and elastin microstructures to the physiological function of the jugular venous valve tissue. This new knowledge of the venous valve tissue-level microstructures is important for advances in basic venous physiology and for future novel approaches to prevent venous valve incompetence.

2.6. Key Considerations for Biomechanical Parameter Estimations of Soft Tissues

Several nonlinear and anisotropic constitutive models have been proposed to describe the biomechanical properties of soft tissues, and reliably estimating the unknown parameters in these models using experimental data is an important step towards developing predictive capabilities. However, the effect of parameter estimation techniques on the resulting biomechanical parameters remains under-analyzed. Aggarwal [18] conducted a thorough study to to analyze the effects of parameter estimation on biomechanical characterizations of soft tissues under planar biaxial testing, including nonlinear preconditioning of the data, selection of the weighted residual, inclusion of the fiber angle as an estimation parameter, and displacement-controlled versus force-controlled testing. Specifically, four invariant-based constitutive models for soft tissues were tested: the Gasser–Ogden–Holzapfel (GOH) model [19], the Humphrey model [20], the Lee-Sacks model [21], and the May–Newman model [22], each with their own set of five or six parameters. It was found that small modifications of weighting the residual by the experimental data and/or taking a log of the parameter in front of the exponential can significantly improve the parameter estimation process. The advantages of the proposed modifications were found not only in terms of convergence speed but also in reducing the possibility of estimating wrong parameter values by getting stuck in a local minima. These results suggest that determining the fiber angles using a non-mechanical test, as in, for example, an optical technique, can greatly help the parameter estimation process.

3. Conclusions

Altogether, the 12 research papers/reviews in this Special Issue on *Advances in Biological Tissue Biomechanics* reflect the importance of experimental and computational biomechanics for improving the fundamental understanding of the function and properties of biological tissues in human body systems. These essential research investigations and technological developments will serve as a foundation for future advances in precision medicine for accurate diagnosis and effective prophylactic management. Finally, the guest editors would like to thank all the authors for their appreciated contributions.

Funding: This editorial was not supported by an external funding.

Acknowledgments: Special credit goes to all of the reviewers active in critically analyzing and improving the contributed articles. The guest editing team of this special issue would also like to thank the staff of *Bioengineering* for their valuable assistance.

Conflicts of Interest: The authors declare no conflict of interest.

References

1. Nishimura, R.A.; Carabello, B.A.; Faxon, D.P.; Freed, M.D.; Lytle, B.W.; O'Gara, P.T.; O'Rourke, R.A.; Shah, P.M. ACC/AHA 2008 Guideline Update on Valvular Heart Disease: Focused Update on Infective Endocarditis—A Report of the American College of Cardiology/American Heart Association Task Force on Practice Guidelines: Endorsed by the Society of Cardiovascular Anesthesiologists, Society for Cardiovascular Angiography and Interventions, and Society of Thoracic Surgeons. *Circulation* **2008**, *118*, 887–896.
2. Maganti, K.; Rigolin, V.H.; Sarano, M.E.; Bonow, R.O. Valvular Heart Disease: Diagnosis and Management. *Mayo Clin. Proc.* **2010**, *85*, 483–500. [CrossRef]
3. Lee, C.-H.; Laurence, D.W.; Ross, C.J.; Kramer, K.E.; Babu, A.R.; Johnson, E.L.; Hsu, M.-C.; Aggarwal, A.; Mir, A.; Burkhart, H.M.; et al. Mechanics of the Tricuspid Valve—From Clinical Diagnosis/Treatment, In-Vivo and In-Vitro Investigations, to Patient-Specific Biomechanical Modeling. *Bioengineering* **2019**, *6*, 47. [CrossRef]
4. Hudson, L.T.; Jett, S.V.; Kramer, K.E.; Laurence, D.W.; Ross, C.J.; Towner, R.A.; Baumwart, R.; Lim, K.M.; Mir, A.; Burkhart, H.M.; et al. A Pilot Study on Linking Tissue Mechanics with Load-Dependent Collagen Microstructures in Porcine Tricuspid Valve Leaflets. *Bioengineering* **2020**, *7*, 60. [CrossRef]
5. Salinas, S.D.; Clark, M.M.; Amini, R. Mechanical Response Changes in Porcine Tricuspid Valve Anterior Leaflet Under Osmotic-Induced Swelling. *Bioengineering* **2019**, *6*, 70. [CrossRef]
6. Ross, C.J.; Zheng, J.; Ma, L.; Wu, Y.; Lee, C.-H. Mechanics and Microstructure of the Atrioventricular Heart Valve Chordae Tendineae: A Review. *Bioengineering* **2020**, *7*, 25. [CrossRef]
7. Ross, C.J.; Hsu, M.-C.; Baumwart, R.; Mir, A.; Burkhart, H.M.; Holzapfel, G.A.; Wu, Y.; Lee, C.-H. Quantification of Load-Dependent Changes in the Collagen Fiber Architecture for Strut Chordae Tendineae-Leaflet Insertion of Porcine Atrioventricular Heart Valves. *Biomech. Model. Mechanobiol.* **2020**, in press.
8. Tesfamariam, M.D.; Mirza, A.M.; Chaparro, D.; Ali, A.Z.; Montalvan, R.; Saytashev, I.; Gonzalez, B.A.; Barreto, A.; Ramella-Roman, J.; Hutcheson, J.D.; et al. Elastin-Dependent Aortic Heart Valve Leaflet Curvature Changes During Cyclic Flexure. *Bioengineering* **2019**, *6*, 39. [CrossRef]

9. Khalili, F.; Gamage, P.P.T.; Sandler, R.H.; Mansy, H.A. Adverse Hemodynamic Conditions Associated with Mechanical Heart Valve Leaflet Immobility. *Bioengineering* **2018**, *5*, 74. [CrossRef]
10. Carmo, M.; Colombo, L.; Bruno, A.; Corsi, F.R.M.; Roncoroni, L.; Cuttin, M.S.; Radice, F.; Mussini, E.; Settembrini, P.G. Alteration of Elastin, Collagen and Their Cross-Links in Abdominal Aortic Aneurysms. *Eur. J. Vasc. Endovasc. Surg.* **2002**, *23*, 543–549. [CrossRef]
11. Patnaik, S.S.; Piskin, S.; Pillalamarri, N.R.; Romero, G.; Escobar, G.P.; Sprague, E.; Finol, E.A. Biomechanical Restoration Potential of Pentagalloyl Glucose after Arterial Extracellular Matrix Degeneration. *Bioengineering* **2019**, *6*, 58. [CrossRef]
12. Wayne, R.; Katherine, F.; Karen, F.; Alan, G.; Kurt, G.; Nancy, H.; Susan, M.H.; Michael, H.; Virginia, H.; Brett, K.; et al. Heart Disease and Stroke Statistics—2008 Update. *Circulation* **2008**, *117*, e25–e146.
13. Liu, W.; Wang, Z. Current Understanding of the Biomechanics of Ventricular Tissues in Heart Failure. *Bioengineering* **2020**, *7*, 2. [CrossRef]
14. Langlois, J.A.; Rutland-Brown, W.; Wald, M.M. The Epidemiology and Impact of Traumatic Brain Injury A Brief Overview. *J. Head Trauma Rehabilm* **2006**, *21*, 375–378. [CrossRef]
15. Prabhu, R.K.; Begonia, M.T.; Whittington, W.R.; Murphy, M.A.; Mao, Y.; Liao, J.; Williams, L.N.; Horstemeyer, M.F.; Sheng, J. Compressive Mechanical Properties of Porcine Brain: Experimentation and Modeling of the Tissue Hydration Effects. *Bioengineering* **2019**, *6*, 40. [CrossRef]
16. Chen, J.; Patnaik, S.S.; Prabhu, R.K.; Priddy, L.B.; Bouvard, J.-L.; Marin, E.; Horstemeyer, M.F.; Liao, J.; Williams, L.N. Mechanical Response of Porcine Liver Tissue under High Strain Rate Compression. *Bioengineering* **2019**, *6*, 49. [CrossRef]
17. Benson, A.A.; Huang, H.-Y.S. Tissue Level Mechanical Properties and Extracellular Matrix Investigation of the Bovine Jugular Venous Valve Tissue. *Bioengineering* **2019**, *6*, 45. [CrossRef]
18. Aggarwal, A. Effect of Residual and Transformation Choice on Computational Aspects of Biomechanical Parameter Estimation of Soft Tissues. *Bioengineering* **2019**, *6*, 100. [CrossRef]
19. Gasser, T.C.; Ogden, R.W.; Holzapfel, G.A. Hyperelastic Modelling of Arterial Layers with Distributed Collagen Fibre Orientations. *J. R. Soc. Interface* **2006**, *3*, 15–35. [CrossRef]
20. Humphrey, J.; Yin, F.A. New Constitutive Formulation for Characterizing the Mechanical Behavior of Soft Tissues. *Biophys. J.* **1987**, *52*, 563–570. [CrossRef]
21. Lee, C.H.; Amini, R.; Gorman, R.C.; Gorman, J.H.; Sacks, M.S. An Inverse Modeling Approach for Stress Estimation in Mitral Valve Anterior Leaflet Valvuloplasty for In-Vivo Valvular Biomaterial Assessment. *J. Biomech.* **2014**, *47*, 2055–2063. [CrossRef] [PubMed]
22. May-Newman, K.; Yin, F.C.P. A Constitutive Law for Mitral Valve Tissue. *J. Biomech. Eng.* **1998**, *120*, 38–47. [CrossRef] [PubMed]

bioengineering

MDPI

Review

Mechanics of the Tricuspid Valve—From Clinical Diagnosis/Treatment, In-Vivo and In-Vitro Investigations, to Patient-Specific Biomechanical Modeling

Chung-Hao Lee [1,2,*], Devin W. Laurence [1], Colton J. Ross [1], Katherine E. Kramer [1], Anju R. Babu [1,3], Emily L. Johnson [4], Ming-Chen Hsu [4,*], Ankush Aggarwal [5], Arshid Mir [6], Harold M. Burkhart [7], Rheal A. Towner [8], Ryan Baumwart [9] and Yi Wu [1]

1 Biomechanics and Biomaterials Design Laboratory, School of Aerospace and Mechanical Engineering, The University of Oklahoma, Norman, OK 73019, USA; dwlaur@ou.edu (D.W.L.); cjross@ou.edu (C.J.R.); Katherine.E.Kramer-1@ou.edu (K.E.K.); babua@nitrkl.ac.in (A.R.B.); yiwu@ou.edu (Y.W.)
2 Institute for Biomedical Engineering, Science and Technology (IBEST), The University of Oklahoma, Norman, OK 73019, USA
3 Department of Biotechnology and Medical Engineering, National Institute of Technology Rourkela, Rourkela, Odisha 769008, India
4 Department of Mechanical Engineering, Iowa State University, Ames, IA 50011, USA; johnsel@iastate.edu
5 Glasgow Computational Engineering Centre, School of Engineering, University of Glasgow, Scotland G12 8LT, UK; Ankush.Aggarwal@glasgow.ac.uk
6 Division of Pediatric Cardiology, Department of Pediatrics, The University of Oklahoma Health Sciences Center, Oklahoma City, OK 73104, USA; Arshid-Mir@ouhsc.edu
7 Division of Cardiothoracic Surgery, Department of Surgery, The University of Oklahoma Health Sciences Center, Oklahoma City, OK 73104, USA; Harold-Burkhart@ouhsc.edu
8 Advance Magnetic Resonance Center, MS 60, Oklahoma Medical Research Foundation, Oklahoma City, OK 73104, USA; Rheal-Towner@omrf.org
9 Center for Veterinary Health Sciences, Oklahoma State University, Stillwater, OK 74078, USA; ryan.baumwart@okstate.edu
* Correspondence: ch.lee@ou.edu (C.-H.L.); jmchsu@iastate.edu (M.-C.H.); Tel.: +1-405-325-4842 (C.-H.L.); +1-515-294-4632 (M.-C.H.)

Received: 28 April 2019; Accepted: 17 May 2019; Published: 22 May 2019

Abstract: Proper tricuspid valve (TV) function is essential to unidirectional blood flow through the right side of the heart. Alterations to the tricuspid valvular components, such as the TV annulus, may lead to functional tricuspid regurgitation (FTR), where the valve is unable to prevent undesired backflow of blood from the right ventricle into the right atrium during systole. Various treatment options are currently available for FTR; however, research for the tricuspid heart valve, functional tricuspid regurgitation, and the relevant treatment methodologies are limited due to the pervasive expectation among cardiac surgeons and cardiologists that FTR will naturally regress after repair of left-sided heart valve lesions. Recent studies have focused on (i) understanding the function of the TV and the initiation or progression of FTR using both in-vivo and in-vitro methods, (ii) quantifying the biomechanical properties of the tricuspid valve apparatus as well as its surrounding heart tissue, and (iii) performing computational modeling of the TV to provide new insight into its biomechanical and physiological function. This review paper focuses on these advances and summarizes recent research relevant to the TV within the scope of FTR. Moreover, this review also provides future perspectives and extensions critical to enhancing the current understanding of the functioning and remodeling tricuspid valve in both the healthy and pathophysiological states.

Keywords: the tricuspid valve; functional tricuspid regurgitation; cardiovascular imaging; mechanical characterization; in-vitro experiments; constitutive modeling; geometrical modeling; finite element modeling; isogeometric analysis (IGA); biaxial mechanical characterization; fluid-structure interactions; material anisotropy; sub-valvular components

---

1. Introduction

The tricuspid valve (TV) regulates blood flow on the right side of the heart between the right atrium (RA) and right ventricle (RV) throughout cardiac cycles. Specifically, the TV is responsible for allowing the deoxygenated blood to flow from the RA into the RV during diastole and preventing retrograde blood flow from the RV into the RA as the RV contracts during systole. The TV prevents such backflow into the RA through the closure of the three TV leaflets, namely the *anterior leaflet* (TVAL), *posterior leaflet* (TVPL), and the *septal leaflet* (TVSL). These leaflets are attached to the RA through the ring-like valvular annulus and to the papillary muscles located on the RV by the chordae tendineae. The proper function of these sub-valvular components is critical to overall function of the TV. Alterations to the function or anatomy of the TV can result in a diseased condition called tricuspid regurgitation (TR) that reduces the overall efficiency of the RV function.

TR occurs when the TV leaflets are unable to completely prevent blood backflow into the RA during systole. Tricuspid regurgitation can be classified by etiology into two categories: *primary* (organic) TR and *secondary* or *functional* TR (FTR) [1–3]. On the one hand, TR is considered primary when there is some type of structural abnormality or damage to the TV apparatus as the primary cause of the TR [2]. Congenital diseases, such as Ebstein's anomaly and hypoplastic left heart syndrome (HLHS), and acquired diseases (e.g., tricuspid leaflet flail resulting from chordae rupture) fall into this category. Interested readers may refer to Table 2.2 from Anwar et al. (2018) [2] for a comprehensive list. On the other hand, TR is classified as functional regurgitation when the TV apparatus itself remains *structurally and mechanically intact*, but instead the TR is secondary to a certain alteration in the surrounding heart geometry/component [1,2,4]. Some examples of the causes of FTR include: RV enlargement, TV annulus dilation, or pulmonary hypertension (cf. Table 2.2 from Anwar et al. (2018) [2]).

FTR often progresses from a combination of three interlinked pathologies that typically stem from a pressure overload or a volume overload in the RV (e.g., pulmonary hypertension) [5–7]. First, as a direct result of the pressure or volume overload, the RV will remodel and become enlarged beyond its physiological configuration [5,7]. An early study by Come et al. (1985) [8] observed about a 60% increase in the RV diameter in patients with TR. Consequently, the annulus will begin to dilate away from the septum to form a more circular shape as compared to the healthy elliptical shape. The TV annulus will lose its saddle-like geometry to become more flattened [9,10]. These alterations will continue to progress, resulting in papillary muscle displacement, leaflet tethering, a reduced coaptation of the TV leaflets, and the formation or worsening of FTR [4–6,9,10].

TR has been historically ignored in the clinical setting, despite affecting approximately 1.6 million Americans [11,12]. This may originate from the pervasive expectation among cardiac surgeons and cardiologists that correction of the left-sided cardiac lesions will lead to natural regression of FTR [13]. However, recent studies by Dreyfus et al. (2005) and Anyanwu and Adams (2010) [14,15] showed that this over-conservative practice was invalid, and untreated FTR frequently progresses to late severe TR, further worsening long-term prognosis and quality of life. Since the time of those clinical studies, the TV has received increasingly more attention in both the clinical and basic research fields, although less than the mitral valve (MV) and the aortic valve (AV) (Figure 1). Nevertheless, many clinically significant questions still need to be addressed, including the determination of the optimal timing and therapeutic option for treating FTR [14,16–22], and the understanding of how to mitigate the recurrence of TR after surgical intervention [23–31]. As expected, there has been some progress toward partially answering these questions and improving patient-specific therapeutics. For example, recent clinical studies have focused on (i) examining the progression and proper assessment of FTR; (ii) in-vitro and in-vivo studies have been conducted to quantify relevant mechanical properties of the TV and its sub-valvular components; and (iii) computational models have been recently developed to explore TV biomechanical function. Despite these recent efforts to improve the understanding of the functioning TV in both healthy and pathophysiological states, gaps in knowledge still exist in understanding the *underlying mechanisms and recurrence* of FTR.

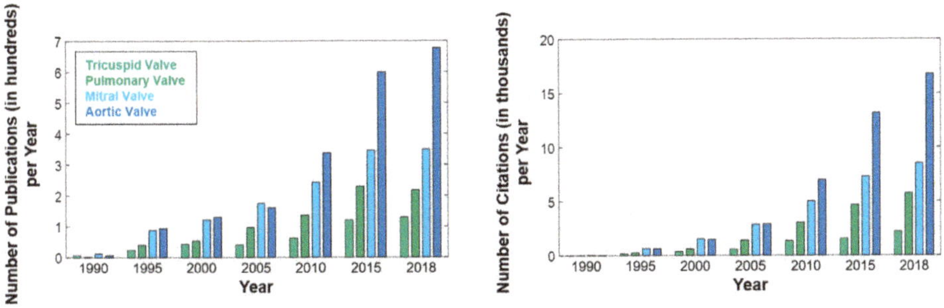

Figure 1. The number of publications (**left**) and the number of total citations (**right**) since 1990 for the four heart valves. Data was adopted from the Web of Science.

In addition to a recent book chapter by Meador et al. (2019) [32], which briefly reviews studies of the biomechanical properties of TV sub-valvular components, this review paper focuses on providing a synopsis and commentary with regard to the recent advances relevant to the TV and FTR. This review paper will also provide an opinion on future perspectives of the TV research for addressing broad topics, such as clinical applications of the presented work and developments critical for patient-specific therapeutics. The remaining sections of this review paper are organized as follows. Section 2 will address the current status of clinical imaging modalities within the scope of the diagnosis and treatment of FTR. Recent advances in understanding the mechanical behaviors and function of the TV using in-vivo and in-vitro methodologies will be discussed in Section 3. In Section 4, current computational modeling tools and related methods will be presented, which aim to enhance the understanding of TV function and disease progression. Concluding remarks, a summary of the key take-away messages, and our perspectives on future TV research will be provided in Section 5.

2. Functional Tricuspid Regurgitation: Diagnosis and Treatment Options

2.1. Sub-Valvular Structures and Components of the TV

The TV regulates the flow of blood between the right atrium and the right ventricle of the heart. Each sub-valvular component of the TV (cf. Figure 2) is critical to the organ-level TV function, and the details of each component are discussed as follows.

2.1.1. TV Annulus

The annulus is a fibromuscular ring that encircles the atrioventricular junction, marking the border between the atrial and the ventricular myocardium. The annulus connects the valve leaflets to the heart chambers. Some studies have found the TV annulus to be pear-shaped [33,34], whereas other studies have predominantly found the annulus to be more saddle-shaped [33,35,36]. The configuration of the TV annulus plays a major role in the coaptation, mobility and the stress distribution in the TV leaflets and chordae tendineae [37,38]. The average diameter of the TV annulus at end systole is 3.15 cm [39]. As the annulus curvature increases, the stress in the TV anterior leaflet decreases, and this alteration in the curvature ultimately results in an increased leaflet strain and abnormal tissue remodeling. During disease conditions, the saddle-shaped annulus enlarges, becoming circular, and the corresponding change in the annulus area typically serves as a predictor of valve disorders such as tricuspid regurgitation [40].

Structurally, the annulus forms the base of the TV leaflets and is composed of two types of discontinuous segments—muscular annulus and collagen-rich fibrous annulus (Figure 2) [41]. The muscular annulus is formed of a circumferentially oriented myofiberous lamina and a second lamina formed of myofibers perpendicular to the circumferential myofibers [42]. Racker et al. (1991) [43] described that the anterior, lateral, and posterior regions of the TV annulus are completely encircled with circumferential myofibers with only a thin muscular connection at the medial region of the TV annulus. The fibrous annulus forms the antero-medial regions and continues with the connective tissues into the TV leaflets. Microscope-based study of the human TV annulus [44] indicated the presence of *myofibers* in the posterior and anterior annulus and *collagen bundles* in the septal annulus.

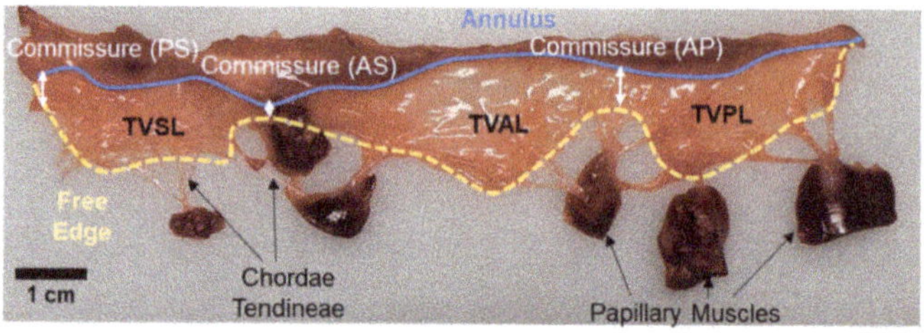

Figure 2. An excised porcine TV tissue sample, showing the three tricuspid valve leaflets, papillary muscle, chordae tendineae, commissures, and the TV annulus.

2.1.2. TV Leaflets

The TV annulus transitions into three leaflets: the TVAL, TVPL, and TVSL (Figure 2). In general, the TV leaflets have a rough zone in the crescentic region where chordae tendineae are attached, a broad basal zone at the apex of the leaflet, and a clear zone [45]. Our recent examination of the porcine valves leaflets demonstrated that the TV leaflets are more translucent and thinner than their MV counterparts as a result of fewer collagen proteins [46]. Histological analysis also revealed the difference in the layered structure between the TV and MV leaflets (cf. Figure 10 from Jett et al. [46]). The TV leaflet tissue layers are composed of extracellular matrix proteins—elastin, collagen, proteoglycans (PGs), and glycosaminoglycans (GAGs) populated with dynamic valvular interstitial cells (VICs). The connective tissue structure is organized into four morphologically and biomechanically distinct layers known as the *atrialis* (A), *spongiosa* (S), *fibrosa* (F), and *ventricularis* (V) (Figure 3) [47,48].

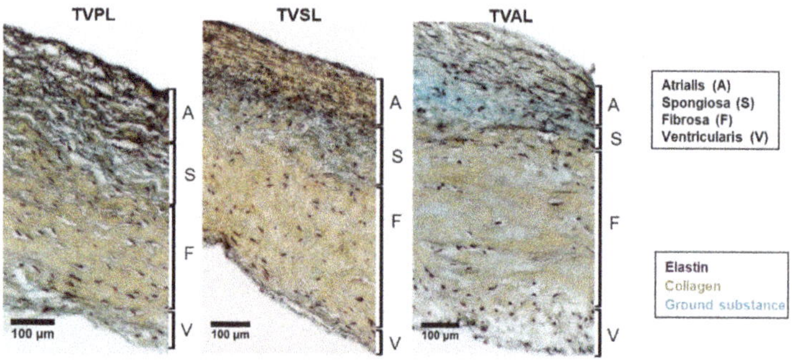

Figure 3. Histological images revealing the porcine TV leaflet microstructure using Movat's Pentachrome straining to emphasize the elastin, collagen, and non-fibrous ground substance. The four morphologically distinct layers are also illustrated in each image, i.e., A: atrialis, S: spongiosa, F: fibrosa, and V: ventricularis.

A dense, collagenous fiber network distinguishes the main load bearing layer of the leaflets, the fibrosa, from the surrounding tissue. The ventricularis, anatomically situated below the fibrosa and facing the ventricular side of the heart, is rich in circumferentially oriented elastin fibers that assist in the stretching and recoiling of the valve tissue. The spongiosa layer is rich in hydrophilic GAGs and PGs that act as a dampening mechanism during rapid leaflet bending [49,50]. The atrialis layer—on the atrial side of the leaflet—is composed of elastin, collagen and GAGs, and this layer of the TV leaflets is reported to have a high innervation density [51]. In addition, VICs are heterogeneous, dynamic cells distributed throughout the leaflets' layers [52]. VICs play a major role in maintaining the structural integrity of the leaflet tissues by regulating the extracellular matrix (ECM) scaffold remodeling. Different VIC phenotypes express molecular markers found in myofibroblast and smooth muscle cells (SMCs). The activated VICs produce myofibroblasts and express smooth muscle α-actin as well as other contractile proteins commonly found in the vascular SMCs [53]. It has also been shown that the MV leaflet VICs are stiffer than the cells in the TV leaflets, implying a correlation between the VIC-regulated collagen biosynthesis and transvalvular pressure loading [52].

2.1.3. TV Chordae Tendineae

The chordae tendineae are fibrous strings that originate from the ventricular papillary muscles or from the ventricle wall and transmit tensile force to the leaflets (Figure 2). The chordae split into three segments either soon after their origin or just before their attachment to the leaflets or the commissural region [54]. Chordae tendineae are composed of elastin, GAGs, collagen fibers, and endothelial cells [55]. The chordae tendineae are typically categorized as basal, marginal, strut, or commissural based on their leaflet attachment location [54,56]. Each category is associated with varying length, cross-sectional area, and mechanical properties of the chordae tendineae. For example, the marginal chordae that are connected to the free edges of the leaflets are stiffer than the basal chordae that are attached to the TV annulus [57].

2.2. Imaging Modalities for Assessing FTR

High-resolution imaging modalities have greatly advanced our understanding of TR and other cardiac abnormalities. Non-invasive imaging techniques used to assess TR include computed tomography (CT), cardiac magnetic resonance imaging (CMRI), and echocardiography. Echocardiography is most frequently employed for diagnosing FTR, but CMRI and CT are increasingly used as a complement [53]. Clinicians use these advanced imaging techniques as a surgical intervention timing-indicator and for preoperative surgery planning.

2.2.1. Echocardiography

Echocardiography, an imaging technique that uses ultrasound waves to image anatomical structures, is the principal modality used to diagnose TR. In the clinic, physicians assess preoperative, intraoperative, and post-operative states of TR generally by two-dimensional echocardiography (2DE). Echocardiography relies on transducer probes to emit "ultra" sound waves at a frequency inaudible to humans (>20,000 Hz) that rebound off inhomogeneities before "echoing" back to the transducer probe. Higher density structures exhibit greater impedance to the propagation of sound.

Echocardiography is relatively inexpensive, widely available, and capable of evaluating the TV both functionally and morphologically. This technique can be performed at a patient's bedside, so it is popular for imaging hemodynamically unstable patients. However, the operator-dependent interface of echocardiography causes certain restrictions. Individual sonographers must adapt conventional probe positioning for patients possessing anatomical variance, such as obesity or emphysema, further altering the uniformity of the data between imaging sessions and in comparative patient studies. Furthermore, consistent and reliable landmarks on the right side of the heart are not as common compared to the left side of the heart. More recent echocardiogram advancements have resulted in the development of real-time three-dimensional echocardiography (3DE), better contrasting agents, and multimodal imaging.

Two major methods exist for performing both 2DE and 3DE: (i) transthoracic echocardiography (TTE) and (ii) transesophageal echocardiography (TEE). For TTE, the transducer probe is positioned noninvasively over the heart. Conversely, in TEE, the probe is inserted down the esophagus to access the heart more directly. While TTE continues to be the cornerstone of diagnostic cardiac ultrasound, TEE offers value as a supplementary tool due to the close probe proximity, decreased signal attenuation, and absence of impedance from intervening lung and bones (Figure 4).

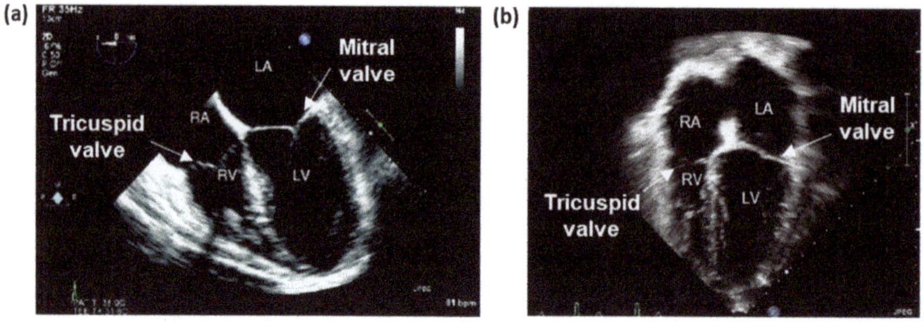

Figure 4. Echocardiographic imaging modalities: (**a**) a four-chamber mid-esophageal view using TEE (image modified from [58]), and (**b**) an apical four-chamber view (A4C) using TTE courtesy of Dr. Mir and Dr. Burkhart from the Children's Heart Center at the University of Oklahoma Health Sciences Center (OUHSC).

Two-Dimensional Imaging Modalities

A 2DE device monitors cardiac image data in B-mode, M-mode, or Doppler. B-mode ultrasonography, or brightness mode, provides 2D grayscale images about a cross-sectional area. Structure brightness can be defined, in decreasing order, as hyperechoic, hypoechoic, and anechoic. High-density structures, such as a calcified valve, reflect most of the sound, resulting in a hyperechoic appearance. In contrast, fluid-filled structures possess low impedance, appearing anechoic. B-mode only provides the most basic image data and, consequently, it is used in conjunction with M-mode or Doppler to convey further information. M-mode ultrasonography, an abbreviation for "motion" mode, uses a rapid succession of pulses along a single ultrasound beam to produce a video-like illustration. By positioning the transducer in a fixed location, B-mode images are recorded at each pulses and changes in the corresponding echo are displayed as a function of time. Valve leaflet coaptation or myocardium movement afford physicians quantifiable time-based data to better interpret the current state of cardiac functionality. Conversely, Doppler ultrasonography depicts the blood velocity, typically showing blood flow toward the device in red and flow away from the transducer in blue. This technique allows physicians to screen patients for cardiac abnormalities, such as TR, by visualizing the regurgitant jet.

Standard Echocardiography Imaging Windows

Due to the complex, multi-component structure of the TV, 2DE requires the acquisition of images from multiple locations to capture the valve's overall 3D geometry and function comprehensively. The right side of the heart is viewed from the mid-esophageal (ME) (30–40 cm) or transgastric (40–45 cm) windows. The views generally used to image the TV include the right ventricular inflow-outflow ME and four-chamber ME (Figure 4a) (transducer angle: 0–20° and 60–90°, respectively) and the basal short-axis and RV-inflow trangastric views (transducer angle: 0–20° and 100–120°, respectively). Also notable in assessing FTR are the views that delineate the RA and RV.

The traditional approaches for TTE include right ventricular inflow (RVIF), parasternal short-axis (PSAX), parasternal long-axis (PLAX) (Figure 5a), apical four-chamber (A4C) (Figures 4b and 5b), and more recently, right ventricular-focused (RVF). Addetia et al. (2016) [59] analyzed the efficacy of these traditional views compared to six nonstandard 2D views devised by their group. Using multiplanar reconstruction of three-dimensional data sets, they showed that their novel 2D views accurately identify the TV leaflets based on defined landmarks and anatomical clues. Such nonconventional imaging protocols may be beneficial for further evaluating TV leaflet pathologies.

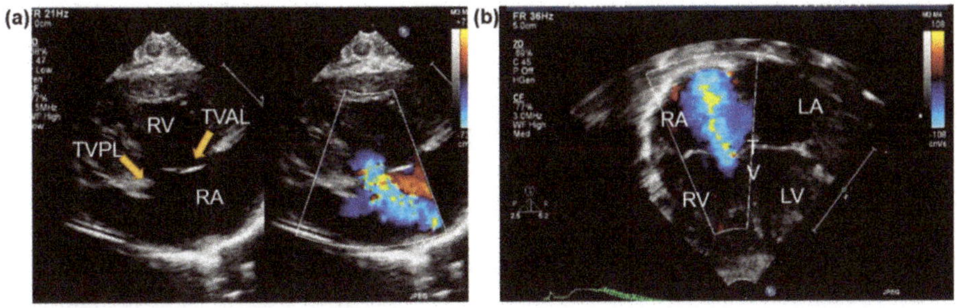

Figure 5. Visualization of severe tricuspid reguritation using two different 2DTTE views and color flow Doppler: (**a**) a parasternal long-axis inflow view, and (**b**) an apical four-chamber view of a newborn with a severe pulmonary hypertension due to diaphragmatic hernia. Images courtesy of Dr. Mir and Dr. Burkhart from the Children's Heart Center at the University of Oklahoma Health Sciences Center (OUHSC).

Three-Dimensional Imaging Modalities

Unlike the MV or AV, the complex, nonplanar structure of the TV makes simultaneously capturing the three TV leaflets in one cross-sectional view nearly impossible using only 2DE imaging. Real-time three-dimensional echocardiography (RT3DE) supplements 2DE with detailed anatomical measurements in 90% of patients [60], and allows for concomitant visualization of the opening and coaptation of the three leaflets through the cardiac cycle [61]. Moreover, in a comparison of 2D TTE and RT3DE, Anwar et al. (2007) [62] concluded that RT3DE more reliably assessed the tricuspid valve annulus (TA) size and function. Accordingly, the advent of RT3DE has prompted numerous studies to elucidate the specific valve geometry and anatomy in healthy patients [23,63].

3DE acquires volume data via transducer probes containing a special matrix array of 2500 piezoelectric crystals that can be independently activated, focused, and steered to scan a pyramidal volume of tissue in three dimensions [64]. In addition to the real-time live mode, the 3DE transducer can also obtain full-volume data, which are the merging of information over four consecutive cardiac cycles using a wide angle to cover a larger region of interest. This allows the ability to view both atrioventricular valves simultaneously. However, full-volume imaging possesses limitations in the potential for poor image and spatial resolution due to physiologically-based artifacts. Additionally, the need to suspend respiration for four cardiac cycles during imaging excludes patients with atrial fibrillation or dyspnea.

The three main RT3DE imaging views are *parasternal*, *apical*, and *subcostal*. In selecting one of these views, it is important to consider the response of an imaging system to a point object, known as the point-spread function of the system, which varies in degree according to the system dimensions in use. Standard RT3DE systems employ a dimension of approximately 0.5 mm (axial), 2.5 mm (lateral), and 3 mm (elevation), and thus, the best images (i.e., least distortion or blurring) are acquired in the axial dimension. Conversely, the elevation dimension produces the greatest degree of spreading [61].

2.2.2. CMRI

MRI technology applies a strong magnetic field to align the body's protons (i.e., hydrogen ions), and radio waves are then generated to disrupt the proton alignment. As the protons realign themselves to the magnetic field, they emit radio signals that the device computer reads and converts into detailed spatiotemporal images [65]. High-resolution, multiplanar images obtained using cine-MRI provide doctors comprehensive information about the morphology and function of the scanned structure. Thus, CMRI may be recommended as a complementary modality to echocardiography in cases of poor image resolution or disqualification from TEE. CMRI is considered the "gold standard" for reliably and accurately measuring the ventricular volumes, ejection fraction, and the myocardial mass [66], which are useful in assessing pre- and post-operative ventricular function in patients with FTR. While MRI produces high-resolution images, the modality is restrictive in practical applications. MRI capabilities are limited in evaluating cases of severe TR due to its inability to be performed on hemodynamically unstable patients. Moreover, patients with permanent pacemakers, implantable cardiac defibrillators, or metal prosthetic heart valves cannot undergo an MRI scan unless they have a newer MRI-compatible system.

2.2.3. Cardiac CT

CT scans use X-ray measurements to create two-dimensional radiographic cross-sections of the heart taken around an axis of rotation. Digital processing yields multiplanar three-dimensional reconstructions of the area of interest with desirable spatial and temporal resolution. Thin image slices allow for detection of distinct valvular boundaries and useful spatial information for assessing RV function [67]. Studies support the prognostic value of cardiac CT for indexing FTR—such parameters include the RA and RV volumes, the leaflet tethering angles and height, and the annular diameter and area [68–70]. CT has also been used in the post-operative assessment of annuloplasty ring dislodgment and quantification of the spatial relationship between pacemaker leading in the RV and the associated TR [67,68,71]. Despite its attractive capabilities and applications to the heart valve leaflets, CT exposure must be monitored and limited due to the potential adverse effects of the radiation during the X-ray measurements.

2.3. *Parameters for Grading TR Severity*

Imaging modalities afford qualitative, quantitative, or semi-quantitative analysis of the right side of the heart. Such analyses allow surgeons to index the degree of TR. The most common parameters used to evaluate TR severity include the annular diameter, the size of the backflow jet, the coaptation mode of the leaflets, the width of vena contracta, and the relative size of the TR jet when compared to the RA and dimensions of the right side of the heart, among others. FTR has traditionally been classified into three categories: *mild*, *moderate*, and *severe*, based on the progression of the disease. Presently, the threshold for severe is an effective regurgitant orifice area (EROA) $\geq$ 40 mm^2, a regurgitant jet volume (R Vol) $\geq$ 45 mL, and a vena contracta (VC) width $\geq$ 7 mm [72]. The establishment of the recommended values for grading TR aid in pre-surgical planning, although, patient-specific variances in the anatomy and pathology, as well as limitations in the current grading scale, confound such recommendations in practice [72].

2.3.1. Regurgitant Jet Area

The degree of TR severity is frequently graded according to the jet area (planimetry of maximal jet area in cm^2). The regurgitant jet area may be assessed qualitatively or semi-quantitatively. Qualitatively, mild TR displays a small central jet, moderate TR displays an intermediate jet, and severe TR displays a very large central or eccentric wall-impinging jet [73]. Quantitative practices for grading severity of TR are not well-established in contrast to those used for MR. Conventional guides to quantitatively grade TR severity suggest a jet area < 5 cm^2 to be considered mild, 5–10 cm^2 to be moderate, and >10 cm^2 to be severe. However, color flow echocardiography may distort the jet size, prompting inaccurate estimations. Therefore, the European Association for Echocardiography does not recommend the use of the regurgitant jet area to grade TR severity [73].

2.3.2. VC Width

Generally imaged in the A4C view, the VC width (mm) describes the narrowest point of the regurgitant flow before the turbulent jet swells outward, as found immediately distal to the regurgitant orifice. To quantify the VC, the components of the regurgitant jet are identified with pulse wave Doppler (at a Nyquist limit of 50–60 cm/s), and the VC width is measured perpendicular to the jet direction [73]. To obtain a reliable measurement, the VC width is averaged over two to three consecutive heart beats. Recommendations by the European Association for Echocardiography suggest that clinicians rely on the VC measurements when quantifying TR severity: severe TR with a VC width ≥7 mm, while VC width values < 7 mm are considered more difficulty to interpret [73,74]. Major benefits of using VC to grade TR severity include: (i) the measurement is independent of hemodynamic and instrumentation factors; and (ii) it can be used on eccentric jets.

2.3.3. Proximal Isovelocity Surface Area (PISA)

Based on the principles of flow dynamics and continuity (conservation of mass), PISA is another useful parameter for estimating valvular insufficiency [73,75,76]. Aptly named, this measurement attempts to quantify the orifice area through which blood flows. Isovelocity refers to the regurgitant region on the color flow map where the color reverses, and the PISA radius stretches from this point, i.e., the edge of the blue hemisphere, to the center of the valve. The surface area (PISA) is determined using $A = 2\pi r^2$. For assessing TR severity, Lancellotti et al. (2010) [73] recommended a PISA radius ≤ 5 mm as mild, 6–9 mm as moderate, and >9 mm as severe. From the semi-quantitative PISA analysis, quantitative parameters, such as the area of defect, EROA and R Vol, can also be derived [76]. Recommendations for EROA and R Vol are defined only for severe tricuspid regurgitation as ≥40 mm^2 and ≥45 mL, respectively [72,73].

2.3.4. TA Diameter

Studies have shown that the pathology of the TV varies by patient, and anatomical parameter changes, such as tricuspid annular dilation, may occur in the absence of considerable TR [19]. However, the converse is not true. In a study consisting of 311 patients undergoing MV surgical repair and 148 of who also receiving concomitant TV repair, Dreyfus et al. (2005) [14] determined that FTR does not occur without a pronounced tricuspid annular dilation. Furthermore, the assessment of tricuspid orifices in each of the study's participants revealed the tricuspid annular dimension to be the only universal feature regarding FTR pathology. Thus, the current threshold for moderate or severe *TR* based on the annular dimension is a diameter > 40 mm; however, this index should be applied conservatively.

2.3.5. Proposed Revisions to The Current TR Severity's Grading Recommendations

To address the limitations of the current grading system, studies have proposed two additional levels of TR and the inclusion of added assessment parameters, such as the right ventricular early inflow-outflow index. For instance, Go et al. (2018) [72] and Hahn and Zamorano (2017) [77] call for two additional grades of severity, "*massive*" and "*torrential*", to better describe TR that is already defined as "severe." The new grading system thresholds "massive" TR as having an EORA of 60–79 mm^2, an R Vol of 60–74 mL, and a VC width of 14–20 mm. On the other hand, "torrential" TR is defined as having an EROA ≥ 80 mm^2, an R Vol ≥ 75 mL, and a VC width ≥ 21 mm.

A recent study [78] retrospectively evaluated the VC width, jet area, EROA, right ventricular early inflow-outflow (RVEIO) index, and the RA and RV volumes using routine TTE data from patients with moderate and severe *TR* ($n = 395$). The RVEIO index was calculated as an integral of the early diastolic filling velocity and the RV outflow velocity during the period of systolic ejection. An RVEIO index ≥ 10, a VC width ≥ 7 cm, a jet area > 10 cm^2, and an EROA ≥ 0.4 cm^2 were shown to be independent predictors of TR. The RVEIO index increased incrementally in relation to TR severity, making the RVEIO index another useful parameter for indexing the severity of TR.

2.4. Surgical Interventions

Heart valve surgery restores proper leaflet function through one of two methods: (i) surgical valve repair, or (ii) surgical valve replacement. TV repair typically uses one of the following surgical techniques: bicuspidization, classic De Vega, flexible band, or rigid ring. Bicuspidization and De Vega annuloplasty are affordable, simple, and present minimal risk of heart block. However, incidence of residual and recurrent TR is high for bicuspidization and moderate for De Vega annuloplasty. Conversely, ring annuloplasty minimizes residual and recurrent TR but is expensive and more difficult to perform [79]. Valve replacement is generally reserved for patients suffering from comorbidities on the right side of the heart that are unamendable using TV surgical repair [80,81].

2.4.1. Repair Methods for Surgical Treatment of FTR

Because patients classified as having *functional* TR, i.e., regurgitation that occurs due to annular dilation resulting from increased pulmonary or right ventricular pressure, compose 70–80% of cases, most surgeries involve repair of the native geometry [14]. Kay et al. (1965) described the first valve repair technique to treat FTR [82], which involved bicuspidization of the TV (i.e., the complete exclusion of the posterior leaflet) using a suture. More modern suture-based techniques include the De Vega procedure [83], which reduces the annulus diameter while maintaining the tri-leaflet structure, effectively treating the annular dilation at the base of the TV anterior and posterior leaflets that occurs in over 80% of cases [75]. One major limitation of the De Vega suture annuloplasty is that the sutures would tear from the fragile annular tissue resulting in suture dehiscence and recurrent TR. Antunes and Girdwood (1983) [84] proposed a variant of this technique using Teflon pledges to reinforce the annuloplasty suture. De Vega suture annuloplasty is relatively safe and effective for treating minor TR when RV dilation is absent [85]. Both types of suture-based annuloplasty are difficult to reproduce, despite superseding bicuspidization, and the results are unpredictable in comparison to ring annuloplasty [79].

Carpentier et al. (1971) [86] proposed the first ring annuloplasty device, a rigid C-shaped prosthetic apparatus, which included a better distribution of tension and a more standardized annular reduction. During surgery, physicians measure the distance between the antero-septal and postero-septal commissures with calipers and select a ring size accordingly. Eight to ten stitches, starting at the midpoint of the septal leaflet and ending at the antero-septal commissure, secure the device to the orifice (Figure 6). Prosthetic rings were adapted from the original idea to include semi-rigid, rigid, and flexible rings. Each type of ring poses advantages and disadvantages. For example, rigid and semi-rigid bands effectively restore and fix the 3D shape of the TV annulus in its native configuration. The most notable advantages of the rigid rings versus the flexible rings are the stabilization and normalization of the septal leaflet dimensions [79]. However, the installation of a rigid annuloplasty ring may increase the forces exerted on the native annulus more as opposed to the forces resulting from a flexible ring. This increase in force may induce annular dehiscence, a well-documented form of valve repair failure [16,30,87,88]. Edwards developed an improved, 3D saddle-shaped, rigid annuloplasty ring to decrease the instances of dehiscence (Figure 6a,b) and improve replicable implantation [89]. The device has demonstrated incredible short-term efficacy, and evaluations are ongoing to assess long-term efficacy. Surgeons must also elect to use partial or complete bands. Partial bands reduce the occurrence of post-operative conduction block by not placing a suture in the region of conduction tissues, whereas complete bands negate the risk of future annular dilation although the septal portion of the annulus is thought not to dilate.

The clinical popularity of both the De Vega suture annuloplasty and the ring annuloplasty has prompted several studies to assess their effectiveness in managing TR during immediate and long-term follow-ups [90]. For example, Rivera et al. (1985) [75] evaluated a randomized study of 159 patients, 76 of whom received Carpentier rings and 83 of whom underwent a modified De Vega procedure. They reported similar results for both techniques in their rates of six-year freedom from 2 + TR. However, the Carpentier ring revealed a more reliable guarantee of long-term freedom from TR (TR recurrence in the De Vega group = 9/19 versus the Carpentier group = 1/16; $p < 0.01$). Another, more recent, study by Huang et al. (2014) [88] compared the treatment outcomes between the suture annuloplasty and the ring annuloplasty. The study observed a significant improvement via post-operative echocardiography assessment of valve function and TR grade (3.4 to 0.6) with no statistical difference between the methods of repair at 1- and 5-year follow-ups (97% and 84% for the De Vega, and 96% and 82% for the ring annuloplasty, respectively). Recurrence-free survival was superior for the ring annuloplasty but not significantly (78.8% vs. 74.5%; $p < 0.62$). More recently, Charfeddine et al. (2017) [91] further documented comparable intermediate survival rates for correcting FTR. In a comparison of ring-based annuloplasty devices, Pfannmüller et al. (2012) [87] demonstrated that patients with rigid bands were at a significantly higher risk of annuloplasty dehiscence. Furthermore, patients suffering from dehiscence exhibited greater residual TR as compared to those without dehiscence.

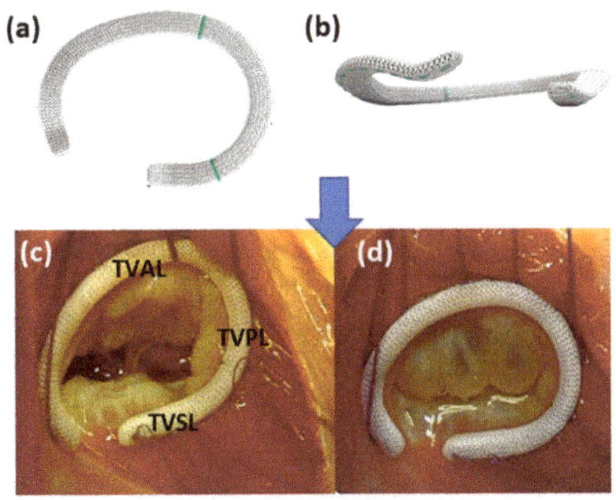

Figure 6. (**a**) Edwards physio-partial annuloplasty ring (**b**) mimics the three-dimensional geometry of the TV annulus. (**c**) The implanted annuloplasty ring is shown during diastole, and (**d**) systole. Images were modified from [92].

Less commonly used techniques for repair include enlargement of the anterior leaflet and "the Clover Technique." Dreyfus et al. (2008) [93] proposed a procedure to remove the anterior leaflet and replace it with a small tissue section of autologous pericardium. This technique is employed when annuloplasty does not sufficiently correct FTR, such as cases of severe leaflet tethering. On the other hand, "the Clover Technique", named for the post-operative shape of the valve, was developed by Lapenna et al. (2010) and Belluschi et al. (2018) [94,95] to correct severe FTR and significant leaflet tethering. The procedure uses both suture and ring annuloplasty to restore the annular geometry and a polypropylene suture to fasten each leaflet at the midpoint of the free edge.

2.4.2. Replacement Methods for Surgical Treatment of FTR

Because annular dilation is a consistent pathological feature of FTR, surgical repair of the annular geometry and TV annuloplasty is widely accepted as the preferred method for restoring proper valve function. However, in the event of severe FTR as a secondary pathology, surgeons may replace the diseased valve tissue. Replacement valves may be either mechanical or bioprosthetic. Mechanical valves come in various configurations including disc valves, bi-leaflet valves, and ball valves. Patients with any implanted mechanical valve will depend on anticoagulant medications for the remainder of their lives. This long-term reliance on medicine to viably sustain mechanical valve implants as well as issues with mechanical valve thrombosis led to the biological prostheses as a secondary approach. Bioprosthetic valves may be an autograft (composed of patient's own tissue), allograft (taken from a donor of the same species), or a xenolog (derived from another species) [96]. Xenograft valves, either porcine or bovine pericardial, are by far the most common valves used in TV replacement. Comparing TV repair and replacement for treating patients with severe TR, no statistical difference was found in early mortality, ten-year overall survival, and ten-year freedom from cardiac death between the patient populations. Thus, TV replacement was concluded to be a viable option for those unsuited for TV repair [81].

3. In-Vivo and In-Vitro Investigations

3.1. In-Vivo Dynamics and Strains of the TV Annulus

Understanding of the TV annulus has been of high focus in earlier literature, whereas, recently, attention has turned toward understanding the mechanisms of FTR. In TV annulus studies, important measurements include: (i) diameter, measured as an antero-posterior (AP) diameter and a septo-lateral (SL) distance due to the annulus' elliptical shape, (ii) height, (iii) area, (iv) circumference, and (v) eccentricity, defined as the ratio between the AP and SL diameters.

In the case of in-vivo studies of the TV annulus, measurements are generally made using MRI, 2DE/3DE, or CT (Figure 7a) [97,98]. A study by Hammarström et al. (1991) [97] used 2DE to measure human annulus parameters with three primary observations: (i) an average annular diameter (between diastole and systole) of 22.5 mm, (ii) the greatest motion occurring along the lateral point of the TV annulus, and (iii) a hinge-point of the annulus movement occurring on the septal side. These observations were later reaffirmed, and more details of the annulus movement were provided through other human in-vivo studies. For example, Ring et al. (2012) [98] used 3DE to quantify: (i) AP and SL average diameters of 41.15 mm and 33.75 mm, respectively, (ii) heights from diastole to systole of 4.2 mm to 5.5 mm, respectively, (iii) areas from diastole to systole of 1145 mm^2 to 1049 mm^2, respectively, (iv) perimeters from diastole to systole of 124 mm to 120 mm, respectively, and (v) eccentricity values from diastole to systole of 1.20 to 1.29, respectively. Regarding the annular movement, during diastole the annulus has a more circular shape, while during systole the annulus becomes more elliptical, as shown through the observed eccentricity values.

The exception to the use of imaging modalities in patients for in-vivo assessment is through open-heart surgery in ovine animal studies in conjunction with sonomicrometry [33,99–101]. To briefly elaborate, sonomicrometry uses piezoelectric transducers, or crystals that are fixed to the anesthetized animal's valve structures by sutures through an open-heart surgery. Once the animal's heart is restored to its healthy hemodynamic profile, the transducer is then energized using a short electrical pulse. This pulse then generates an acoustic wave that is captured through ultrasound technologies. Sonomicrometry has its limitations, however, in that it can be difficult to ensure that the same crystal locations are used between animal specimen due to intraspecies heart valve geometry variations. Additionally, echocardiography is generally used to ensure a proper crystal placement after surgical operation, although shadowing from the crystals can lead to improper evaluations [102].

Rausch et al. (2018) [101] used such techniques to retrieve relevant clinical parameters of the ovine TV annulus, as well as establish engineering metrics (e.g., strain and curvature along the TV annulus). Specifically, they determined that the annular motion is asymmetric in contraction. This asymmetry is elucidated by the deformations originating at the antero-septal and postero-septal vertices, as demonstrated by the contraction concentrations (tangential strain increases up to +0.10 mm/mm), the elongations of the antero-postero vertex, and the mid-septal region of the annulus (relative curvature increase up to +0.03 1/mm). Moreover, Malinowski et al. (2015) [99] used the ovine model to determine the effect of pulmonary hypertension on the TV annular dynamics and found that the disease causes a 12% increase in the total annular area. Employing the same ovine model, Malinowski et al. (2018) [100] also observed the effect of annuloplasty suture on the TV annular mechanics, noting that the tricuspid valve surgical repair resulted in an increase in the compressive TV annular strains. More recently, Mathur et al. (2019) [103] refined the approach by placing four sonomicrometery crystals on each of the three TV leaflets and six on the surrounding TV annulus. Through this approach, they observed that the TVAL and TVPL had similar closing patterns, while the TVSL appeared to have a smaller range of motion and closing velocity. Additionally, their animal study showed that the magnitude of the strain in the belly region of the leaflets was larger than the free edge region and the leaflet strain throughout the cardiac cycle was qualitatively different for both regions. Similar studies have been performed to analyze other pathology-induced changes in the TV annular dynamics [104,105] and the effects of surgical interventions [106].

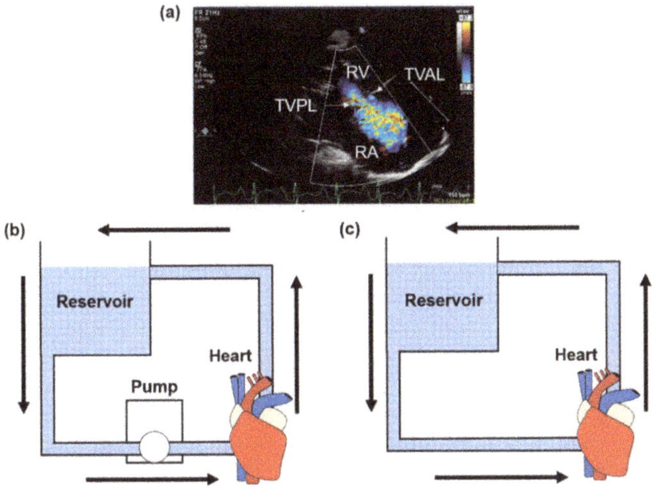

Figure 7. Example methodologies for investigating TV annulus dynamics: (**a**) clinical 3DE in-vivo imaging (image courtesy of Dr. Mir and Dr. Burkhart from the Children's Heart Center at the University of Oklahoma Health Sciences Center), (**b**) an in-vitro pump-driven fluid flow loop paired with sonomicrometry, or (**c**) a Langendorff model (or a working heart model) paired with sonomicrometry.

Although In-vivo studies are useful as they allow for non-invasive, accurate clinical assessment, they are generally limited by the use of imaging techniques that are unable to capture the fine details of valve apparatus movements (usually due to interference with other bodily structures) or fine resolution dynamic information due to limited capture rates. In-vitro studies address this need for higher resolution and dynamic information of the sub-valvular structures of the tricuspid valve.

3.2. In-Vitro Flow and Pressure Systems

In-vitro studies are useful for obtaining high-resolution and dynamic information about the annulus movement, usually through a flow system or pressure system (Figure 7b). Flow systems use a pulsatile fluid flow through the valve apparatus to induce TV motions as they would occur in-vivo. However, this system is limited as it can only create passively beating hearts. In the case of passively beating hearts, there is no active ventricle contraction. As such, ventricular pressures applied by an in-vitro flow system cause the inverse annulus action, with expansion occurring rather than contraction during systole. Pressure systems, on the other hand, operate by applying a pressure to the ventricles to force full leaflet closure, allowing analyses of the pressurized valve statically. There is an exception to pressure or flow systems for in-vitro analysis of heart operation to create active beating hearts, known as the Langendorff or the working heart model (Figure 7c) [107]. In these models, the human/ovine/porcine heart is retrieved immediately preceding the human or animal death and the organ is placed into a flow system and perfused with a solution to provide cell nutrients and maintain cell life. The heart is resuscitated and actively beats ex-vivo with the same TV annulus movement that would be found in-vivo. Regardless of the system used for creating ex-vivo beating hearts, the system is paired with a certain imaging modality, such as sonomicrometry, for collecting high-resolution data.

3.2.1. In-Vitro Flow Systems

For studies involving an in-vitro flow system, measurements were usually made using sonomicrometry or 2DE/3DE [100,101,108–110]. These studies were generally performed on ovine or porcine hearts because of their availability and lower cost. Imaging analysis of hearts that have been excised from the surrounding bodily structures produces higher-fidelity imaging and information about the TV dynamics compared to non-invasive imaging techniques. For example, a porcine in-vitro study by Khoiy et al. (2018) [108] used a fluid flow system to investigate the effects of chordae rupture on the TV geometry and function. With regard to non-ruptured valves, they found a change in the average diameter for the TV annulus that agrees with the clinical observations for human hearts by Ring et al. (2012) [98]. Focusing on the geometrical changes of the TV with ruptured chordae tendineae, Khoiy et al. (2018) [108] observed that the TV annulus area increased by ~9% during the cardiac cycle.

Moreover, Malinowski et al. (2018) [100] performed a characterization of the TV annular strain using human hearts perfused with donor blood to restore normal beating function ex-vivo (i.e., the Langendorff system). From this study, a mean compressive annular strain of 4% ± 2% was quantified, with the greatest compression occurring at the anterior side and the smallest compression at the septal side. Similar studies have been performed to derive the human tricuspid annulus geometry [111].

In-vitro flow systems could also be useful in measuring the TV leaflet dynamics, as was performed by Khoiy et al. (2016) [109]. Specifically, the porcine TVSL was fixed with sonomicrometry crystals, and the heart was placed in a flow system to emulate human physiologic conditions. The TVSL was shown to have a peak areal strain of 9.8%, a peak circumferential strain of 5.6%, and a peak radial strain of 4.3% at the maximum right ventricular pressure. Furthermore, the areal strain distributions were shown to be *non-uniform* across the TV leaflet, while the principal strains were more spatially uniform. Additionally, the circumferential and radial strains were also shown to exhibit some heterogeneity, with higher circumferential strains at the TVPL and increased radial strains near the TVAL.

3.2.2. In-Vitro Pressure Systems

Pressure systems are primarily useful for analyzing the deformations of the pressurized valve structures during systolic closure. Pant et al. (2017) [112] used a pressure system to examine the change in the alignment of the microstructure of the TV leaflets from a non-pressurized to a pressurized state. For this study, porcine hearts were placed into a system that hydrostatically pressurized the leaflets to force coaptation, and the leaflets were fixed using glutaraldehyde. The primary finding of the study was that a higher alignment of the collagen fibers exists in the pressurized TV leaflets, as opposed to the relaxed, free tissues. Additionally, the leaflet area was found to increase in the pressurized state for the TVAL (405 ± 31 mm^2 to 479 ± 63 mm^2) and decrease for both the TVPL (413 ± 23 mm^2 to 337 ± 26 mm^2) and the TVSL (429 ± 19 mm^2 to 374 ± 40 mm^2). The anisotropy indices of the leaflets were also found to increase by 2 to 3 times in the pressurized leaflets, with the largest increase observed in the TVAL. Similar results have been found in another study by Basu et al. [113].

3.3. *Chordae Tendineae Force Measurements*

Other important dynamic structures of the TV are the chordae tendineae. In a study by Troxler et al. (2012) [114], porcine TVs were excised for use in the Georgia Tech right heart simulator to measure the forces the chordae tendineae experience during the cardiac cycle. In this study, miniature C-ring force transducers were attached to the strut chordae tendineae, and the chordal forces were measured under the normal and emulated pathology conditions. In the normal condition, strut chordal forces ranged from 0.1 to 0.4 N, depending on the papillary muscle and leaflet insertion points. TV annular dilation (100% dilation) caused the chordal forces to nearly double, ranging from 0.2 N to 0.7 N, whereas papillary muscle dysplasia only led to an increase in the chordal force if all three insertion regions were displaced. For combined pathologies, a greater increase in chordal forces was observed when the papillary muscles were moved apically, rather than laterally.

3.4. Biomechanical Quantifications of the Subvalvlar Structures of the TV

With the recent growing interest in the TV's functions, several in-vitro studies have been performed to analyze the deformations and mechanics of the TV leaflet and chordae tendineae tissues. Some recent studies have assumed the leaflets to be thin membranes and subsequently used Laplace's Law to estimate the in-vivo mechanical response [46,115,116]. On the other hand, some investigations have determined a parametric spline representation for the leaflets and subsequently used an inverse modeling approach for approximating the leaflets' mechanical responses [117–120]. Nevertheless, in-vitro biomechanical analyses using a biaxial mechanical testing technique have been well developed for characterizing soft biological tissues that have distinct transversely isotropic material properties. Within the scope of heart valves, most of the previous literature has been focused on the analysis of the MV or AV leaflet tissues, whereas limited research exists on the TV counterparts (cf. Figure 1). However, as emphasized in Section 1, the TV has become of increasing interest in the past decade since the seminal clinical paper from Dreyfus et al. (2005) [14]. In-vitro studies of the TV leaflets usually focus on biaxial mechanical testing of the tissue under various loading conditions to retrieve the stress-strain responses that may be representative of the physiological deformations of the TV leaflets.

3.4.1. Biaxial and Uniaxial Mechanical Properties of the TV Leaflets

Previous studies have elucidated the leaflet tissue's mechanical properties by performing biaxial mechanical testing of the central, belly regions of the TV leaflets (Figure 8a,b) [46,116,121–123]. Because of the complex microstructure of the fibrous tissues, biaxial mechanical testing is generally performed at various loading ratios of a targeted stress in each tissue direction (i.e., circumferential and radial directions). Coupling between the two tissue directions plays a critical role in the overall stress-stretch responses. Moreover, the TV leaflet tissues exhibit repeatable cyclic mechanical response after the tissue has been preconditioned to restore their in-vivo functionality.

Several key findings from these biaxial mechanical testing studies of the TV leaflets were summarized as follows. First, the TV leaflets have an anisotropic and non-linear mechanical response, with the radial direction being generally more extensible than the circumferential direction. For example, in Jett et al. (2018) [46], the porcine TVAL tested to a 115 kPa equibiaxial stress was found to have a circumferential peak strain of ~22% and a radial peak strain of ~73%. Second, the mechanics differs between each of the three TV leaflets, with the greatest extensibility and the greatest anisotropy generally observed in the TVPL (Figure 8c). Moreover, Pham et al. (2017) [124] tested *human* TV leaflets to a 70 kPa equibiaxial stress, and they reported circumferential strains of: TVAL, 10%; TVPL, 13%; and TVSL, 10%; as well as radial strains as: TVAL, 17%; TVPL, 22%; and TVSL, 20%. Third, studies have also demonstrated that the TV leaflets' mechanical responses have a slight dependence on the loading rate. In Jett et al. (2018) [46], the TV leaflets were tested at 2.29 N/m, 4.42 N/m, and 7.92 N/min loading rates, and only a ~7% difference was observed in the tissue's peak strains. Fourth, the TV leaflets also possess a significant stress–relaxation behavior [46,121,125]. For example, in the study by Laurence et al. (2019) [125], 6 delimited regions of the porcine TVAL were stretched to a 50 N/m equibiaxial membrane tension and allowed to relax for 900 seconds, with a 20–30% decay of the initial membrane tension reported.

In addition, the above-mentioned anisotropic, non-linear nature of the TV leaflet's mechanical response stems from the gradual recruitment of crimped collagen fibers preferentially oriented in the circumferential direction. Since the crimped collagen fibers do not contribute to the mechanical behavior, the tissue's mechanical response initially has a long toe-region in which there are large deformations corresponding to low stresses. As the collagen fibers are recruited and straightened, the tissue's mechanical response stiffens, resulting in the distinct non-linearity. The orientation, recruitment, and re-orientation of the collagen and elastin fibers throughout the TV leaflet tissue under loading has been determined by imaging techniques [126–128]. For instance, Alavi et al. (2015) [129] used second harmonic imaging techniques to understand the distribution and orientation of collagen fibers during uniaxial and biaxial loading. Two key findings from their study are: (i) collagen fibers in the superficial layers were aligned in between the radial and circumferential directions when the TV leaflets were in their relaxed state, whereas collagen fibers were oriented in the circumferential direction in deeper layers; (ii) the collagen fibers were reoriented according to the applied biaxial loading.

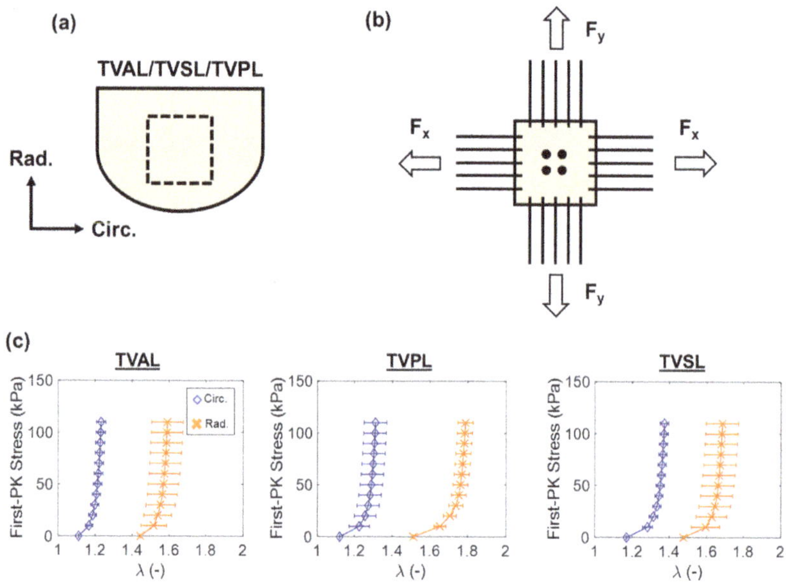

Figure 8. (**a**) Schematic of the excised leaflet tissue and the central bulk region taken from the TVAL, TVPL, or TVSL. (**b**) An illustration of the mounted TV leaflet tissue specimen onto the biaxial testing system with the collagen fiber orientations (circumferential and radial directions) aligned with the testing x- and y-directions. (**c**) Mean ± SEM of the first-PK stress versus stretch results of the porcine TVAL, TVPL, and TVSL tissues (n = 6) under equibiaxial loading protocol (F_x:F_y = 1:1) at room temperature (22 °C). Images were modified from Jett et al. (2018) [46].

Another interesting recent study by Basu et al. (2018) [130] analyzed the mechanics of the TV leaflet-annulus transition regions through uniaxial mechanical testing. In their study, they found that the largest Young's modulus occurred at the septal leaflet-annulus transition (208.7 ± 67.2 kPa), followed by the posterior (136.8 ± 56.9 kPa) and anterior transitions (92.0 ± 66.8 kPa). However, the extensibility of the transition regions was found to be similar across all three leaflet-annulus zones.

3.4.2. Bending Properties of the TV Leaflets

In-vivo, the TV leaflets experience large flexure between the pressure changes of diastole and systole. Biaxial mechanical testing can shed some light on the leaflet mechanical behaviors; however, bending mechanical tests can provide additional insight into the leaflet mechanics. For example, Fu et al. (2018) [131] observed that for the MV anterior leaflet, the leaflet bending angle can be a useful quantitative metric for when clinicians should perform surgical interventions. Thus, it is important to perform such quantifications of the healthy TV leaflets' bending mechanical behaviors. Bending mechanical testing is typically performed using a custom-made device, based on the device made by Merryman et al. (2006) [132]. This device fixes the leaflet on each end and then allows for a moment to be applied either with or against the natural curvature of the leaflet, and the corresponding displacement is measured using fiduciary markers and digital image correlation techniques. Brazile et al. (2015) [133] used such a device to quantify the bending mechanical behaviors of the TV leaflets. From this study, they observed that when the bending moment was applied with the natural leaflet curvature there was a non-linear momentum-curvature relationship; when the bending moment was applied against the natural leaflet curvature a stiffer behavior was observed. Quantitatively, at a peak change of curvature of 0.075 mm^{-1} the flexural rigidity of the leaflet in the circumferential direction was 54% higher against the curvature than with the curvature, whereas in the radial direction there was a 97% difference. With regard to the instantaneous effective bending modulus similar trends were observed, with the percent difference between the curvatures being 47% in the circumferential direction and 93% in the radial direction. The results of this study can be useful in the development of polymeric replacement leaflets where the materials can be tailored to specific bending mechanical properties [134].

3.4.3. Spatial Variations in Tissue Mechanics of TV Leaflets

With recent advancements in computational power, models have been refined for the heart valves to consider the leaflets' complex microstructure, structural heterogeneity, and distinct layers [135–142]. However, these models are typically based on investigations of bulk, central regions of the MV or AV leaflets (cf. Subsection 3.4.1) rather than the spatially varied mechanical properties of the TV leaflets. Our group sought to fill this gap in knowledge by providing a mechanical characterization of the TV leaflet specimens over their spatial domain. In Laurence et al. (2019) [125], we first sectioned the TVAL into six smaller regions (labeled A–F, Figure 9a) that were mechanically tested under our established biaxial mechanical testing protocols [46].

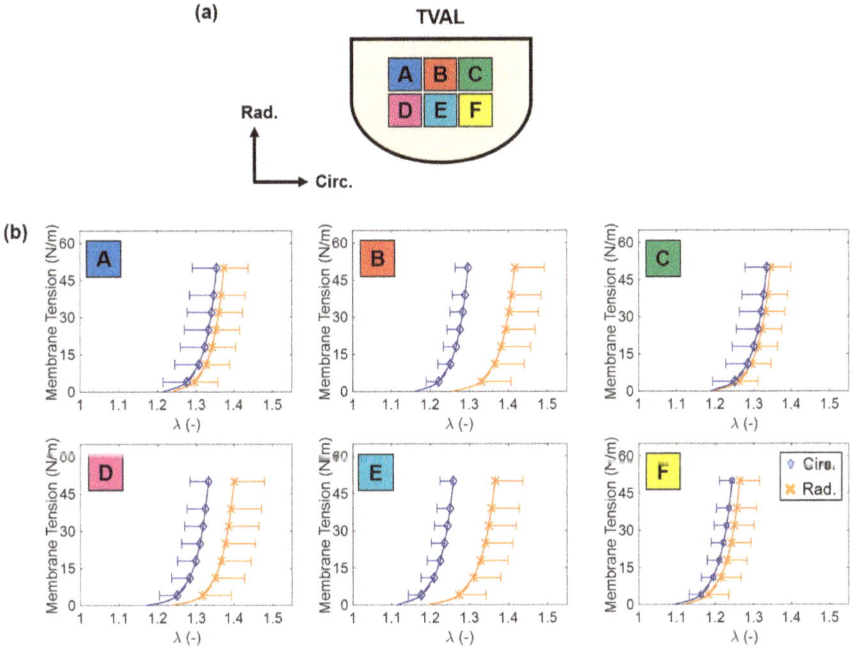

Figure 9. (**a**) Schematic of the TVAL with six smaller tested regions for investigating the regional variance in the tissue's mechanical properties. (**b**) Mean ± SEM (n = 10~13) of membrane tension versus total tissue stretch results for the 6 tissue regions under the equibiaxial loading protocol. Images were modified from Laurence et al. (2019) [125].

This study, which is the first of its kind, demonstrated the spatial variability of the TV leaflet's mechanical properties (Figure 9b). Specifically, from equibiaxial mechanical testing of the tissues to a 50 N/m membrane tension, central regions of the leaflet (B and E) were observed to exhibit a greater material anisotropy than those edge regions (A, C, D, and F). With regard to the tissue's extensibility, higher extensibility was observed in regions near the TV annulus (A–C) as compared to regions near the free edge (D–F). In addition, the regions near the TV annulus had circumferential peak stretch values approximately 4% higher than those in the edge regions, whereas a similar, but less pronounced trend was observed for the radial peak stretches (i.e., ~2.5% higher in regions bordering the TV annulus). The results of this study are useful in better understanding the leaflet mechanics and how the stress could vary and be distributed spatially over the TV leaflet.

3.4.4. Microstructural Constituent's Contributions to Tissue Mechanics of the TV Leaflet

While our study on the regional variance in the TV leaflet tissue's mechanical properties provides useful insight into the overall biomechanical behaviors of the TV, more detailed biomechanical characterizations could be made to investigate the mechanical contributions associated with the underlying microstructural constituents. For the AV leaflets, the mechanical contributions of each constituent have been comprehensively quantified using a biaxial mechanical testing procedure in which tissues are tested before and after enzymatic removal of the constituent of interest [49,50,143,144]. However, this has not been done for the TV leaflets. Therefore, we sought to fill this knowledge gap by applying this enzymatic-treatment procedure to the TV leaflets [145].

In our study on the GAG contributions to the TV leaflet tissue's mechanical properties, we retrieved porcine TVALs and followed a three-step procedure: (1) TVAL tissue specimens were biaxially mechanically tested; (2) enzyme treatment was performed to remove the GAGs; and (3) treated tissues were biaxially mechanically tested using the same procedure as in Step (1). From this study, it was observed that the GAG-removed tissues (treated, T−) experienced greater stretch than those intact (control, C−) tissues (Figure 10). Specifically, in the loading to a 75 N/m equibiaxial membrane tension, the GAG-removed leaflets were 4.7% and 7.6% more extensible in the circumferential and radial directions, respectively, compared to the native leaflet tissues. In addition, stress–relaxation testing at a 75 N/m equibiaxial membrane tension revealed a lesser relaxation behavior of the leaflets after GAG removal treatment in both the circumferential relaxation (C−: 17.1% relaxation, T−: 15.0% relaxation) and the radial relaxation (C−: 16.4% relaxation, T−: 14.5% relaxation). Ongoing investigations are currently being conducted by our group to determine the contributions of *collagen fibers* and *elastin fibers* to the overall biomechanical behaviors of the TV leaflets. This unique study further enhances our understanding of the leaflet microstructural constituents, which can be useful in the fields of tissue-engineered heart valves where the recreation of the microstructure remains an emerging challenge.

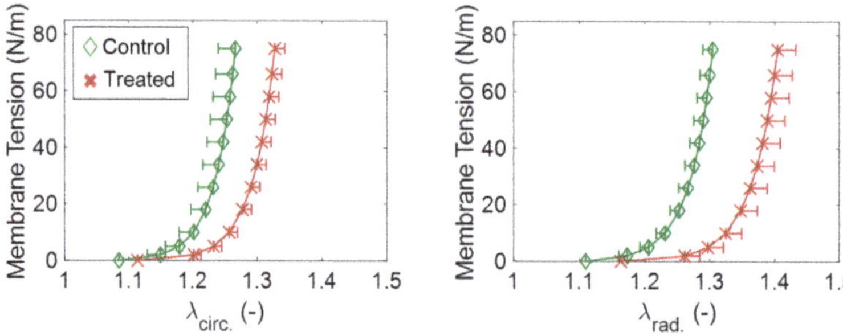

Figure 10. Mean ± SEM (n = 6) of the membrane tension versus total tissue stretch results (**left**: circumferential direction, **right**: radial direction) for the TVAL between the control and enzyme-treated groups under equibiaxial loading (F_X:F_Y = 1:1). Figures were modified from Ross et al. (2019) [145].

3.4.5. Mechanics of TV Chordae Tendineae

The TV chordae tendineae are essential to proper leaflet movement and have been of recent focus in TV literature. In-vivo studies for the CT are useful for obtaining mappings of the chordae distributions and geometries in the valve, which can be helpful for clinical assessments or computational simulations.

In-vitro studies of the chordae tendineae are performed under uniaxial mechanical testing, or using strain gauges and a flow/pressure system. Uniaxial mechanical testing is generally performed on chordae tendineae that have been separated from the leaflet and papillary muscles through clamping to a servo-hydraulic tensile testing machine [56,122,146–152]. Testing protocol then follows either as cyclic loading to a force or stress that is representative of physiologic loading, or as loading until chordal rupture. Our group has expanded on this by excising and mechanically testing a functioning chordae "group" with preservation of papillary muscles and leaflet points of attachment (Figure 11a).

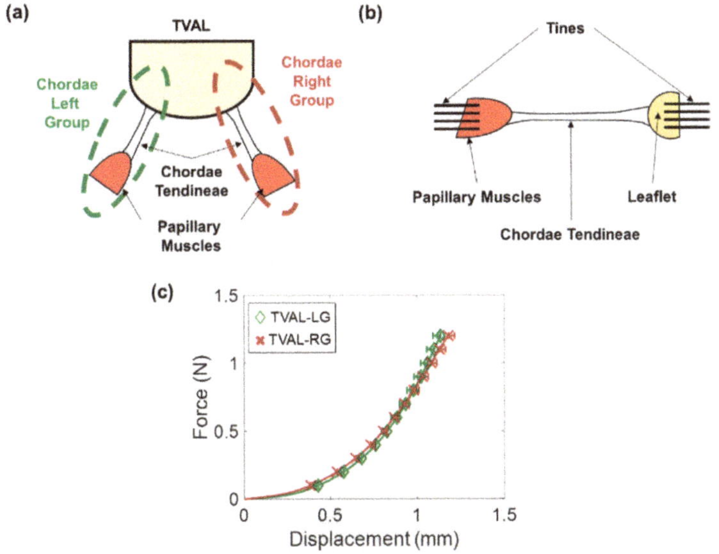

Figure 11. (**a**) The TVAL left (LG) and right (RG) strut chordae tendineae were excised as a group structure, preserving points of attachment to the papillary muscles and leaflet for (**b**) uniaxial mechanical testing to observe (**c**) the mechanical properties of the chordae tendineae as a tissue group.

Lim (1980) [146] characterized non-linear mechanical responses for the TV chordae, as well as an increased tissue extensibility with an increased thickness. Comparing the chordae tendineae between the mitral valve and tricuspid valve, the TV chordae had a lower fiber density, resulting in a lower extensibility [146]. More recently, Pokutta-Paskaleva et al. (2018) [122] characterized the mechanical properties of the three porcine chordal subsets (strut, basal, and marginal). In their study, the chordae tendineae were uniaxially loaded until rupture (Figure 11b). Of the three subsets, it was observed that different chordal subsets exhibit different mechanical properties, with basal chords being the most extensible in the TVPL and TVSL (a Green strain of ~15% at the 3 MPa Cauchy stress). In terms of leaflet attachment, the chordae tendineae that attach to the TVAL were observed to have a lower extensibility than their TVPL and TVSL counterparts (TVAL, 0–5%; TVPL, 6–15%; TVSL, 13–16%) [122]. In our group's study, which analyzed functional strut chordae tendineae groups as opposed to individual chordal segments as done in previous studies, a greater extensibility was observed (Figure 11). At a 1MPa stress, we found that the porcine strut chordae tendineae groups have a peak strain of approximately 11–14%. Our findings (Figure 11c) could provide useful information about the tissue mechanics of the TV chordae tendineae by mimicking the in-vivo functioning environment, which can be implemented in computational models to refine therapeutics related to chordal pathologies, such as chordae rupture.

4. Computational Biomechanical Modeling of the TV

Computational modeling allows for in-silico investigations of TV function that provide unique insight that would be otherwise unobtainable considering conventional in-vitro or clinical methods. The fundamental building blocks of the TV biomechanical modeling framework are selecting the appropriate numerical scheme (e.g., structural modeling, fluid modeling, or fluid-structural modeling), defining the geometry (solid or fluid) and the corresponding computational mesh, specifying the material properties of the tissue and/or fluid domains, prescribing boundary conditions that mimic the clinical scenario of interest, and defining other essential parameters such as contact between leaflet surfaces as well as between blood flow and the TV structure. The selected numerical scheme will use these inputs to provide an approximation of the modeling scenario, such as the closure of the TV. The accuracy of the approximation depends largely on how well the user can describe the complexity of the material, the geometry, and the boundary conditions. Two modeling schemes are typically used in the context of heart valve biomechanics. The first is bio-solid modeling, which describes the behavior of a bio-solid (e.g., heart valves) under specified loading and displacement conditions. The second is fluid-structure interaction (FSI) modeling, which predicts how a fluid and solid will behave and interact through coupled structural dynamics and computational fluid dynamics. Finite element (FE) modeling is typically employed to solve the bio-solid subproblem, while the finite volume method or FE method are both common choices for solving the fluid subproblem. Some research groups elected to use commercially available software packages such as ABAQUS (Dassault Systèmes), LS-DYNA (Livermore Software Technology Corporation), or FEBio (open source) while others opt to develop their own in-house solver. The remainder of this section will emphasize the disparity between the developed computational models for the left-sided and right-sided heart valves, followed by a brief description of recent work for the TV geometrical modeling, constitutive modeling, and computational modeling.

4.1. *Disparity of Computational Models for the Left-Sided and Right-Sided Heart Valves*

The first 3D FE model for the MV was reported over two decades ago in a study by Kunzelman et al. (1993) [153], whereas the first AV FE model was developed over three decades ago by Hamid et al. (1987) [65]. These early numerical studies contained many assumptions regarding the leaflets' material and geometrical properties that have since been addressed. For example, Kunzelman's paper for the MV revealed the anisotropy of the tissue by increasing the Young's modulus in one direction; in contrast, more recent MV computational models have fully mapped the regionally varying fiber architectures with complex constitutive models to more realistically represent the valve's function [127,154–157]. Kunzelman's early study also used an idealized representation of a porcine MV, while a recent study by Wang and Sun (2013) [158] modeled a patient-specific MV geometry from clinical CT slices. Additionally, FSI models have since been used to obtain more accurate predictions of the MV [159,160], AV [161], or bioprosthetic valve [162,163] dynamics. These advances are closely mirrored with the development of a structural constitutive model [164], patient-specific modeling [165,166], and FSI models [161,167,168]. However, limited research has gone into modeling the TV and pulmonary valve (PV) due to the clinical oversight previously discussed (cf. Section 1). The first study developing a computational model for the TV did not arrive until 2010 in a numerical study by Stevanella et al. (2010) [169] as a direct extension of their earlier MV modeling work [170], which is a stark difference compared to the multiple decades of MV and AV modeling. This delayed development of the TV models has limited the progression toward the refined modeling methods used for the MV and AV, leaving considerable opportunities for future extensions and developments.

4.2. *Geometrical Modeling of the TV*

4.2.1. Modeling the TV Geometry

One important step toward enhanced TV computational models is accurately modeling the intricate TV geometry. Two common approaches have been used for the TV leaflet geometry: (i) manual or semi-automated segmentation of medical imaging data and (ii) representation of the geometry using cubic splines. Aversa and Carredu (2017) [171] was the first group to perform manual segmentation of the TV geometry for later use in a computational model. Specifically, they performed real-time 3DE imaging of an in-vitro beating heart and then manually segmented the images using a MATLAB framework. Aversa and Carredu (2017) [171] had difficulties accurately capturing the commissures of the TV leaflets, which was likely due to the lack of image clarity in this region. Kong et al. (2018) [172] later manually segmented clinical CT imaging data for three human patients' TV geometries using the Avizo (Zuse Institute Berlin) software suite. Due to the nature of the in-vivo imaging method, Kong et al. (2018) [172] were unable to directly compare the segmented geometry to the real geometry; however, the three segmented TV geometries were later used in their computational modeling framework and, as discussed in Section 4.4, they obtained excellent closing behavior of the TV. As for semi-automated image segmentation, Pouch et al. (2017) [173] explored the use of semi-automated and multi-atlas methods for modeling the TV geometry for a heart with HLHS, a congenital heart disease.

The use of cubic splines to represent the complex TV geometry can allow for faster geometry acquisition and the use of the TV geometry within a parametric framework (cf. Section 4.2.2). Stevanella et al. (2010) [169] was the first group to use this method by modeling the TV geometry using cubic splines based on anatomical measurements (e.g., leaflet height, commissure height, etc.) of ex-vivo hearts. Stevanella et al. relied on existing information from literature [33] about the TV annulus shape to create the 3D geometry of the TV apparatus. The later work by Kamensky et al. (2018) [174] also used the spline method, except cubic basis-splines (B-splines) were used rather than cubic splines. Additionally, Kamensky et al. used the annulus and chordae information obtained from the micro-CT imaging data of a formalin-fixed ovine heart to construct a 3D geometry of the TV. We have since expanded on the work by Kamensky et al., (cf. more details in the Section 4.2.2).

4.2.2. Parametric Design of Heart Valve Geometries

Parametric design is typically performed using parameterized computer-aided design (CAD) geometries that are constructed from several selected design variables. Thubrikar et al. (1990) [175] first used a parametric design for heart valve leaflet applications. They considered the AV and introduced a 3D geometry that used a parametric description to search for an optimal prosthetic valve design with improved performance. More recent studies incorporating parametric valve designs include Labrosse et al. (2006) [176], Haj-Ali et al. (2012) [168], Kouhi and Morsi (2013) [177], Fan et al. (2013) [178], Li and Sun (2017) [179], and Xu et al. (2018) [180]. These existing studies have provided guidelines for suitable aortic prosthetic valve designs. However, the overall absence of similar developments in parametric design and modeling of the TV has limited understanding of TV geometry and function.

In this work, we present an extended version of the TV modeling capabilities in Kamensky et al. (2018) [174] that includes a more comprehensive range of valve and chordae configurations. The flexibility of this updated geometry-modeling framework encompasses a variety of possible TV designs, making it effective for many types of TV applications. The highly adaptable framework employs a parametric definition of the valve geometry that enables precise control of the valve. The TV surface and chordae were constructed from a combination of patient data and parameter inputs to obtain the full model of the TV, as shown in Figures 12 and 13. The valve and chordae were then parameterized by the illustrated input parameters, including the leaflet and commissure heights and the distance parameters that control the chordae configuration. The versatility of this framework allows for the modeling of healthy, diseased, and patient-specific TVs while maintaining geometries that are not overly complex. The CAD-based B-spline models produced with this framework can be easily converted into analysis-suitable meshes for FE analysis or directly used for isogeometric analysis (IGA) [181]. The use of these CAD geometries for computational modeling applications will be discussed in Section 4.4.

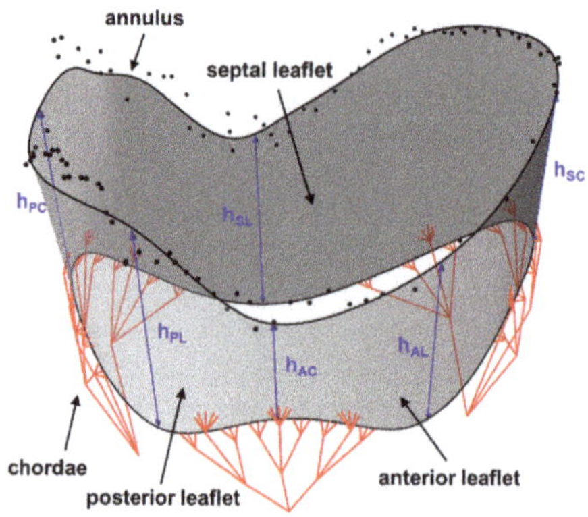

Figure 12. Parametric definition of a B-spline TV surface and chordae. Each leaflet and commissure height is indicated as *h*, where the subscript S, P, or A denotes the septal, posterior, or anterior leaflet, and C or L denotes a commissure or a leaflet location.

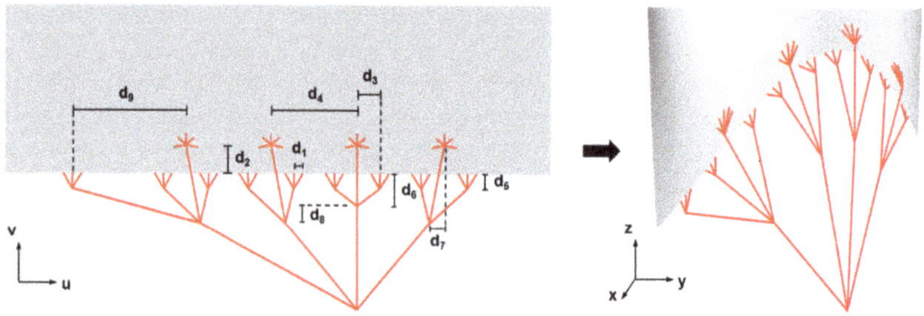

Figure 13. Chordae mapping from the 2D parametric space to the 3D topology.

Each step of the geometry construction process within the proposed framework is illustrated in Figure 14. First, the TV annulus scan data from a patient was fit using a B-spline curve that generates the initial annulus shape (Figure 14a). The scan data points that divide the three leaflets were translated onto the fitted annular curve and used to define the location of the commissure and leaflet heights. These input height parameters determined the offset distances of the division points (Figure 14b). The offset points were then interpolated to define the bottom edge of the leaflet geometry (Figure 14c). The offset direction was defined as the normal of the best-fit plane of the annulus data. The set of cubic B-spline curves, two describing the leaflet edges and six defining the leaflet dimensions, generated a bidirectional curve network that was subsequently used to interpolate the valve geometry as a Gordon surface [182] (Figure 14d). The resulting surface can be easily re-parameterized to ensure that the valve surface elements remain suitable for analysis. Finally, the constructed valvular surface was connected to a corresponding set of structured chordae that were parametrically constructed based on patient data (Figure 14e). The chordae distance inputs were set in the parametric space of the B-spline surface, as shown in Figure 13. Using this approach, the attachment locations in the 2D parametric space were naturally mapped into the physical space to determine the actual attachments and spacing of the chordae on the 3D surface (Figure 14f).

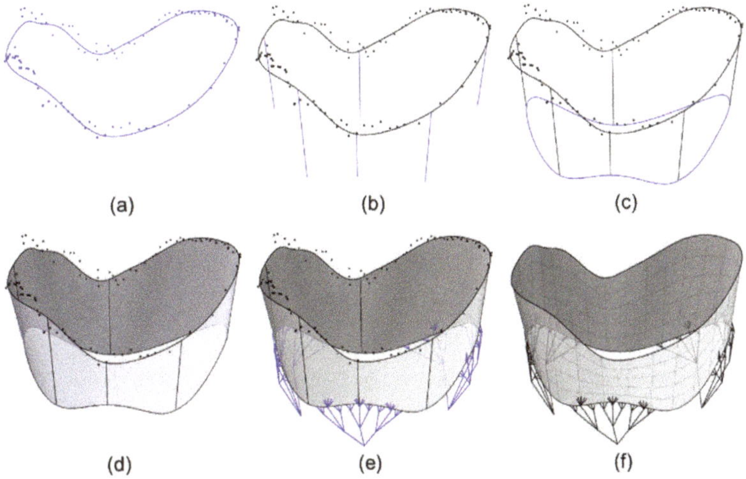

Figure 14. Geometry construction process for the parametric TV: (**a**) B-spline fit of the annulus from the scanned micro-CT data, (**b**) the definitions of the commissure height and the leaflet height, (**c**) a determination of the TV leaflet free edge, (**d**) an interpolation of the TV leaflet surface, (**e**) the attachment of chordae tendineae to the TV leaflets, and (**f**) the final geometry model.

The versatility of the geometry-modeling framework is further demonstrated in Figures 15 and 16. Specifically, Figure 15 shows examples of varying leaflet geometric parameters, whereas Figure 16 illustrates a comparison of the healthy TV with a flattened and dilated TV annulus. As exhibited, the modeling framework can be flexibly adjusted to accommodate different valve geometry configurations and applications. With this enhanced framework, there are clear prospective developments that would initiate substantial progress in moving TV modeling toward the existing MV and AV modeling capacity. The improved models also have a significant potential to enhance clinical understanding of the TV geometry and provide insight into the function of both healthy and diseased TVs. Besides clinical and surgical applications, the wide range of capabilities offered by this framework also extends the feasibility of developing and analyzing prosthetic valve designs that closely mimic the native TV geometry.

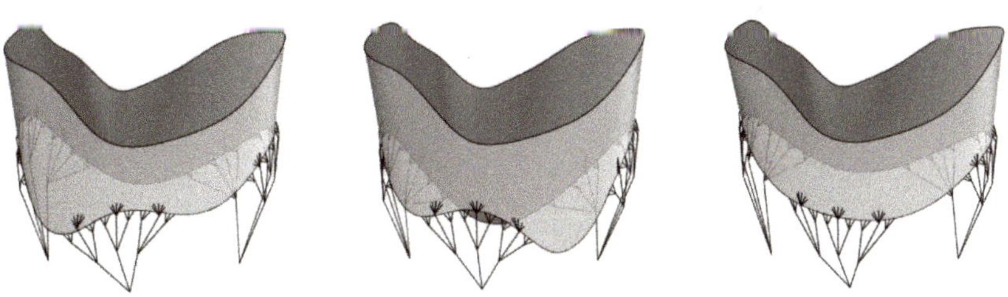

Figure 15. TV geometries with varying leaflet and commissure heights.

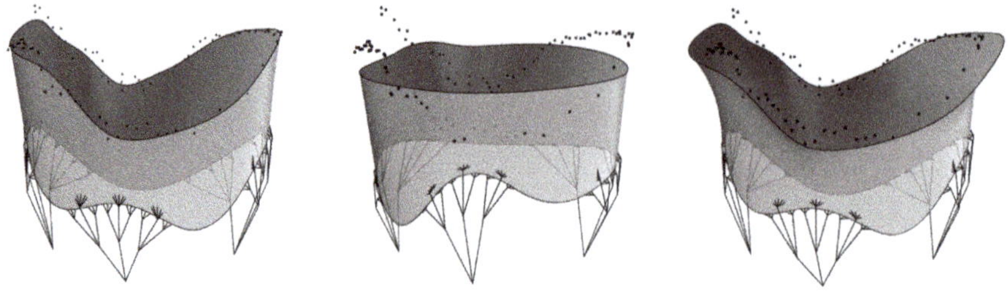

Figure 16. TV geometries of healthy, flattened, and dilated valves.

4.3. Constitutive Modeling of the TV Leaflets

Despite extensive research in developing constitutive models for the functioning MV, there have been very few similar works for the TV. Much of this gap is primarily due to the limited mechanical characterizations for the TV leaflets until Khoiy et al. (2016) [116] studying porcine TV leaflets, Pham et al. (2017) [124] characterizing human tricuspid leaflets, and subsequently Jett et al. (2018) [46,121] performing extensive biaxial testing on both porcine and ovine atrioventricular heart valve leaflets. Since then, three studies in particular have determined constitutive model parameters for the TV leaflets. However, there is still a significant gap in development of most representative constitutive models for the TV, especially models that are structurally based rather than phenomenological.

The first study was from Aversa and Careddu (2017) [171]. In this study, the previous human TV leaflet data for equibiaxial mechanical testing from Pham et al. (2017) [124] was used to estimate parameters of an invariant-based strain energy density function W with the form [157]:

$$W = C_{10}\left[e^{C_{01}(I_1-3)} - 1\right] + \frac{c_0}{2}\left[(1-\beta)e^{c_1(I_1-3)^2} + \beta e^{c_2(I_4-1)^2} - 1\right], \tag{1}$$

where C_{10}, C_{01}, c_0, c_1, and c_2 are the material model constants, β is the parameter describing the material anisotropy (β = 0: purely isotropic, and β = 1: anisotropic), and I_1 and I_4 are the first and fourth invariants of the right Cauchy-Green tensor **C**. Pham et al. (2017) [124] used this model to fit only the equibiaxial data, which provides an excellent fit as used in their subsequent computational modeling.

The second study is from Kong et al. (2018) [172] and uses the same human data from Aversa and Careddu (2017) except with a different strain energy density form. In this study, a strain energy density form from Holzapfel et al. (2000) [183] was used, which represents a fiber-reinforced material with two families of fibers (denoted by i = 1,2):

$$W = C_{10}\left[e^{C_{01}(I_1-3)} - 1\right] + \frac{k_1}{2k_2}\sum_{i=1}^{2}\left\{e^{k_2[\kappa I_1 + (1-3\kappa)I_{4i} - 1]^2} - 1\right\}, i = 1, 2. \tag{2}$$

Herein, C_{10}, C_{01}, k_1, and k_2 are material properties, κ is the model parameter defining the distribution of the family of fibers (κ = 0: well-aligned fibers; κ = 1/3 randomly aligned fibers), and I_i and I_{4i} are the first and fourth invariants of **C**. Similar to the study by Aversa and Careddu (2017), they only fit the data from the equibiaxial protocol from each TV leaflet, demonstrating a very good fit to the experimental data.

The third study for constitutive modeling of the TV leaflets is from Khoiy et al. (2018) [116]. This extensive study was devoted specifically to constitutive modeling of the TV leaflets, whereas the previous two studies used it in their FE simulations. They used the biaxial mechanical data from their quantifications of porcine TV leaflet mechanical properties [116] to determine the appropriate constitutive parameters for the Fung-type strain energy density form:

$$W = \frac{c}{2}\left(e^{(a_1 E_{CC}^2 + a_2 E_{RR}^2 + 2a_3 E_{CC} E_{RR})} - 1\right), \tag{3}$$

where c, a_1, a_2, and a_3 are material constants, and E_{CC} and E_{RR} are the Green strain in the circumferential and radial directions, respectively. They were able to obtain excellent fits ($R^2 > 0.85$) for the model considering five different loading protocols (T_{Circ}:T_{Rad} = 1:1, 1:0.75, 0.75:1, 1:0.5, 0.5:1).

Furthermore, Khoiy et al. (2018) [116] sought to understand how well the material model performed when attempting to predict the TV mechanical behaviors. They fit the Fung-type constitutive model (Equation (3)) to the data of only four of the five loading protocols (T_{Circ}:T_{Rad} = 1:1, 1:0.75, 1:0.5, 0.5:1) and predicted the TV's stress-strain behavior using the fifth protocol (T_{Circ}:T_{Rad} = 0.75:1). The results from this prediction of the fifth protocol agreed well with the experimental data ($R^2 > 0.88$), illustrating the potential for this strain energy density form to be able to fully capture the TV mechanical behavior. In the same study, Khoiy et al. also explored different measures for averaging experimental data prior to performing the constitutive model fitting: the membrane tension, the Cauchy stress, and the first Piola-Kirchhoff stress. They found that membrane tension is sufficient if the tissue thickness does not vary significantly. Otherwise, either the Cauchy stress or first Piola-Kirchhoff (1st-PK) stress must be used. This is currently the most thorough study for determining the constitutive model parameters for the TV leaflets. However, more extensive studies exist for the MV [142,184,185] and AV [141,164] and similar studies are necessary for the TV to move toward a more realistic representation of the TV mechanical response.

In summary, there are limited studies for determining constitutive parameters for the TV leaflets. These studies are not nearly as extensive as those for the MV leaflet tissue where the leaflets have been modeled using detailed structural constitutive models [142,185], or where the effect of selected constitutive model on biomechanical function has been determined using a FE framework [127]. While the phenomenological constitutive model forms used for the TV have shown a reasonable representation of the TV mechanical function, it is essential to not limit TV constitutive modeling to those few forms. Countless other constitutive model forms, both phenomenological and structural, exist in the literature for the other heart valve leaflets, and should be thoroughly examined to determine if they effectively capture the mechanical behavior of the three TV leaflets. These studies should be complemented with investigations to determine the effect of constitutive model form on the biomechanical function of the TV using a patient-specific computational modeling framework.

4.4. Computational Models of the TV

Computational models combine the topics discussed in this section to predict the TV biomechanical function. Approaches using bio-solid modeling or the FSI framework have been currently employed for these investigations. Nevertheless, as previously described in Section 4.1, the limited number of computational models is in stark contrast to the numerous present for the MV or AV. While it is evident that more computational modeling investigations are needed, the handful of existing and ongoing studies provide the essential foundation for cutting-edge advancements of TV computational models. For example, the study by Stevanella et al. (2010) [169] provides the fundamental work behind parametric representations of the TV for use in bio-solid models. The work by Kong et al. (2018) [172] is a critical step in modeling patient-specific conditions using a patient's clinical imaging data. Even more recently, Singh-Gryzbon et al. (2019) [186] have employed FSI modeling of the TV biomechanical function.

4.4.1. Bio-Solid Models of the TV

The first bio-solid model published for the TV was from Stevanella et al. (2010) [169], in which a parametric representation of the ex-vivo TV geometry (cf., Section 4.2) was used and was subsequently discretized into an FE-suitable mesh. The leaflets were represented by 40,300 three-node triangular plane-stress shell elements and the chordae tendineae were represented by a series of two-node truss elements. To simulate valvular closure, a time-dependent physiological pressure traction was prescribed to the ventricular side of the leaflets while the annulus contracted in accordance with previous values [33]. These simulations were performed using the commercial ABAQUS/Explicit dynamic FE software package. Results from the FE model showed a stress values less than 100 kPa, peak strains of ~52%, and papillary muscle forces ranging from 0.37–0.75 N. It should be noted that these simulations used material properties for the MV leaflets, which may not be the most representative of the TV tissue mechanics.

The next study for computationally modeling the TV was Aversa and Careddu (2017) [171]. In this study, Aversa and Careddu took a different approach than Stevanella's study and used 3DE imaging data to inform their bio-solid model geometry (cf., Section 4.2). The FE mesh consisted of three-node triangular plane-stress shell elements for the TV leaflets and a series of truss elements to represent the chordae. For simulating TV closure, a time-dependent pressure traction matching was applied to the ventricular surface, the annulus was prescribed to contract based on their in-vitro measurements, and the papillary muscles were assumed to have no rigid body motion. Results of their FE model showed an incomplete closure of the TV leaflets at peak systole. Furthermore, the stress and strain results were irregular with belly stress values near 300 kPa and peak strains of 0.40 occurring near the chordal attachments, which may be due to the inability to properly define the free margin and commissures of the leaflet, which caused the chordae to not behave incorrectly.

Kamensky et al. (2018) [174] were the next group to perform bio-solid modeling of the TV. Within this study, Kamensky et al. used B-spline surfaces and curves to represent the TV and chordae geometry within an isogeometric computational framework. To simulate the TV closure, they applied a constant transvalvular pressure of 25 mmHg to the ventricular surface. Using this isogeometric framework, two scenarios were simulated: (i) the healthy TV geometry with no modifications, and (ii) rupturing one of the chordae attached to the TVAL. The results from the chordae rupture scenario showed a significant prolapse of the anterior leaflet into the RA. Although this study is brief in its application, it is the first scenario simulating the TV closure within an isogeometric framework and considering a pathological modification.

More recently, Kong et al. (2018) reported a more clinically oriented study [172] that aimed to develop a patient-specific FE framework using clinical CT imaging data. The geometries of three patients who were suspected of having coronary artery disease were incorporated into their FE modeling framework, in which eight-node hexahedral elements and two-node truss elements were used for the TV leaflets and the chordae tendineae, respectively. Kong et al. implemented a chordae configuration based on ex-vivo measurements of the chordae tendineae due to the limited spatial and temporal resolutions of the CT imaging modality. For simulating the TV closure, applied a pressure traction of 23.7 mmHg was applied to the ventricular side of the TV leaflets, and dynamic motions of the TV annulus and papillary muscles as obtained from the medical imaging data were prescribed. The results of the FE simulations varied between the three patients' geometries in comparison to previous simulation studies. The average stress in each TV leaflet ranged from 24–91 kPa, which relatively agrees with the values from Stevanella et al. (2010) [169] but is much lower than the values found in Aversa and Careddu (2017) [171]. Furthermore, the average TV leaflet's strains (0.12–0.32) we were much lower than previous studies (Stevanella et al. ~0.52; Aversa and Carredu 0.40).

In addition, bio-solid modeling has recently been used by our group to better understand the influence of common TV pathologies associated with FTR on the TV mechanical function [187]. Minor modifications to the TV geometry from our previous study [174] were made to investigate five scenarios: (i) healthy (no modifications), (ii) pulmonary hypertension, (iii) annulus dilation, (iv) papillary muscle displacement resulting from left ventricle enlargement, and (v) chordae rupture. Both the mechanics-based quantities, such as the von Mises stress (Figure 17) or Green strain, and the clinically relevant metrics, including the tenting height, tenting area, and the coaptation height, were compared among those study scenarios. Systematic quantifications of these mechanical and/or geometrical changes for each scenario are expected to enhance the current understanding of FTR, especially for situations that may not be possibly replicated under clinical settings.

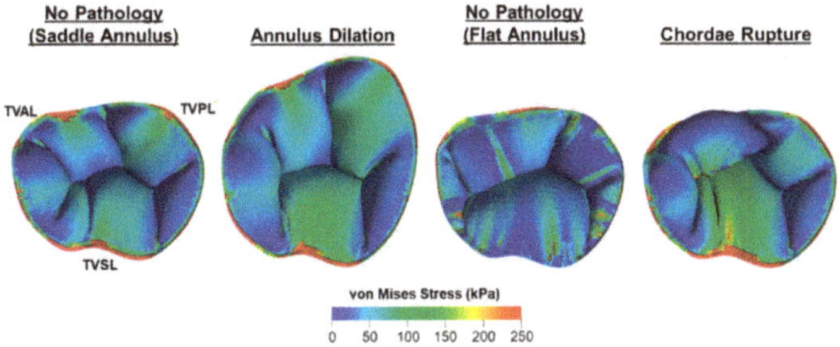

Figure 17. Representative FE results adopted from an ongoing study by Laurence et al. (2019) [187].

4.4.2. FSI Model of the TV

Singh-Gryzbon et al. (2019) [186] has recently extended computational modeling of the TV to the consideration of bio-solid/fluid interactions under an FSI framework that would allow evaluations of regurgitant states and other pathological conditions associated with FTR. Specifically, for the bio-solid subproblem, they reconstructed the TV geometry from segmentation of the micro-CT imaging data of an excised, glutaraldehyde-fixed porcine TV at the unloaded (diastolic) and pressure-loaded (systolic) states. The corresponding computational mesh, composed of tetrahedral finite elements ($n = 1.20 \times 10^5$) with a characteristic length of 0.01–0.02 mm, was then created. Constitutive model forms and parameters previously described by Khoiy et al. (2018) [108] and Toma et al. (2016) [188] were adopted to describe the material properties of the TV leaflets and chordae, respectively. A smooth particle hydrodynamics (SPH) approach [189–192] rather than the typical FE method under the immersed boundary [193–195] or arbitrary Lagrangian-Eulerian [161,196] formulations was employed for modeling the fluid subproblem. This choice was primarily due to the ability of SPH to more accurately capture the separation of flow throughout valve closure and the relatively quick simulation completion time compared to the typical mesh-based methods. The interaction between the bio-solid and fluid subproblems was enforced through a classical penalty-based contact algorithm for the contact between SPH particles and the TV mesh. The boundary conditions considered in their FSI simulations included fixed boundaries for the TV annulus and the base of papillary muscles and a prescribed physiological velocity waveform to the inlet of the fluid domain. Simulations were performed for the healthy scenario with no regurgitation and the regurgitant scenario in which the papillary muscles were displaced to induce FTR.

When comparing the simulation results from the healthy and regurgitant scenarios, there was an increase in the average leaflet belly stress from 12.45 ± 6.25 kPa to 13.38 ± 7.96 kPa and an increase in the corresponding average principle strain from 0.169 ± 0.099 to 0.1848 ± 0.113. The values presented in this study were markedly lower than previous computational studies that only modeled the bio-solid component of the TV, with the closest comparison being to the study by Kong et al. (2018) [172] (Table 1). Nevertheless, the developed FSI model from this study is a critical first step in moving toward holistic modeling of the dynamic TV biomechanical function.

Table 1. Nomenclature for all the abbreviations adopted in this review paper.

	Abbreviation	Description
Anatomy	AP	Antero-posterior
	AV	Aortic valve
	MV	Mitral valve
	PV	Pulmonary valve
	RA	Right atrium
	RV	Right ventricle
	SL	Septo-lateral
	TA	Tricuspid valve annulus
	TV	Tricuspid valve
	TVAL	Tricuspid valve anterior leaflet
	TVPL	Tricuspid valve posterior leaflet
	TVSL	Tricuspid valve septal leaflet
	VC	Vena contracta
Computational Modeling	CAD	Computer-aided design
	FE	Finite element
	FSI	Fluid-structure interaction
	IGA	Isogeometric analysis
	SPH	Smooth particle hydrodynamics
Disease/Pathology	FTR	Functional tricuspid regurgitation
	HLHS	Hypoplastic left heart syndrome
	ToF	Tetralogy of Fallot
	TR	Tricuspid regurgitation
Imaging and Grading of the TR Severity	2DE	Two-dimensional echocardiography
	3DE	Three-dimensional echocardiography
	A4C	Apical four-chamber view
	CT	Computed tomography
	CMRI	Cardiac magnetic resonance imaging
	EROA	Effective regurgitant orifice area
	ME	Mid-esophageal
	PISA	Proximal isovelocity surface area
	PLAX	Parasternal long axis
	PSAX	Parasternal short axis
	RT3DE	Real-time three-dimensional echocardiography
	RVEIO	Right ventricular early inflow-outflow
	RVF	Right ventricular-focused
	RVIF	Right ventricular inflow
	R Vol	Regurgitant jet volume
	TEE	Transesophageal echocardiography
	TTE	Transthoracic echocardiography
Mechanics	$\mathbf{C}$	Right Cauchy-Green tensor
	E_{CC}	Green strain in the tissue's circumferential direction
	E_{RR}	Green strain in the tissue's radial direction
	F_x	Force in the x-direction
	F_y	Force in the y-direction
	I_1	First invariant of the right Cauchy-Green tensor $\mathbf{C}$
	I_4	Fourth invariant of the right Cauchy-Green tensor $\mathbf{C}$
	λ	Stretch ratio
	T_{circ}	Membrane tension in the circumferential direction
	T_{rad}	Membrane tension in the radial direction

Table 1. *Cont.*

	Abbreviation	Description
Microstructure	A	Atrialis layer
	ECM	Extracellular matrix
	F	Fibrosa layer
	GAGs	Glycosaminoglycans
	PGs	Proteoglycans
	S	Spongiosa layer
	SMC	Smooth muscle cell
	V	Ventricularis layer
	VIC	Valvular interstitial cell
Specimen Labels	-C	Control specimen
	-T	Treated specimen
	A/S	Atrialis/spongiosa layer
	F/V	Fibrosa/ventricularis layer
	LG	Left chordae group
	RG	Right chordae group

5. Closing Remarks and Future Perspectives

This review paper has summarized recent advances within the scope of assessing or understanding FTR regarding TV biomechanics. First, current image modalities were discussed as well as recent clinical studies that have used the various modalities to gain insight into the development and progression of FTR. Typical therapeutic options for treating patients with FTR were also discussed. Then, recent in-vitro and in-vivo studies involving the TV were surveyed. These studies were primarily focused on understanding the kinematics and/or mechanical behaviors of components of the TV apparatus, including the TV leaflets, annulus, or chordae tendineae. Lastly, several recent TV computational simulation studies were reviewed with key aspects highlighted, such as the constitutive modeling and geometrical modeling. While the existing computational models are broad in the TV biomechanical modeling applications, the stark difference in the number and complexity of computational models for the TV, as compared to the MV/AV counterparts, underscores the emerging need for advances in personalized TV modeling. Such advances would open new avenues for predictive computer simulations to be directly used in clinical settings.

When understanding the initiation and progression of FTR through in-silico methods, one can look at TV bio-solid modeling as our group has done previously. However, certain aspects of FTR, such as right ventricular enlargement, require interactions between the TV with the RV to completely comprehend the underlying mechanisms. Thus, one future extension within the context of understanding disease initiation and progression is to develop a coupled TV-RV model for fully capturing such essential complex interactions. If developed properly within an FSI modeling framework, a deeper connection between the increase in transvalvular pressure (i.e., pulmonary hypertension) and the subsequent enlargement of the RV or other aspects of FTR could be established. Moreover, this coupled model could be extended to consider both the LVs and RVs with the MV and TVs to explore congenital heart diseases. Some applicational examples include the HLHS where the left ventricle is typically underdeveloped [173,197–199], or the Tetralogy of Fallot (ToF) with the four distinct defects affecting flow through the LV and RV [200,201].

Computational models such as those outlined in Section 4.4 or the coupled TV-RV model could also be tailored to clinical applications to provide recommendations concerning proper timing or methodology of clinical intervention. Such clinical extensions would require specific advancements to ensure the timeliness and accuracy of diagnosis or recommendation. For example, a framework would need to be developed that incorporates automated (or semi-automated) segmentation of the patient's imaged heart geometry, uses this geometry in an inverse bio-solid modeling scheme to quantify the mechanical properties of the TV leaflets, and finally performs in-silico simulations to determine a suggested timing for clinical intervention. These analyses could involve connections made through in-vivo or in-vitro studies, as outlined in Section 3, or through an incorporation of a machine learning model [202,203] trained to recognize key features correlated with necessary or impending clinical intervention. Aspects of this proposed framework could be first tested using an animal model with direct validations using methods such as those described in Section 3 for comparing the mechanical properties of sub-valvular components of the TV.

Certain perspectives must be enhanced or developed for the above-mentioned clinically applicable extensions. First, non-invasive methods for quantifying TV leaflet properties are necessary to capture patient-specific TV mechanical properties. Secondly, one must be able to determine patient-specific boundary conditions derived from clinical imaging data. Lastly, current computational models must be enhanced to realistically model complex FSIs and other bio-solid components of the heart, such as the RV. While these extensions are non-trivial and may take time to properly develop and validate, their collective impacts to the clinical setting will be profound.

Author Contributions: Writing—original draft, D.W.L., C.J.R., K.E.K., E.L.J., C.-H.L., and A.R.B.; Writing—review and editing, A.A., A.M., H.M.B., R.A.T., R.B., M.-C.H., Y.W., and C.-H.L.; Funding acquisition, C.-H.L.; Project administration, C.-H.L.

Funding: Support from the American Heart Association Scientist Development Grant (SDG) Award (16SDG27760143) is gratefully acknowledged. CHL was in part supported by the institutional start-up funds from the School of Aerospace and Mechanical Engineering (AME), the IBEST-OUHSC Funding for Interdisciplinary Research, and the research funding through the Faculty Investment Program from the Research Council at the University of Oklahoma (OU).

Conflicts of Interest: The authors of this paper have no financial or personal relationships with other people or organizations that could inappropriately influence (bias) our work.

References

1. Badano, L.P.; Muraru, D.; Enriquez-Sarano, M. Assessment of functional tricuspid regurgitation. *Eur. Heart J.* **2013**, *34*, 1875–1885. [CrossRef]
2. Anwar, A.M.; Folkert, J.; Soliman, O.I. Clinical recognition of tricuspid valve disease. In *Practical Manual of Tricuspid Valve Diseases*; Springer: Berlin, Germany, 2018; pp. 25–48.
3. Nishimura, R.A.; Otto, C.M.; Bonow, R.O.; Carabello, B.A.; Erwin, J.P.; Guyton, R.A.; O'Gara, P.T.; Ruiz, C.E.; Skubas, N.J.; Sorajja, P. 2014 AHA/ACC guideline for the management of patients with valvular heart disease: Executive summary: A report of the American College of Cardiology/American Heart Association Task Force on Practice Guidelines. *J. Am. Coll. Cardiol.* **2014**, *63*, 2438–2488. [CrossRef] [PubMed]
4. Sun, Y.-P.; O'Gara, P.T. Epidemiology, anatomy, pathophysiology and clinical evaluation of functional tricuspid regurgitation. *Minerva Cardioangiol.* **2017**, *65*, 469–479. [PubMed]
5. Badano, L.P.; Ginghina, C.; Easaw, J.; Muraru, D.; Grillo, M.T.; Lancellotti, P.; Pinamonti, B.; Coghlan, G.; Marra, M.P.; Popescu, B.A. Right ventricle in pulmonary arterial hypertension: Haemodynamics, structural changes, imaging, and proposal of a study protocol aimed to assess remodelling and treatment effects. *Eur. J. Echocardiogr.* **2009**, *11*, 27–37. [CrossRef]
6. Spinner, E.M.; Lerakis, S.; Higginson, J.; Pernetz, M.; Howell, S.; Veledar, E.; Yoganathan, A.P. Correlates of tricuspid regurgitation as determined by 3D echocardiography: Pulmonary arterial pressure, ventricle geometry, annular dilatation, and papillary muscle displacement. *Circ. Cardiovasc. Imaging* **2012**, *5*, 43–50. [CrossRef]
7. Hinderliter, A.L.; Willis, P.W.; Long, W.A.; Clarke, W.R.; Ralph, D.; Caldwell, E.J.; Williams, W.; Ettinger, N.A.; Hill, N.S.; Summer, W.R. Frequency and severity of tricuspid regurgitation determined by Doppler echocardiography in primary pulmonary hypertension. *Am. J. Cardiol.* **2003**, *91*, 1033–1037. [CrossRef]
8. Come, P.C.; Riley, M.F. Tricuspid anular dilatation and failure of tricuspid leaflet coaptation in tricuspid regurgitation. *Am. J. Cardiol.* **1985**, *55*, 599–601. [CrossRef]
9. Ton-Nu, T.-T.; Levine, R.A.; Handschumacher, M.D.; Dorer, D.J.; Yosefy, C.; Fan, D.; Hua, L.; Jiang, L.; Hung, J. Geometric determinants of functional tricuspid regurgitation: Insights from 3-dimensional echocardiography. *Circulation* **2006**, *114*, 143–149. [CrossRef]

10. 1Song, J.-M.; Jang, M.-K.; Kim, Y.-J.; Kim, D.-H.; Kang, D.-H.; Song, J.-K. Right ventricular remodeling determines tricuspid valve geometry and the severity of functional tricuspid regurgitation: A real-time 3-dimensional echocardiography study. *Korean Circ. J.* **2010**, *40*, 448–453.
11. Stuge, O.; Liddicoat, J. Emerging opportunities for cardiac surgeons within structural heart disease. *J. Thorac. Cardiovasc. Surg.* **2006**, *132*, 1258–1261. [CrossRef]
12. Taramasso, M.; Pozzoli, A.; Guidotti, A.; Nietlispach, F.; Inderbitzin, D.T.; Benussi, S.; Alfieri, O.; Maisano, F. Percutaneous tricuspid valve therapies: The new frontier. *Eur. Heart J.* **2016**, *38*, 639–647. [CrossRef]
13. Braunwald, N.S.; Ross, J., Jr.; Morrow, A.G. Conservative management of tricuspid regurgitation in patients undergoing mitral valve replacement. *Circulation* **1967**, *35*, I-63–I-69. [CrossRef]
14. Dreyfus, G.D.; Corbi, P.J.; Chan, K.M.J.; Bahrami, T. Secondary tricuspid regurgitation or dilatation: Which should be the criteria for surgical repair? *Ann. Thorac. Surg.* **2005**, *79*, 127–132. [CrossRef]
15. Anyanwu, A.C.; Adams, D.H. Functional tricuspid regurgitation in mitral valve disease: Epidemiology and prognostic implications. *Semin. Thoracic Cardiovasc. Surg.* **2010**, *22*, 69–75. [CrossRef]
16. Navia, J.L.; Nowicki, E.R.; Blackstone, E.H.; Brozzi, N.A.; Nento, D.E.; Atik, F.A.; Rajeswaran, J.; Gillinov, A.M.; Svensson, L.G.; Lytle, B.W. Surgical management of secondary tricuspid valve regurgitation: Annulus, commissure, or leaflet procedure? *J. Thorac. Cardiovasc. Surg.* **2010**, *139*, 1473–1482. [CrossRef] [PubMed]
17. Navia, J.L.; Brozzi, N.A.; Klein, A.L.; Ling, L.F.; Kittayarak, C.; Nowicki, E.R.; Batizy, L.H.; Zhong, J.; Blackstone, E.H. Moderate tricuspid regurgitation with left-sided degenerative heart valve disease: To repair or not to repair? *Ann. Thorac. Surg.* **2012**, *93*, 59–69. [CrossRef]
18. Di Mauro, M.; Bezante, G.P.; Di Baldassarre, A.; Clemente, D.; Cardinali, A.; Acitelli, A.; Salerni, S.; Penco, M.; Calafiore, A.M.; Gallina, S. Functional tricuspid regurgitation: An underestimated issue. *Int. J. Cardiol.* **2013**, *168*, 707–715. [CrossRef]
19. Chan, V.; Burwash, I.G.; Lam, B.-K.; Auyeung, T.; Tran, A.; Mesana, T.G.; Ruel, M. Clinical and echocardiographic impact of functional tricuspid regurgitation repair at the time of mitral valve replacement. *Ann. Thorac. Surg.* **2009**, *88*, 1209–1215. [CrossRef]
20. Calafiore, A.M.; Gallina, S.; Iacò, A.L.; Contini, M.; Bivona, A.; Gagliardi, M.; Bosco, P.; Di Mauro, M. Mitral valve surgery for functional mitral regurgitation: Should moderate-or-more tricuspid regurgitation be treated? A propensity score analysis. *Ann. Thorac. Surg.* **2009**, *87*, 698–703. [CrossRef] [PubMed]
21. Rogers, J.H.; Bolling, S.F. The tricuspid valve: Current perspective and evolving management of tricuspid regurgitation. *Circulation* **2009**, *119*, 2718–2725. [CrossRef]
22. Antunes, M.J.; Barlow, J.B. Management of tricuspid valve regurgitation. *Heart* **2007**, *93*, 271–276. [CrossRef]
23. Fukuda, S.; Song, J.-M.; Gillinov, A.M.; McCarthy, P.M.; Daimon, M.; Kongsaerepong, V.; Thomas, J.D.; Shiota, T. Tricuspid valve tethering predicts residual tricuspid regurgitation after tricuspid annuloplasty. *Circulation* **2005**, *111*, 975–979. [CrossRef]
24. Matsuyama, K.; Matsumoto, M.; Sugita, T.; Nishizawa, J.; Tokuda, Y.; Matsuo, T. Predictors of residual tricuspid regurgitation after mitral valve surgery. *Ann. Thorac. Surg.* **2003**, *75*, 1826–1828. [CrossRef]
25. Tang, G.H.L.; David, T.E.; Singh, S.K.; Maganti, M.D.; Armstrong, S.; Borger, M.A. Tricuspid valve repair with an annuloplasty ring results in improved long-term outcomes. *Circulation* **2006**, *114*, I-577–I-581. [CrossRef]
26. Kim, Y.-J.; Kwon, D.-A.; Kim, H.-K.; Park, J.-S.; Hahn, S.; Kim, K.-H.; Kim, K.-B.; Sohn, D.-W.; Ahn, H.; Oh, B.-H. Determinants of surgical outcome in patients with isolated tricuspid regurgitation. *Circulation* **2009**, *120*, 1672–1678. [CrossRef] [PubMed]
27. Onoda, K.; Yasuda, F.; Takao, M.; Shimono, T.; Tanaka, K.; Shimpo, H.; Yada, I. Long-term follow-up after Carpentier-Edwards ring annuloplasty for tricuspid regurgitation. *Ann. Thorac. Surg.* **2000**, *70*, 796–799. [CrossRef]
28. Fukuda, S.; Gillinov, A.M.; McCarthy, P.M.; Matsumura, Y.; Thomas, J.D.; Shiota, T. Echocardiographic follow-up of tricuspid annuloplasty with a new three-dimensional ring in patients with functional tricuspid regurgitation. *J. Am. Soc. Echocardiogr.* **2007**, *20*, 1236–1242. [CrossRef] [PubMed]
29. Kim, J.B.; Jung, S.-H.; Choo, S.J.; Chung, C.H.; Lee, J.W. Clinical and echocardiographic outcomes after surgery for severe isolated tricuspid regurgitation. *J. Thorac. Cardiovasc. Surg.* **2013**, *146*, 278–284. [CrossRef]
30. Fukuda, S.; Gillinov, A.M.; Song, J.-M.; Daimon, M.; Kongsaerepong, V.; Thomas, J.D.; Shiota, T. Echocardiographic insights into atrial and ventricular mechanisms of functional tricuspid regurgitation. *Am. Heart J.* **2006**, *152*, 1208–1214. [CrossRef] [PubMed]

31. Min, S.-Y.; Song, J.-M.; Kim, J.-H.; Jang, M.-K.; Kim, Y.-J.; Song, H.; Kim, D.-H.; Lee, J.W.; Kang, D.-H.; Song, J.-K. Geometric changes after tricuspid annuloplasty and predictors of residual tricuspid regurgitation: A real-time three-dimensional echocardiography study. *Eur. Heart J.* **2010**, *31*, 2871–2880. [CrossRef] [PubMed]
32. Meador, W.D.; Mathur, M.; Rausch, M.K. Tricuspid valve biomechanics: A brief review. In *Advances in Heart Valve Biomechanics;* Springer: Berlin, Germany, 2018; pp. 105–114.
33. Hiro, M.E.; Jouan, J.; Pagel, M.R.; Lansac, E.; Lim, K.H.; Lim, H.-S.; Duran, C.M.G. Sonometric study of the normal tricuspid valve annulus in sheep. *J. Heart. Valve Dis.* **2004**, *13*, 452–460.
34. Deloche, A.; Guérinon, J.; Fabiani, J.; Morillo, F.; Caramanian, M.; Carpentier, A.; Maurice, P.; Dubost, C. Anatomical study of rheumatic tricuspid valvulopathies. Applications to the critical study of various methods of annuloplasty. *Arch. Mal. Coeur Vaiss.* **1974**, *67*, 497.
35. Chandra, S.; Powell, K.; Breburda, C.; Mikic, I.; Shekhar, R.; Morehead, A.; Cosgrove, D.; Thomas, J. Three dimensional reconstruction (shape and motion) of tricuspid annulus in normals and in patients after tricuspid annuloplasty with a flexible ring. *Comput. Cardiol.* **1996**, 693–696. [CrossRef]
36. Jouan, J.; Pagel, M.R.; Hiro, M.E.; Lim, K.H.; Lansac, E.; Duran, C.M. Further information from a sonometric study of the normal tricuspid valve annulus in sheep: geometric changes during the cardiac cycle. *J. Heart. Valve Dis.* **2007**, *16*, 511.
37. Jensen, M.O.; Jensen, H.; Levine, R.A.; Yoganathan, A.P.; Andersen, N.T.; Nygaard, H.; Hasenkam, J.M.; Nielsen, S.L. Saddle-shaped mitral valve annuloplasty rings improve leaflet coaptation geometry. *J. Thorac. Cardiovasc. Surg.* **2011**, *142*, 697–703. [CrossRef]
38. Salgo, I.S.; Gorman, J.H., III; Gorman, R.C.; Jackson, B.M.; Bowen, F.W.; Plappert, T.; St John Sutton, M.G.; Edmunds, L.H., Jr. Effect of annular shape on leaflet curvature in reducing mitral leaflet stress. *Circulation* **2002**, *106*, 711–717. [CrossRef]
39. Dwivedi, G.; Mahadevan, G.; Jimenez, D.; Frenneaux, M.; Steeds, R.P. Reference values for mitral and tricuspid annular dimensions using two-dimensional echocardiography. *Echo Res. Pr.* **2014**, *1*, 43–50. [CrossRef]
40. Spinner, E.M.; Buice, D.; Yap, C.H.; Yoganathan, A.P. The effects of a three-dimensional, saddle-shaped annulus on anterior and posterior leaflet stretch and regurgitation of the tricuspid valve. *Ann. Biomed. Eng.* **2012**, *40*, 996–1005. [CrossRef]
41. Angelini, A.; Ho, S.; Anderson, R.H.; Davies, M.J.; Becker, A.E. A histological study of the atrioventricular junction in hearts with normal and prolapsed leaflets of the mitral valve. *Heart* **1988**, *59*, 712–716. [CrossRef]
42. Keith, A.; Flack, M. The form and nature of the muscular connections between the primary divisions of the vertebrate heart. *J. Anat. Physiol.* **1907**, *41*, 172.
43. Racker, D.K.; Ursell, P.C.; Hoffman, B.F. Anatomy of the tricuspid annulus. Circumferential myofibers as the structural basis for atrial flutter in a canine model. *Circulation* **1991**, *84*, 841–851. [CrossRef] [PubMed]
44. Dudziak, M.; Skwarek, M.; Hreczecha, J.; Jerzemowski, J.; Grzybiak, M. Microscopic study of right fibrous annulus. *Folia Morphol. (Praha)* **2009**, *68*, 32–35.
45. Silver, M.D.; Lam, J.H.C.; Ranganathan, N.; Wigle, E.D. Morphology of the human tricuspid valve. *Circulation* **1971**, *43*, 333–348. [CrossRef]
46. Jett, S.; Laurence, D.; Kunkel, R.; Babu, A.R.; Kramer, K.; Baumwart, R.; Towner, R.; Wu, Y.; Lee, C.-H. An investigation of the anisotropic mechanical properties and anatomical structure of porcine atrioventricular heart valves. *J. Mech. Behav. Biomed. Mater.* **2018**, *87*, 155–171. [CrossRef]
47. Gross, L.; Kugel, M.A. Topographic anatomy and histology of the valves in the human heart. *Am. J. Pathol.* **1931**, *7*, 445. [PubMed]
48. Sacks, M.S.; Yoganathan, A.P. Heart valve function: A biomechanical perspective. *Philos. Trans. R. Soc. Lond. B Biol. Sci.* **2007**, *362*, 1369–1391. [CrossRef] [PubMed]
49. Eckert, C.E.; Fan, R.; Mikulis, B.; Barron, M.; Carruthers, C.A.; Friebe, V.M.; Vyavahare, N.R.; Sacks, M.S. On the biomechanical role of glycosaminoglycans in the aortic heart valve leaflet. *Acta Biomater.* **2013**, *9*, 4653–4660. [CrossRef] [PubMed]
50. Lovekamp, J.J.; Simionescu, D.T.; Mercuri, J.J.; Zubiate, B.; Sacks, M.S.; Vyavahare, N.R. Stability and function of glycosaminoglycans in porcine bioprosthetic heart valves. *Biomaterials* **2006**, *27*, 1507–1518. [CrossRef] [PubMed]

51. Marron, K.; Yacoub, M.H.; Polak, J.M.; Sheppard, M.N.; Fagan, D.; Whitehead, B.F.; de Leval, M.R.; Anderson, R.H.; Wharton, J. Innervation of human atrioventricular and arterial valves. *Circulation* **1996**, *94*, 368–375. [CrossRef]
52. Merryman, W.D.; Youn, I.; Lukoff, H.D.; Krueger, P.M.; Guilak, F.; Hopkins, R.A.; Sacks, M.S. Correlation between heart valve interstitial cell stiffness and transvalvular pressure: Implications for collagen biosynthesis. *Am. J. Physiol. Heart Circ. Physiol.* **2006**, *290*, H224–H231. [CrossRef]
53. Messier, R.H., Jr.; Bass, B.L.; Aly, H.M.; Jones, J.L.; Domkowski, P.W.; Wallace, R.B.; Hopkins, R.A. Dual structural and functional phenotypes of the porcine aortic valve interstitial population: Characteristics of the leaflet myofibroblast. *J. Surg. Res.* **1994**, *57*, 1–21. [CrossRef]
54. Lam, J.H.C.; Ranganathan, N.; Wigle, E.D.; Silver, M.D. Morphology of the human mitral valve: I. Chordae tendineae: A new classification. *Circulation* **1970**, *41*, 449–458. [CrossRef]
55. Millington-Sanders, C.; Meir, A.; Lawrence, L.; Stolinski, C. Structure of chordae tendineae in the left ventricle of the human heart. *J. Anat.* **1998**, *192*, 573–581. [CrossRef]
56. Casado, J.A.; Diego, S.; Ferreño, D.; Ruiz, E.; Carrascal, I.; Méndez, D.; Revuelta, J.M.; Pontón, A.; Icardo, J.M.; Gutiérrez-Solana, F. Determination of the mechanical properties of normal and calcified human mitral chordae tendineae. *J. Mech. Behav. Biomed. Mater.* **2012**, *13*, 1–13. [CrossRef]
57. Kunzelman, K.S.; Cochran, K.P. Mechanical properties of basal and marginal mitral valve chordae tendineae. *ASAIO J.* **1990**, *36*, M405–M407.
58. 5Humphries, J.A.; Kramer, C.J.; Sengupta, P.P.; Khandheria, B.K. Transesophageal echocardiography. In *Case Based Echocardiography;* Springer: Berlin, Germany, 2010; pp. 85–101.
59. Addetia, K.; Yamat, M.; Mediratta, A.; Medvedofsky, D.; Patel, M.; Ferrara, P.; Mor-Avi, V.; Lang, R.M. Comprehensive two-dimensional interrogation of the tricuspid valve using knowledge derived from three-dimensional echocardiography. *J. Am. Soc. Echocardiogr.* **2016**, *29*, 74–82. [CrossRef]
60. Anwar, A.M.; Geleijnse, M.L.; Soliman, O.I.; McGhie, J.S.; Frowijn, R.; Nemes, A.; van den Bosch, A.E.; Galema, T.W.; Folkert, J. Assessment of normal tricuspid valve anatomy in adults by real-time three-dimensional echocardiography. *Int. J. Cardiovasc. Imaging* **2007**, *23*, 717–724. [CrossRef]
61. Badano, L.P.; Agricola, E.; de Isla, L.P.; Gianfagna, P.; Zamorano, J.L. Evaluation of the tricuspid valve morphology and function by transthoracic real-time three-dimensional echocardiography. *Eur. J. Echocardiogr.* **2009**, *10*, 477–484. [CrossRef]
62. Anwar, A.M.; Geleijnse, M.L.; ten Cate, F.J.; Meijboom, F.J. Assessment of tricuspid valve annulus size, shape and function using real-time three-dimensional echocardiography. *Interact. Cardiovasc. Thorac. Surg.* **2006**, *5*, 683–687. [CrossRef]
63. Kwan, J.; Kim, G.-C.; Jeon, M.-J.; Kim, D.-H.; Shiota, T.; Thomas, J.D.; Park, K.-S.; Lee, W.-H. 3D geometry of a normal tricuspid annulus during systole: A comparison study with the mitral annulus using real-time 3D echocardiography. *Eur. J. Echocardiogr.* **2007**, *8*, 375–383. [CrossRef]
64. Vegas, A. Three dimensional transoesophageal echocardiography: Principles and clinical applications. *Ann. Card. Anaesth.* **2016**, *19*, S35. [CrossRef] [PubMed]
65. Rivera, R.; Duran, E.; Ajuria, M. Carpentier's flexible ring versus De Vega's annuloplasty. A prospective randomized study. *J. Thorac. Cardiovasc. Surg.* **1985**, *89*, 196–203.
66. Pennell, D.J.; Sechtem, U.P.; Higgins, C.B.; Manning, W.J.; Pohost, G.M.; Rademakers, F.E.; van Rossum, A.C.; Shaw, L.J.; Kent Yucel, E. Clinical indications for cardiovascular magnetic resonance (CMR): Consensus Panel report. *J. Cardiovasc. Magn. Reson.* **2004**, *6*, 727–765. [CrossRef]
67. Van Praet, K.M.; Stamm, C.; Starck, C.T.; Sündermann, S.; Meyer, A.; Montagner, M.; Nazari Shafti, T.Z.; Unbehaun, A.; Jacobs, S.; Falk, V. An overview of surgical treatment modalities and emerging transcatheter interventions in the management of tricuspid valve regurgitation. *Expert Rev. Cardiovasc. Ther.* **2018**, *16*, 75–89. [CrossRef]
68. Huttin, O.; Voilliot, D.; Mandry, D.; Venner, C.; Juillière, Y.; Selton-Suty, C. All you need to know about the tricuspid valve: Tricuspid valve imaging and tricuspid regurgitation analysis. *Arch. Cardiovasc. Dis.* **2016**, *109*, 67–80. [CrossRef]
69. Kabasawa, M.; Kohno, H.; Ishizaka, T.; Ishida, K.; Funabashi, N.; Kataoka, A.; Matsumiya, G. Assessment of functional tricuspid regurgitation using 320-detector-row multislice computed tomography: Risk factor analysis for recurrent regurgitation after tricuspid annuloplasty. *J. Thorac. Cardiovasc. Surg.* **2014**, *147*, 312–320. [CrossRef]

70. Nemoto, N.; Lesser, J.R.; Pedersen, W.R.; Sorajja, P.; Spinner, E.; Garberich, R.F.; Vock, D.M.; Schwartz, R.S. Pathogenic structural heart changes in early tricuspid regurgitation. *J. Thorac. Cardiovasc. Surg.* **2015**, *150*, 323–330. [CrossRef] [PubMed]
71. Gopalan, D. Right heart on multidetector CT. *Br. J. Radiol.* **2011**, *84*, S306–S323. [CrossRef]
72. Go, Y.Y.; Dulgheru, R.; Lancellotti, P. The conundrum of tricuspid regurgitation grading. *Front. Cardiovasc. Med.* **2018**, *5*, 164. [CrossRef] [PubMed]
73. Lancellotti, P.; Moura, L.; Pierard, L.A.; Agricola, E.; Popescu, B.A.; Tribouilloy, C.; Hagendorff, A.; Monin, J.-L.; Badano, L.; Zamorano, J.L. European Association of Echocardiography recommendations for the assessment of valvular regurgitation. Part 2: Mitral and tricuspid regurgitation (native valve disease). *Eur. J. Echocardiogr.* **2010**, *11*, 307–332. [CrossRef]
74. Tribouilloy, C.M.; Enriquez-Sarano, M.; Bailey, K.R.; Tajik, A.J.; Seward, J.B. Quantification of tricuspid regurgitation by measuring the width of the vena contracta with Doppler color flow imaging: A clinical study. *J. Am. Coll. Cardiol.* **2000**, *36*, 472–478. [CrossRef]
75. Lambert, A.S. Proximal isovelocity surface area should be routinely measured in evaluating mitral regurgitation: A core review. *Anesth. Analg.* **2007**, *105*, 940–943. [CrossRef] [PubMed]
76. Hahn, R.T.; Zamorano, J.L. The need for a new tricuspid regurgitation grading scheme. *Eur. Heart J. Cardiovasc. Imaging* **2017**, *18*, 1342–1343. [CrossRef] [PubMed]
77. Izgi, I.A.; Acar, E.; Kilicgedik, A.; Guler, A.; Cakmak, E.O.; Demirel, M.; Izci, S.; Yilmaz, M.F.; Inanir, M.; Kirma, C. A new and simple method for clarifying the severity of tricuspid regurgitation. *Echocardiography* **2017**, *34*, 328–333. [CrossRef]
78. Chikwe, J.; Anyanwu, A.C. Surgical strategies for functional tricuspid regurgitation. *Semin. Thorac. Cardiovasc. Surg.* **2010**, *22*, 90–96. [CrossRef] [PubMed]
79. Pinney, S.P. The role of tricuspid valve repair and replacement in right heart failure. *Curr. Opin. Cardiol.* **2012**, *27*, 288–295. [CrossRef]
80. Chang, H.W.; Jeong, D.S.; Cho, Y.H.; Sung, K.; Kim, W.S.; Lee, Y.T.; Park, P.W. Tricuspid valve replacement vs. repair in severe tricuspid regurgitation. *Circ. J.* **2017**, *81*, 330–338. [CrossRef]
81. Kay, J.H.; Maselli-Campagna, G.; Tsuji, H.K. Surgical treatment of tricuspid insufficiency. *Ann. Surg.* **1965**, *162*, 53–58. [CrossRef]
82. De Vega, N.; De Rabago, G.; Castillon, L.; Moreno, T.; Azpitarte, J. A new tricuspid repair. Short-term clinical results in 23 cases. *J. Cardiovasc. Surg. (Torino)* **1973**, *14*, 384–386.
83. Antunes, M.J.; Girdwood, R.W. Tricuspid annuloplasty: A modified technique. *Ann. Thorac. Surg.* **1983**, *35*, 676–678. [CrossRef]
84. Raja, S.G.; Dreyfus, G.D. Basis for intervention on functional tricuspid regurgitation. *Semin. Thorac. Cardiovasc. Surg.* **2010**, *22*, 79–83. [CrossRef] [PubMed]
85. Carpentier, A. A new reconstructive operation for correction of mitral and tricuspid insufficiency. *J. Thorac. Cardiovasc. Surg.* **1971**, *61*, 1–13. [PubMed]
86. Pfannmüller, B.; Doenst, T.; Eberhardt, K.; Seeburger, J.; Borger, M.A.; Mohr, F.W. Increased risk of dehiscence after tricuspid valve repair with rigid annuloplasty rings. *J. Thorac. Cardiovasc. Surg.* **2012**, *143*, 10501055. [CrossRef]
87. Huang, X.; Gu, C.; Men, X.; Zhang, J.; You, B.; Zhang, H.; Wei, H.; Li, J. Repair of functional tricuspid regurgitation: Comparison between suture annuloplasty and rings annuloplasty. *Ann. Thorac. Surg.* **2014**, *97*, 1286–1292. [CrossRef] [PubMed]
88. Filsoufi, F.; Salzberg, S.P.; Coutu, M.; Adams, D.H. A three-dimensional ring annuloplasty for the treatment of tricuspid regurgitation. *Ann. Thorac. Surg.* **2006**, *81*, 2273–2277. [CrossRef] [PubMed]
89. Mas, P.T.; Rodríguez-Palomares, J.F.; Antunes, M.J. Secondary tricuspid valve regurgitation: A forgotten entity. *Heart* **2015**, *101*, 1840–1848.
90. Bernal, J.M.; Gutiérrez-Morlote, J.; Llorca, J.; San José, J.M.; Morales, D.; Revuelta, J.M. Tricuspid valve repair: An old disease, a modern experience. *Ann. Thorac. Surg.* **2004**, *78*, 2069–2074. [CrossRef] [PubMed]
91. Charfeddine, S.; Hammami, R.; Triki, F.; Abid, L.; Hentati, M.; Frikha, I.; Kammoun, S. Plastic repair of tricuspid valve: Carpentier's ring annuloplasty versus De VEGA technique. *Pan Afr. Med. J.* **2017**, *27*, 119. [CrossRef] [PubMed]
92. Dreyfus, G.D.; Raja, S.G.; John Chan, K.M. Tricuspid leaflet augmentation to address severe tethering in functional tricuspid regurgitation. *Eur. J. Cardiothorac. Surg.* **2008**, *34*, 908–910. [CrossRef]

93. Lapenna, E.; De Bonis, M.; Verzini, A.; La Canna, G.; Ferrara, D.; Calabrese, M.C.; Taramasso, M.; Alfieri, O. The clover technique for the treatment of complex tricuspid valve insufficiency: Midterm clinical and echocardiographic results in 66 patients. *Eur. J. Cardiothorac. Surg.* **2010**, *37*, 1297–1303. [CrossRef]
94. Belluschi, I.; Del Forno, B.; Lapenna, E.; Nisi, T.; Iaci, G.; Ferrara, D.; Castiglioni, A.; Alfieri, O.; De Bonis, M. Surgical techniques for tricuspid valve disease. *Front. Cardiovasc. Med.* **2018**, *5*, 118. [CrossRef]
95. Bloomfield, P. Choice of heart valve prosthesis. *Heart* **2002**, *87*, 583–589. [CrossRef]
96. Hammarström, E.; Wranne, B.; Pinto, F.J.; Puryear, J.; Popp, R.L. Tricuspid annular motion. *J. Am. Soc. Echocardiogr.* **1991**, *4*, 131–139. [CrossRef]
97. Ring, L.; Rana, B.S.; Kydd, A.; Boyd, J.; Parker, K.; Rusk, R.A. Dynamics of the tricuspid valve annulus in normal and dilated right hearts: a three-dimensional transoesophageal echocardiography study. *Eur. Heart J. Cardiovasc. Imaging* **2012**, *13*, 756–762. [CrossRef]
98. Malinowski, M.; Wilton, P.; Khaghani, A.; Langholz, D.; Hooker, V.; Eberhart, L.; Hooker, R.L.; Timek, T.A. The effect of pulmonary hypertension on ovine tricuspid annular dynamics. *Eur. J. Cardiothorac. Surg.* **2015**, *49*, 40–45. [CrossRef]
99. Malinowski, M.; Schubert, H.; Wodarek, J.; Ferguson, H.; Eberhart, L.; Langholz, D.; Jazwiec, T.; Rausch, M.K.; Timek, T.A. Tricuspid annular geometry and strain after suture annuloplasty in acute ovine right heart failure. *Ann. Thorac. Surg.* **2018**, *106*, 1804–1811. [CrossRef]
100. Rausch, M.K.; Malinowski, M.; Wilton, P.; Khaghani, A.; Timek, T.A. Engineering analysis of tricuspid annular dynamics in the beating ovine heart. *Ann. Biomed. Eng.* **2018**, *46*, 443–451. [CrossRef]
101. Fawzy, H.; Fukamachi, K.; Mazer, C.D.; Harrington, A.; Latter, D.; Bonneau, D.; Errett, L. Complete mapping of the tricuspid valve apparatus using three-dimensional sonomicrometry. *J. Thorac. Cardiovasc. Surg.* **2011**, *141*, 1037–1043. [CrossRef]
102. Mathur, M.; Jazwiec, T.; Meador, W.D.; Malinowski, M.; Goehler, M.; Ferguson, H.; Timek, T.A.; Rausch, M.K. Tricuspid valve leaflet strains in the beating ovine heart. *Biomech. Model. Mechanobiol.* **2019**, 1–11. [CrossRef]
103. Malinowski, M.; Proudfoot, A.G.; Eberhart, L.; Schubert, H.; Wodarek, J.; Langholz, D.; Rausch, M.K.; Timek, T.A. Large animal model of acute right ventricular failure with functional tricuspid regurgitation. *Int. J. Cardiol.* **2018**, *264*, 124–129. [CrossRef]
104. Rausch, M.K.; Malinowski, M.; Meador, W.D.; Wilton, P.; Khaghani, A.; Timek, T.A. The Effect of Acute Pulmonary Hypertension on Tricuspid Annular Height, Strain, and Curvature in Sheep. *Cardiovasc. Eng. Technol.* **2018**, *9*, 365–376. [CrossRef] [PubMed]
105. Malinowski, M.; Jaźwiec, T.; Goehler, M.; Bush, J.; Quay, N.; Ferguson, H.; Rausch, M.K.; Timek, T.A. Impact of tricuspid annular size reduction on right ventricular function, geometry and strain. *Eur. J. Cardiothorac. Surg.* **2019**. [CrossRef]
106. Schechter, M.A.; Southerland, K.W.; Feger, B.J.; Linder, D., Jr.; Ali, A.A.; Njoroge, L.; Milano, C.A.; Bowles, D.E. An isolated working heart system for large animal models. *J. Vis. Exp.* **2014**. [CrossRef] [PubMed]
107. Khoiy, K.A.; Pant, A.D.; Amini, R. Quantification of material constants for a phenomenological constitutive model of porcine tricuspid valve leaflets for simulation applications. *J. Biomech. Eng.* **2018**, *140*, 094503. [CrossRef] [PubMed]
108. Khoiy, K.A.; Biswas, D.; Decker, T.N.; Asgarian, K.T.; Loth, F.; Amini, R. Surface strains of porcine tricuspid valve septal leaflets measured in ex vivo beating hearts. *J. Biomech. Eng.* **2016**, *138*, 111006. [CrossRef]
109. Spinner, E.M.; Shannon, P.; Buice, D.; Jimenez, J.H.; Veledar, E.; del Nido, P.J.; Adams, D.H.; Yoganathan, A.P. In-vitro characterization of the mechanisms responsible for functional tricuspid regurgitation. *Circulation* **2011**, *124*, 920–929. [CrossRef]
110. Malinowski, M.; Jazwiec, T.; Goehler, M.; Quay, N.; Bush, J.; Jovinge, S.; Rausch, M.K.; Timek, T.A. Sonomicrometry-derived 3-dimensional geometry of the human tricuspid annulus. *J. Thorac. Cardiovasc. Surg.* **2019**, *157*, 1452–1461. [CrossRef]
111. Pant, A.D.; Thomas, V.S.; Black, A.L.; Verba, T.; Lesicko, J.G.; Amini, R. Pressure-induced microstructural changes in porcine tricuspid valve leaflets. *Acta Biomater.* **2018**, *67*, 248–258. [CrossRef]
112. Basu, A.; He, Z. Annulus tension on the tricuspid valve: an in-vitro study. *Cardiovasc. Eng. Technol.* **2016**, *7*, 270–279. [CrossRef]
113. Troxler, L.G.; Spinner, E.M.; Yoganathan, A.P. Measurement of strut chordal forces of the tricuspid valve using miniature C ring transducers. *J. Biomech.* **2012**, *45*, 1084–1091. [CrossRef]

114. Arts, T.; Meerbaum, S.; Reneman, R.; Corday, E. Stresses in the closed mitral valve: A model study. *J. Biomech.* **1983**, *16*, 539–547. [CrossRef]
115. Khoiy, K.A.; Amini, R. On the biaxial mechanical response of porcine tricuspid valve leaflets. *J. Biomech. Eng.* **2016**, *138*, 104504–104506. [CrossRef]
116. Aggarwal, A.; Ferrari, G.; Joyce, E.; Daniels, M.J.; Sainger, R.; Gorman, J.H., III; Gorman, R.; Sacks, M.S. Architectural trends in the human normal and bicuspid aortic valve leaflet and its relevance to valve disease. *Ann. Biomed. Eng.* **2014**, *42*, 986–998. [CrossRef]
117. Aggarwal, A.; Sacks, M.S. A framework for determination of heart valves' mechanical properties using inverse-modeling approach. In Proceedings of the International Conference on Functional Imaging and Modeling of the Heart, Maastricht, The Netherlands, 25–27 June 2015; pp. 285–294.
118. Aggarwal, A.; Sacks, M.S. An inverse modeling approach for semilunar heart valve leaflet mechanics: Exploitation of tissue structure. *Biomech. Model. Mechanobiol.* **2016**, *15*, 909–932. [CrossRef]
119. Abbasi, M.; Barakat, M.S.; Dvir, D.; Azadani, A.N. A non-invasive material characterization framework for bioprosthetic heart valves. *Ann. Biomed. Eng.* **2019**, *47*, 97–112. [CrossRef]
120. Jett, S.; Laurence, D.; Kunkel, R.; Babu, A.R.; Kramer, K.; Baumwart, R.; Towner, R.; Wu, Y.; Lee, C.-H. Biaxial mechanical data of porcine atrioventricular valve leaflets. *Data in Brief* **2018**, *21*, 358–363. [CrossRef]
121. Pokutta-Paskaleva, A.; Sulejmani, F.; DelRocini, M.; Sun, W. Comparative mechanical, morphological, and microstructural characterization of porcine mitral and tricuspid leaflets and chordae tendineae. *Acta Biomater.* **2019**, *85*, 241–252. [CrossRef]
122. Ross, C.; Laurence, D.; Wu, Y.; Lee, C.-H. Biaxial mechanical characterizations of atrioventricular heart valves. *J. Vis. Exp.* **2019**, *146*, e59170. [CrossRef]
123. Pham, T.; Sulejmani, F.; Shin, E.; Wang, D.; Sun, W. Quantification and comparison of the mechanical properties of four human cardiac valves. *Acta Biomater.* **2017**, *54*, 345–355. [CrossRef]
124. Laurence, D.; Ross, C.; Jett, S.; Johns, C.; Echols, A.; Baumwart, R.; Towner, R.; Liao, J.; Bajona, P.; Wu, Y.; et al. An investigation of regional variations in the biaxial mechanical properties and stress relaxation behaviors of porcine atrioventricular heart valve leaflets. *J. Biomech.* **2019**, *83*, 16–27. [CrossRef]
125. Kunzelman, K.S.; Cochran, R.P.; Murphree, S.S.; Ring, W.S.; Verrier, E.D.; Eberhart, R.C. Differential collagen distribution in the mitral valve and its influence on biomechanical behaviour. *J. Heart. Valve Dis.* **1993**, *2*, 236–244.
126. Lee, C.-H.; Rabbah, J.-P.; Yoganathan, A.P.; Gorman, R.C.; Gorman, J.H., III; Sacks, M.S. On the effects of leaflet microstructure and constitutive model on the closing behavior of the mitral valve. *Biomech. Model. Mechanobiol.* **2015**, *14*, 1281–1302. [CrossRef]
127. Goth, W.; Potter, S.; Allen, A.C.B.; Zoldan, J.; Sacks, M.S.; Tunnell, J.W. Non-destructive reflectance mapping of collagen fiber alignment in heart valve leaflets. *Ann. Biomed. Eng.* **2019**, *47*, 1250–1264. [CrossRef]
128. Alavi, S.H.; Sinha, A.; Steward, E.; Milliken, J.C.; Kheradvar, A. Load-dependent extracellular matrix organization in atrioventricular heart valves: Differences and similarities. *Am. J. Physiol. Heart Circ. Physiol.* **2015**, *309*, H276–H284. [CrossRef]
129. Basu, A.; Lacerda, C.; He, Z. Mechanical Properties and Composition of the Basal Leaflet-Annulus Region of the Tricuspid Valve. *Cardiovasc. Eng. Technol.* **2018**, *9*, 217–225. [CrossRef]
130. Fu, J.-T.; Popal, M.S.; Jiao, Y.-Q.; Zhang, H.-B.; Zheng, S.; Hu, Q.-M.; Han, W.; Meng, X. A predictor for mitral valve repair in patient with rheumatic heart disease: the bending angle of anterior mitral leaflet. *J. Thorac. Dis.* **2018**, *10*, 2908. [CrossRef]
131. Merryman, W.D.; Engelmayr, G.C., Jr.; Liao, J.; Sacks, M.S. Defining biomechanical endpoints for tissue engineered heart valve leaflets from native leaflet properties. *Prog. Pediatr. Cardiol.* **2006**, *21*, 153–160. [CrossRef]
132. Brazile, B.; Wang, B.; Wang, G.; Bertucci, R.; Prabhu, R.; Patnaik, S.S.; Butler, J.R.; Claude, A.; Brinkman-Ferguson, E.; Williams, L.N. On the bending properties of porcine mitral, tricuspid, aortic, and pulmonary valve leaflets. *J. Long Term Effects Med. Implants* **2015**, *25*, 41–53. [CrossRef]
133. Amoroso, N.J.; D'Amore, A.; Hong, Y.; Rivera, C.P.; Sacks, M.S.; Wagner, W.R. Microstructural manipulation of electrospun scaffolds for specific bending stiffness for heart valve tissue engineering. *Acta Biomater.* **2012**, *8*, 4268–4277. [CrossRef]

134. Amini, R.; Eckert, C.E.; Koomalsingh, K.; McGarvey, J.; Minakawa, M.; Gorman, J.H., III; Gorman, R.C.; Sacks, M.S. On the in-vivo deformation of the mitral valve anterior leaflet: effects of annular geometry and referential configuration. *Ann. Biomed. Eng.* **2012**, *40*, 1455–1467. [CrossRef]
135. Arzani, A.; Mofrad, M.R.K. A strain-based finite element model for calcification progression in aortic valves. *J. Biomech.* **2017**, *65*, 216–220. [CrossRef]
136. Kunzelman, K.S.; Reimink, M.S.; Cochran, R.P. Annular dilatation increases stress in the mitral valve and delays coaptation: A finite element computer model. *Cardiovasc. Surg.* **1997**, *5*, 427–434. [CrossRef]
137. Lee, C.-H.; Carruthers, C.A.; Ayoub, S.; Gorman, R.C.; Gorman, J.H., III; Sacks, M.S. Quantification and simulation of layer-specific mitral valve interstitial cells deformation under physiological loading. *J. Theor. Biol.* **2015**, *373*, 26–39. [CrossRef]
138. Li, J.; Luo, X.Y.; Kuang, Z.B. A nonlinear anisotropic model for porcine aortic heart valves. *J. Biomech.* **2001**, *34*, 1279–1289. [CrossRef]
139. Prot, V.; Skallerud, B. Nonlinear solid finite element analysis of mitral valves with heterogeneous leaflet layers. *Comput. Mech.* **2009**, *43*, 353–368. [CrossRef]
140. Rego, B.V.; Sacks, M.S. A functionally graded material model for the transmural stress distribution of the aortic valve leaflet. *J. Biomech.* **2017**, *54*, 88–95. [CrossRef]
141. Sacks, M.S. Incorporation of experimentally-derived fiber orientation into a structural constitutive model for planar collagenous tissues. *J. Biomech. Eng.* **2003**, *125*, 280–287. [CrossRef]
142. Shah, S.R.; Vyavahare, N.R. The effect of glycosaminoglycan stabilization on tissue buckling in bioprosthetic heart valves. *Biomaterials* **2008**, *29*, 1645–1653. [CrossRef]
143. Rodriguez, K.J.; Piechura, L.M.; Porras, A.M.; Masters, K.S. Manipulation of valve composition to elucidate the role of collagen in aortic valve calcification. *BMC Cardiovasc. Disord.* **2014**, *14*, 29. [CrossRef]
144. Kramer, K.; Ross, C.; Laurence, D.; Babu, A.; Wu, Y.; Towner, R.; Mir, A.; Burkhart, H.M.; Holzapfel, G.A.; Lee, C.-H. An investigation of layer-specific tissue biomechanics of porcine atrioventricular heart valve leaflets. *Acta Biomater.*. Under review.
145. Lim, K.O. Mechanical properties and ultrastructure of normal human tricuspid valve chordae tendineae. *Jpn. J. Physiol.* **1980**, *30*, 455–464. [CrossRef]
146. Lim, K.O.; Boughner, D.R. Mechanical properties of human mitral valve chordae tendineae: Variation with size and strain rate. *Can. J. Physiol. Pharmacol.* **1975**, *53*, 330–339. [CrossRef]
147. Ritchie, J.; Jimenez, J.; He, Z.; Sacks, M.S.; Yoganathan, A.P. The material properties of the native porcine mitral valve chordae tendineae: An in-vitro investigation. *J. Biomech.* **2006**, *39*, 1129–1135. [CrossRef]
148. Roberts, N.; Morticelli, L.; Jin, Z.; Ingham, E.; Korossis, S. Regional biomechanical and histological characterization of the mitral valve apparatus: Implications for mitral repair strategies. *J. Biomech.* **2016**, *49*, 2491–2501. [CrossRef]
149. Sedransk, K.L.; Grande-Allen, K.J.; Vesely, I. Failure mechanics of mitral valve chordae tendineae. *J. Heart. Valve Dis.* **2002**, *11*, 644–650. [PubMed]
150. Wilcox, A.G.; Buchan, K.G.; Espino, D.M. Frequency and diameter dependent viscoelastic properties of mitral valve chordae tendineae. *J. Mech. Behav. Biomed. Mater.* **2014**, *30*, 186–195. [CrossRef] [PubMed]
151. Zuo, K.; Pham, T.; Li, K.; Martin, C.; He, Z.; Sun, W. Characterization of biomechanical properties of aged human and ovine mitral valve chordae tendineae. *J. Mech. Behav. Biomed. Mater.* **2016**, *62*, 607–618. [CrossRef]
152. Kunzelman, K.S.; Cochran, R.P.; Chuong, C.; Ring, W.S.; Verrier, E.D.; Eberhart, R.D. Finite element analysis of the mitral valve. *J. Heart. Valve Dis.* **1993**, *2*, 326–340.
153. Hamid, M.S.; Sabbah, H.N.; Stein, P.D. Vibrational analysis of bioprosthetic heart valve leaflets using numerical models: Effects of leaflet stiffening, calcification, and perforation. *Circul. Res.* **1987**, *61*, 687–694. [CrossRef]
154. Lee, C.-H.; Amini, R.; Gorman, R.C.; Gorman, J.H., 3rd; Sacks, M.S. An inverse modeling approach for stress estimation in mitral valve anterior leaflet valvuloplasty for in-vivo valvular biomaterial assessment. *J. Biomech.* **2014**, *47*, 2055–2063. [CrossRef]
155. Prot, V.; Skallerud, B.; Sommer, G.; Holzapfel, G.A. On modelling and analysis of healthy and pathological human mitral valves: Two case studies. *J. Mech. Behav. Biomed. Mater.* **2010**, *3*, 167–177. [CrossRef]
156. Lee, C.-H.; Amini, R.; Sakamoto, Y.; Carruthers, C.A.; Aggarwal, A.; Gorman, R.C.; Gorman, J.H., III; Sacks, M.S. Mitral valves: A computational framework. In *Multiscale Modeling in Biomechanics and Mechanobiology*; Springer: Berlin, Germany, 2015; pp. 223–255.

157. Lee, C.-H.; Oomen, P.J.A.; Rabbah, J.P.; Yoganathan, A.; Gorman, R.C.; Gorman, J.H., III; Amini, R.; Sacks, M.S. A high-fidelity and micro-anatomically accurate 3D finite element model for simulations of functional mitral valve. In Proceedings of the International Conference on Functional Imaging and Modeling of the Heart, London, UK, 20–22 June 2013; pp. 416–424.
158. Wang, Q.; Sun, W. Finite element modeling of mitral valve dynamic deformation using patient-specific multi-slices computed tomography scans. *Ann. Biomed. Eng.* **2013**, *41*, 142–153. [CrossRef]
159. Kunzelman, K.S.; Einstein, D.R.; Cochran, R.P. Fluid–structure interaction models of the mitral valve: function in normal and pathological states. *Philos. Trans. R. Soc. B: Biol. Sci.* **2007**, *362*, 1393–1406. [CrossRef]
160. Lau, K.D.; Diaz, V.; Scambler, P.; Burriesci, G. Mitral valve dynamics in structural and fluid–structure interaction models. *Med. Eng. Phys.* **2010**, *32*, 1057–1064. [CrossRef]
161. De Hart, J.; Peters, G.W.M.; Schreurs, P.J.G.; Baaijens, F.P.T. A three-dimensional computational analysis of fluid–structure interaction in the aortic valve. *J. Biomech.* **2003**, *36*, 103–112. [CrossRef]
162. Hsu, M.-C.; Kamensky, D.; Bazilevs, Y.; Sacks, M.S.; Hughes, T.J. Fluid–structure interaction analysis of bioprosthetic heart valves: Significance of arterial wall deformation. *Comput. Mech.* **2014**, *54*, 1055–1071. [CrossRef]
163. Hsu, M.-C.; Kamensky, D.; Xu, F.; Kiendl, J.; Wang, C.; Wu, M.C.; Mineroff, J.; Reali, A.; Bazilevs, Y.; Sacks, M.S. Dynamic and fluid–structure interaction simulations of bioprosthetic heart valves using parametric design with T-splines and Fung-type material models. *Comput. Mech.* **2015**, *55*, 1211–1225. [CrossRef]
164. Billiar, K.L.; Sacks, M.S. Biaxial mechanical properties of the native and glutaraldehyde-treated aortic valve cusp: Part II—A structural constitutive model. *J. Biomech. Eng.* **2000**, *122*, 327–335. [CrossRef]
165. Morganti, S.; Conti, M.; Aiello, M.; Valentini, A.; Mazzola, A.; Reali, A.; Auricchio, F. Simulation of transcatheter aortic valve implantation through patient-specific finite element analysis: Two clinical cases. *J. Biomech.* **2014**, *47*, 2547–2555. [CrossRef]
166. Wang, Q.; Sirois, E.; Sun, W. Patient-specific modeling of biomechanical interaction in transcatheter aortic valve deployment. *J. Biomech.* **2012**, *45*, 1965–1971. [CrossRef]
167. Nicosia, M.A.; Cochran, R.P.; Einstein, D.R.; Rutland, C.J.; Kunzelman, K.S. A coupled fluid-structure finite element model of the aortic valve and root. *J. Heart. Valve Dis.* **2003**, *12*, 781–789. [PubMed]
168. Haj-Ali, R.; Marom, G.; Zekry, S.B.; Rosenfeld, M.; Raanani, E. A general three-dimensional parametric geometry of the native aortic valve and root for biomechanical modeling. *J. Biomech.* **2012**, *45*, 2392–2397. [CrossRef]
169. Stevanella, M.; Votta, E.; Lemma, M.; Antona, C.; Redaelli, A. Finite element modelling of the tricuspid valve: A preliminary study. *Med. Eng. Phys.* **2010**, *32*, 1213–1223. [CrossRef]
170. Votta, E.; Caiani, E.; Veronesi, F.; Soncini, M.; Montevecchi, F.M.; Redaelli, A. Mitral valve finite-element modelling from ultrasound data: A pilot study for a new approach to understand mitral function and clinical scenarios. *Philos. Trans. R. Soc. Lond. A Math. Phys. Eng. Sci.* **2008**, *366*, 3411–3434. [CrossRef]
171. Aversa, A.; Careddu, E. Image-Based Analysis of Tricuspid Valve Biomechanics: Towards a Novel Approach Integrating In-vitro 3D-Echocardiography and Finite Element Modelling. Master's Thesis, Politecnico di Milano, Milan, Italy, 2017.
172. Kong, F.; Pham, T.; Martin, C.; McKay, R.; Primiano, C.; Hashim, S.; Kodali, S.; Sun, W. Finite element analysis of tricuspid valve deformation from multi-slice computed tomography images. *Ann. Biomed. Eng.* **2018**, *46*, 1112–1127. [CrossRef]
173. Pouch, A.M.; Aly, A.H.; Lasso, A.; Nguyen, A.V.; Scanlan, A.B.; McGowan, F.X.; Fichtinger, G.; Gorman, R.C.; Gorman, J.H.; Yushkevich, P.A. Image segmentation and modeling of the pediatric tricuspid valve in hypoplastic left heart syndrome. In Proceedings of the International Conference on Functional Imaging and Modeling of the Heart, Toronto, ON, Canada, 11–13 June 2017; pp. 95–105.
174. Kamensky, D.; Xu, F.; Lee, C.-H.; Yan, J.; Bazilevs, Y.; Hsu, M.-C. A contact formulation based on a volumetric potential: Application to isogeometric simulations of atrioventricular valves. *Comput. Meth. Appl. Mech. Eng.* **2018**, *330*, 522–546. [CrossRef]
175. Thubrikar, M. Geometry of the aortic valve. In *The Aortic Valve*; CRC Press: Boca Raton, FL, USA, 1990; pp. 1–20.
176. Labrosse, M.R.; Beller, C.J.; Robicsek, F.; Thubrikar, M.J. Geometric modeling of functional trileaflet aortic valves: Development and clinical applications. *J. Biomech.* **2006**, *39*, 2665–2672. [CrossRef] [PubMed]

177. Kouhi, E.; Morsi, Y.S. A parametric study on mathematical formulation and geometrical construction of a stentless aortic heart valve. *J. Artificial Organs* **2013**, *16*, 425–442. [CrossRef]
178. Fan, R.; Bayoumi, A.S.; Chen, P.; Hobson, C.M.; Wagner, W.R.; Mayer, J.E., Jr.; Sacks, M.S. Optimal elastomeric scaffold leaflet shape for pulmonary heart valve leaflet replacement. *J. Biomech.* **2013**, *46*, 662–669. [CrossRef] [PubMed]
179. Li, K.; Sun, W. Simulated transcatheter aortic valve deformation: A parametric study on the impact of leaflet geometry on valve peak stress. *Int. J. Numer. Method. Biomed. Eng.* **2017**, *33*, e02814. [CrossRef]
180. Xu, F.; Morganti, S.; Zakerzadeh, R.; Kamensky, D.; Auricchio, F.; Reali, A.; Hughes, T.J.; Sacks, M.S.; Hsu, M.C. A framework for designing patient-specific bioprosthetic heart valves using immersogeometric fluid–structure interaction analysis. *Int. J. Numer. Method. Biomed. Eng.* **2018**, *34*, e2938. [CrossRef]
181. Hughes, T.J.R.; Cottrell, J.A.; Bazilevs, Y. Isogeometric analysis: CAD, finite elements, NURBS, exact geometry and mesh refinement. *Comput. Meth. Appl. Mech. Eng.* **2005**, *194*, 4135–4195. [CrossRef]
182. Piegl, L.; Tiller, W. *The NURBS Book*; Springer Science & Business Media: Berlin, Germany, 2012.
183. Holzapfel, G.A.; Gasser, T.C.; Ogden, R.W. A new constitutive framework for arterial wall mechanics and a comparative study of material models. *J. Elast. Phys. Sci. Solids* **2000**, *61*, 1–48.
184. Prot, V.; Skallerud, B.; Holzapfel, G.A. Transversely isotropic membrane shells with application to mitral valve mechanics. Constitutive modelling and finite element implementation. *Int. J. Numer. Meth. Eng.* **2007**, *71*, 987–1008. [CrossRef]
185. Zhang, W.; Ayoub, S.; Liao, J.; Sacks, M.S. A meso-scale layer-specific structural constitutive model of the mitral heart valve leaflets. *Acta Biomater.* **2016**, *32*, 238–255. [CrossRef]
186. Singh-Gryzbon, S.; Sadri, V.; Toma, M.; Pierce, E.L.; Wei, Z.A.; Yoganathan, A.P. Development of a computational method for simulating tricuspid valve dynamics. *Ann. Biomed. Eng.* **2019**, *47*, 1422–1434. [CrossRef]
187. Laurence, D.; Johnson, E.; Hsu, M.-C.; Mir, A.; Burkhart, H.M.; Wu, Y.; Lee, C.-H. Finite element simulation framework for investigating pathological effects on organ-level tricuspid valve biomechanical function. In Proceedings of the Summer Biomechanics, Bioengineering and Biotransport Conference, Seven Springs, PA, USA, 25–28 June 2019.
188. Toma, M.; Bloodworth, C.H.; Einstein, D.R.; Pierce, E.L.; Cochran, R.P.; Yoganathan, A.P.; Kunzelman, K.S. High-resolution subject-specific mitral valve imaging and modeling: Experimental and computational methods. *Biomech. Model. Mechanobiol.* **2016**, *15*, 1619–1630. [CrossRef]
189. Collé, A.; Limido, J.; Vila, J.-P. An accurate SPH scheme for dynamic fragmentation modelling. In Proceedings of the EPJ Web of Conferences, Arcachon, France, 9–14 September 2018; p. 01030.
190. Lanson, N.; Vila, J.P. Meshless methods for conservation laws. *Math. Comput. Sim.* **2001**, *55*, 493–501. [CrossRef]
191. Monaghan, J.J.; Gingold, R.A. Shock simulation by the particle method SPH. *JCoPh* **1983**, *52*, 374–389. [CrossRef]
192. Sun, P.; Colagrossi, A.; Marrone, S.; Antuono, M.; Zhang, A.M. Multi-resolution Delta-plus-SPH with tensile instability control: Towards high Reynolds number flows. *CoPhC* **2018**, *224*, 63–80. [CrossRef]
193. Borazjani, I. Fluid–structure interaction, immersed boundary-finite element method simulations of bio-prosthetic heart valves. *Comput. Meth. Appl. Mech. Eng.* **2013**, *257*, 103–116. [CrossRef]
194. Griffith, B.E.; Luo, X.; McQueen, D.M.; Peskin, C.S. Simulating the fluid dynamics of natural and prosthetic heart valves using the immersed boundary method. *Int. J. Appl. Mech.* **2009**, *1*, 137–177. [CrossRef]
195. Griffith, B.E. Immersed boundary model of aortic heart valve dynamics with physiological driving and loading conditions. *Int. J. Numer. Method. Biomed. Eng.* **2012**, *28*, 317–345. [CrossRef]
196. Espino, D.M.; Shepherd, D.E.T.; Hukins, D.W.L. Evaluation of a transient, simultaneous, arbitrary Lagrange–Euler based multi-physics method for simulating the mitral heart valve. *Comput. Methods Biomech. Biomed. Eng.* **2014**, *17*, 450–458. [CrossRef] [PubMed]
197. Takahashi, K.; Inage, A.; Rebeyka, I.M.; Ross, D.B.; Thompson, R.B.; Mackie, A.S.; Smallhorn, J.F. Real-time 3-dimensional echocardiography provides new insight into mechanisms of tricuspid valve regurgitation in patients with hypoplastic left heart syndrome. *Circulation* **2009**, *120*, 1091. [CrossRef] [PubMed]
198. Nii, M.; Guerra, V.; Roman, K.S.; Macgowan, C.K.; Smallhorn, J.F. Three-dimensional tricuspid annular function provides insight into the mechanisms of tricuspid valve regurgitation in classic hypoplastic left heart syndrome. *J. Am. Soc. Echocardiogr.* **2006**, *19*, 391–402. [CrossRef]

199. Reyes, A., 2nd; Bove, E.L.; Mosca, R.S.; Kulik, T.J.; Ludomirsky, A. Tricuspid valve repair in children with hypoplastic left heart syndrome during staged surgical reconstruction. *Circulation* **1997**, *96*, II-341-3–II-344-5. [PubMed]
200. Apitz, C.; Webb, G.D.; Redington, A.N. Tetralogy of Fallot. *Lancet* **2009**, *374*, 1462–1471. [CrossRef]
201. Bertranou, E.G.; Blackstone, E.H.; Hazelrig, J.B.; Turner, M.E., Jr.; Kirklin, J.W. Life expectancy without surgery in Tetralogy of Fallot. *Am. J. Cardiol.* **1978**, *42*, 458–466. [CrossRef]
202. Maglogiannis, I.; Loukis, E.; Zafiropoulos, E.; Stasis, A. Support vectors machine-based identification of heart valve diseases using heart sounds. *Comput. Methods Programs Biomed.* **2009**, *95*, 47–61. [CrossRef]
203. Das, R.; Sengur, A. Evaluation of ensemble methods for diagnosing of valvular heart disease. *Expert Syst. Appl.* **2010**, *37*, 5110–5115. [CrossRef]

bioengineering

MDPI

Article

A Pilot Study on Linking Tissue Mechanics with Load-Dependent Collagen Microstructures in Porcine Tricuspid Valve Leaflets

Luke T. Hudson [1], Samuel V. Jett [1], Katherine E. Kramer [1], Devin W. Laurence [1], Colton J. Ross [1], Rheal A. Towner [2], Ryan Baumwart [3], Ki Moo Lim [4], Arshid Mir [5], Harold M. Burkhart [6], Yi Wu [1] and Chung-Hao Lee [1,7,*]

1 Biomechanics and Biomaterials Design Laboratory (BBDL), School of Aerospace and Mechanical Engineering, The University of Oklahoma, Norman, OK 73019, USA; luke.t.hudson-1@ou.edu (L.T.H.); samuel.v.jett-1@ou.edu (S.V.J.); Katherine.E.Kramer-1@ou.edu (K.E.K.); dwlaur@ou.edu (D.W.L.); cjross@ou.edu (C.J.R.); yiwu@ou.edu (Y.W.)

2 Advanced Magnetic Resonance Center, MS 60, Oklahoma Medical Research Foundation, Oklahoma City, OK 73104, USA; Rheal-Towner@omrf.org

3 Department of Veterinary Clinical Sciences, College of Veterinary Medicine, Washington State University, Pullman, WA 99164, USA; ryan.baumwart@wsu.edu

4 Department of Medical IT Convergence Engineering, Kumoh National Institute of Technology, Gumi, Gyeongbuk 39177, Korea; kmlim@kumoh.ac.kr

5 Division of Pediatric Cardiology, Department of Pediatrics, The University of Oklahoma Health Sciences Center, Oklahoma City, OK 73104, USA; Arshid-Mir@ouhsc.edu

6 Division of Cardiothoracic Surgery, Department of Surgery, The University of Oklahoma Health Sciences Center, Oklahoma City, OK 73104, USA; Harold-Burkhart@ouhsc.edu

7 Institute for Biomedical Engineering, Science and Technology (IBEST), The University of Oklahoma, Norman, OK 73019, USA

* Correspondence: ch.lee@ou.edu; Tel.: +1-405-325-4842

Received: 7 April 2020; Accepted: 15 June 2020; Published: 18 June 2020

Abstract: The tricuspid valve (TV) is composed of three leaflets that coapt during systole to prevent deoxygenated blood from re-entering the right atrium. The connection between the TV leaflets' microstructure and the tissue-level mechanical responses has yet to be fully understood in the TV biomechanics society. This pilot study sought to examine the load-dependent collagen fiber architecture of the three TV leaflets, by employing a multiscale, combined experimental approach that utilizes tissue-level biaxial mechanical characterizations, micro-level collagen fiber quantification, and histological analysis. Our results showed that the three TV leaflets displayed greater extensibility in the tissues' radial direction than in the circumferential direction, consistently under different applied biaxial tensions. Additionally, collagen fibers reoriented towards the direction of the larger applied load, with the largest changes in the alignment of the collagen fibers under radially-dominant loading. Moreover, collagen fibers in the belly region of the TV leaflets were found to experience greater reorientations compared to the tissue region closer to the TV annulus. Furthermore, histological examinations of the TV leaflets displayed significant regional variation in constituent mass fraction, highlighting the heterogeneous collagen microstructure. The combined experimental approach presented in this work enables the connection of tissue mechanics, collagen fiber microstructure, and morphology for the TV leaflets. This experimental methodology also provides a new research platform for future developments, such as multiscale models for the TVs, and the design of bioprosthetic heart valves that could better mimic the mechanical, microstructural, and morphological characteristics of the native tricuspid valve leaflets.

Keywords: tricuspid regurgitation; biaxial mechanical testing; polarized spatial frequency domain imaging; spatial alignment; collagen fiber reorientation; material anisotropy

1. Introduction

The tricuspid valve (TV) plays a key role in regulating the unidirectional blood flow within the right side of the heart. The TV is composed of three collagenous leaflets that are attached to the right atrium at the valvular annulus and connected to the papillary muscles of the right ventricular walls through the chordae tendineae. Proper opening and closing of the TV leaflets prevent the backflow of deoxygenated blood from the right ventricle (RV) into the right atrium. Pressure and/or volume overload in the RV can cause alterations in the right ventricular geometry and further result in the development of diseases such as functional tricuspid regurgitation (FTR). These organ-level changes are prevalent among the general population and can be detrimental, potentially leading to heart failure [1–3]. Despite these repercussions, FTR has been largely under-investigated in the last two decades compared to other valvular heart diseases (e.g., left-sided heart valves: the mitral valve and the aortic valve). This may originate from the pervasive expectation among surgeons and cardiologists that correcting those left-sided cardiac lesions will naturally resolve the accompanied FTR [4]. However, recent studies by Dreyfus et al. (2005) and Anyanwu and Adams (2010) [2,5] have shown that this conservative clinical viewpoint and practice may not be valid, and those untreated FTR cases later progress to severe TR that further worsens long-term prognosis.

Apart from these organ-level, geometrical changes, tissue remodeling has been observed on the microscopic level, indicating the importance of the leaflets' microstructure in connection to proper physiological function [6–8]. The microstructure of the TV leaflets can be morphologically categorized into four distinct layers, as described from the atrial to the ventricular surfaces [9–11]: (i) the atrialis, composed primarily of radially-oriented elastin fibers that provide the tissues' low-stress elasticity; (ii) the spongiosa, containing non-fibrous components, such as glycosaminoglycans (GAGs) and proteoglycans (PGs), that act as a dampening mechanism during rapid leaflet bending; (iii) the collagen-rich fibrosa—the primary load bearing layer; and (iv) the ventricularis, which is composed of circumferentially-oriented elastin fibers that facilitate movement and restore the leaflets to their undeformed geometries during TV opening [12,13]. The connection of the tissue-level and microstructural changes to the overall organ-level TV function have not yet been fully understood in the heart valve biomechanics society. Quantifying these microstructural and tissue-level changes can inform accurate constitutive models, that may be used in in-silico finite element simulations for guiding clinical therapeutics.

In spite of the lack of connections between the microstructures and tissue biomechanics to the organ-level TV function, research efforts have been made in the past 3–5 years to characterize the tissue mechanics of the TV leaflets [10,14–17] and other sub-valvular components, such as the chordae tendineae [18–20] and the TV annulus [21–26], in addition to evolving clinical studies on the TV function [27,28] and post-operative outcome [29–31]. Previous studies of the heart valve leaflets have shown that the collagen fiber networks adjust and reorient in response to applied mechanical loading [32–35]. More recently, Laurence et al. (2019) further investigated the regional variations in the biaxial mechanical and biaxial stress relaxation properties of the TV leaflets [36]. The variance between the central (belly) region and the edge region (closer to the commissure) elicits the need for a more in-depth understanding of the microstructural differences within the TV leaflet tissue, that would aid in establishing such a connection to the organ-level TV function. In addition, the mechanical properties of specific tissue layers were also examined by Kramer et al. (2019) [9]. Specifically, in the TV anterior leaflet, different tissue constituents may contribute to the different mechanical behaviors of each layer (i.e., a more compliant stress-strain response in the combined atrialis/spongiosa layer, as opposed to the combined fibrosa/ventricularis layer). Evidently, the role of the collagen fiber architecture in tissue mechanics warrants further investigations to improve our understanding of current TV pathologies and function.

On the other hand, to examine collagen fiber networks, optical techniques, such as second harmonic generation (SHG) imaging [35,37–39] and small angle light scattering (SALS) [40,41], have been employed to provide valuable insight into the microstructural characterization of the heart valve leaflets and other soft tissues. However, these techniques are limited in their ability to capture the spatially varied, load-dependent collagen fiber architecture. For example, SHG has been used to obtain high-resolution images of the collagen fiber architecture of the mitral valve anterior leaflet at a biaxially-loaded state [34]. This investigation was limited to a micron-level field of view (FOV), restricting the SHG modality from effectively examining the spatial variance in the collagen fiber architectures (CFAs) at a larger FOV. In contrast, the SALS imaging modality can capture the CFA of the entire tissue specimen, but this technique requires chemical fixation along with optical clearing solution [42,43], limiting its capability to investigate the adaptive, load-dependent collagen fiber architecture. To addresses these shortcomings, polarized spatial frequency domain imaging (pSDFI), a recently developed optical imaging technique, has been used to observe the CFA of the tissue on a millimeter-scale FOV, while removing the need for fixative solutions. Previously, Goth et al. [44] displayed the collagen microstructural quantification capabilities of pSFDI using ovine aortic heart valve leaflets at an unloaded state; however, they did not yet highlight the load-dependent architectural changes. Recently, our group has developed a combined instrument, which integrated an in-house pSFDI device with a commercial biaxial testing system, to investigate the load-dependent changes in the CFAs for bovine tendon tissues, with highly-aligned CFAs, and a representative porcine mitral valve anterior leaflet with more dispersed CFAs [45].

Thus, the objective of this pilot study is to utilize the above combined instrument to observe the load-dependent changes in the CFAs for three TV leaflets from a representative porcine heart. This investigation will provide key insight into the microstructures and tissue-level mechanics of the TV. The observed load-dependent changes in the CFAs at the microstructural level will aid in providing a better understanding of how the TV diseases influence the tissue mechanics and overall TV function, and in informing TV computational models for guiding clinical therapeutics, such as feasibility or valve repair.

2. Materials and Methods

2.1. Porcine Heart Acquisition and Tissue Preparation

Three fresh, normal porcine hearts were obtained from a local USDA-approved abattoir (Country Home Meat Company, Edmond, OK, USA), and dissected to retrieve the three TV leaflets (Figure 1a): the anterior leaflet (TVAL), the posterior leaflet (TVPL), and the septal leaflet (TVSL) (n = 3 for each of the three TV leaflets). Excess chordae tissue was trimmed from the leaflets, and thickness measurements were made using a digital caliper (Westward Tools 1AAU4—0.01 mm resolution) at three different locations, to determine an average leaflet thickness. Tissues were then stored in phosphate-buffered saline (PBS) at 4 °C, prior to testing within 48 h.

2.2. Biaxial Mechanical Testing

The TV leaflet tissues were mounted to a commercial biaxial testing system (CellScale, Canada, 1.5 N load cells), using four BioRakes to facilitate a 10 × 10 mm effective testing region in the central portion of the leaflet specimens (Figure 1b,c). During mounting, the tissues' circumferential (C) and radial (R) directions were aligned with the x- and y-directions of the biaxial testing device. Following our previously developed displacement-controlled protocols [9,10,36,46,47] for investigating the atrioventricular heart valve tissue's nonlinear mechanical behaviors and material anisotropy, the specimen was next submerged in a PBS bath at 37 °C for the entire duration of mechanical testing. A preconditioning protocol was utilized with six equibiaxial and non-equibiaxial loading/unloading cycles, targeting a membrane tension of 25 N/m to restore the leaflets to their in-vivo physiological configurations [14,48]. After preconditioning, the tissues were subjected to displacement-controlled biaxial testing experiments, emulating biaxial tensions (T) at varying loading ratios (T_C:T_R = 1:1, 1:0.5, 0.5:1, and 2:2), to capture a broad range of potential physiological deformation states [10,17,19]. These biaxial tension protocols were each applied for three loading/unloading cycles, and the forces and displacements were recorded at 5 Hz. The membrane tension-stretch data was extracted from the third loading cycle. Note that the membrane tension was computed from the recorded force divided by the effective edge length (10 mm), and the tissue stretch (λ) was calculated using $d_{tine}^{load}/d_{tine}^{PPC}$ for both the circumferential and radial directions, where d_{tine}^{PPC} is the tine distance after preconditioning, and d_{tine}^{load} is the tine-to-tine distance at the loading state.

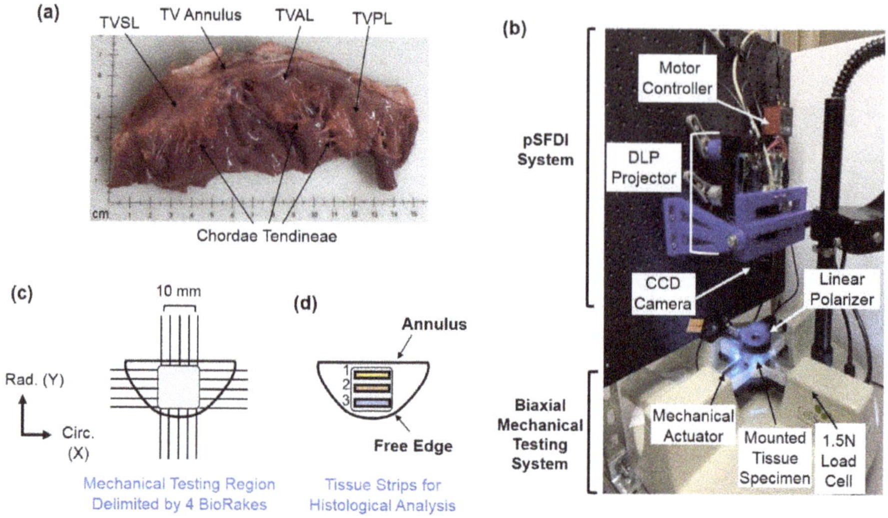

Figure 1. (**a**) The porcine TV was dissected to retrieve the three TV leaflets for use in (**b**) biaxial mechanical testing. (**c**) A side view of the combined pSFDI-biaxial testing system, displaying optical components, mounted tissue specimen, and mechanical testing components. (**d**) TV leaflet strips from the effective testing region were used for histological analyses.

2.3. *pSFDI-Based Collagen Microstructure Quantifications*

Following mechanical testing, the pSFDI system was integrated with the biaxial mechanical tester by vertical placement above the testing sample (Figure 1b). Following the procedure of pSFDI-based collagen fiber quantifications, the incident spatial frequency light patterns were produced from an LED projector (Texas Instruments, Dallas, TX, USA), with a wavelength of 490 nm (cyan). A 5-megapixel CCD camera (Basler, Germany) was used to capture the reflected light intensity responses through a rotating linear polarizer (Thorlabs Inc., Newton, NJ, USA), at 37 distinct polarization states (i.e., 0° to 180°, 5° increments).

The above imaging procedure was repeated for three linear phase shifts (0°, 120°, and 240°) of the spatial frequency pattern, based on spatial frequency domain imaging (SFDI) theory [49–51]. Image processing and data analyses were completed via custom MATLAB (MathWorks, Natick, MA, USA) programs, to examine the collagen fiber architecture of the tissue's region of interest (ROI), as defined by the BioRake tines. The quantified CFA information includes the collagen fiber orientation θ_{fiber} and the degree of optical anisotropy (DOA) at different loading states (i.e., post-preconditioning, and (T_C:T_R = 1:1, 0.5:1, 1:0.1). The DOA stems from the structural anisotropy of the collagen fibers, and therefore, we may attribute the optical anisotropy within the sample to the structural alignment of the CFA (DOA = 1, fully aligned case). Please refer to more details of the pSFDI data analysis in Appendix A, and further information about the pSFDI theory can be found in the recent studies by Goth et al. (2016, 2019) [42,44] and Jett et al. (2020) [45].

In addition to quantifying the load-dependent CFAs of the selected tissue's ROI, a 3 × 3 grid array as shown in Figure 2 was used to further analyze the spatial variations of the changes in the collagen fiber architecture in response to mechanical loads.

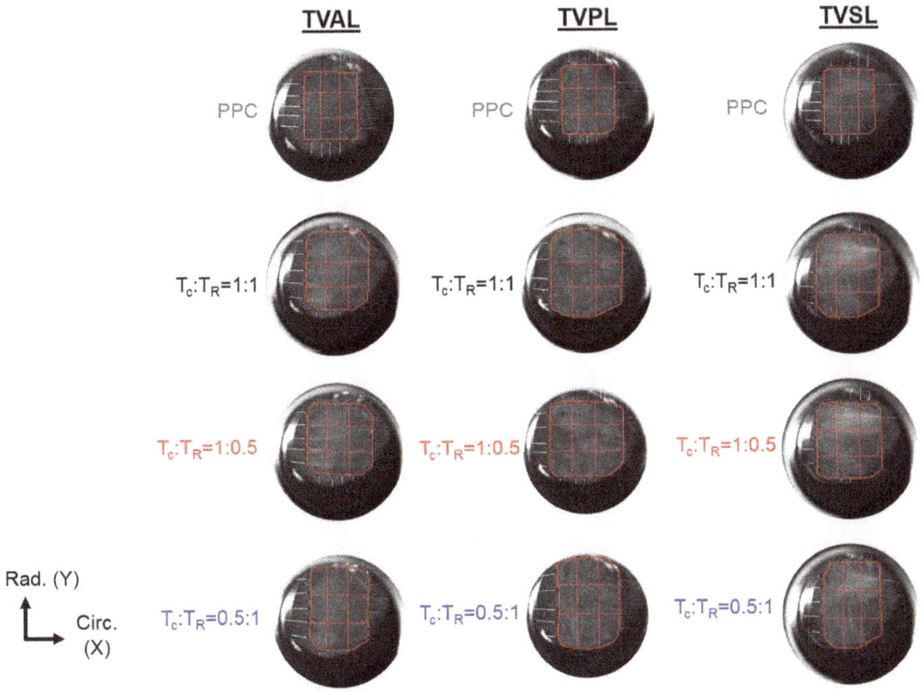

Figure 2. 3 × 3 grid sub-regions used for analyzing the spatial variations in the quantified load-dependent fiber orientation angle and the degree of optical anisotropy (DOA) for the anterior leaflet (TVAL), the posterior leaflet (TVPL), and the septal leaflet (TVSL) tissue specimens. (T_C: circumferential tension, T_R: radial tension, PPC: post-preconditioning).

2.4. Histological Analysis

To assess the distribution of extracellular matrix (ECM) constituents within each TV leaflet tissue, three rectangular samples (1 × 4 mm) were excised from each of the tested leaflets, with Region 1 defined as the strip near the TV annulus, and Region 3 defined as the strip close to the free edge (Figure 1d). Dissected tissue strips were fixed in 10% formalin at room temperature (23 °C) for 48 h, embedded in paraffin wax, and sectioned (5–7 µm) for histological staining with Movat's Pentachrome. For all stained tissue samples, three microscopic images were acquired at a 10× objective lens (Olympus, Shinjuku, Tokyo, Japan), and the images were then analyzed using a color deconvolution plugin [52] in ImageJ (National Institute of Health, Bethesda, MD, USA).

In brief, the color deconvolution method was used to separate the red/green/blue (RGB) images into various ECM components, based upon the Movat's Pentachrome stain: (i) collagen fiber (yellow), (ii) elastin (dark purple), and (iii) non-fibrous ground substance (blue). The staining of these ECM constituents allowed for the determination of the four morphologically distinct layers. The atrialis and ventricularis layers were identified by their outward surfaces towards the atrial and ventricular chambers. The spongiosa can be distinguished from the atrialis layer by the presence of hydrated GAGs and PGs (blue), whereas the fibrosa consists of primarily circumferentially-oriented collagen fibers (yellow). The ventricularis was then defined by the presence of both elastin (dark purple) and collagen (yellow), located below the collagen-rich fibrosa layer [9,10]. The RGB-separated images were next made binary, and thresholding was applied to determine the integrated optical density (*IOD*) of each constituent. Mass fractions (ω_i) of the three morphological components were calculated using $\omega_i = \mathrm{IOD}_i/\left(\mathrm{IOD}_g + \mathrm{IOD}_c + \mathrm{IOD}_e\right)$, where the subscripts g, c, and e denote the non-fibrous ground substance, collagen fibers, and elastin, respectively, and the subscript i signifies the intact tissue, which carries g, c, and e. Additionally, the thicknesses of the whole leaflet and the different tissue layers (i.e., atrialis, spongiosa, fibrosa, and ventricularis, from the atrial to ventricular sides) were measured from each image at three random locations and reported as the mean.

2.5. Statistical Analysis

Statistical analysis was performed in Prism (GraphPad, San Diego, CA, USA) for the TV leaflets of the representative heart #1. Firstly, the mass fraction for the three regionally-varying tissue strips were compared within the same leaflet for each of the three TV leaflets. Secondly, the variations in the mass fractions for the different tissue strips in the same relative locations across different leaflets were compared. These observations were compared based on the null hypothesis that the mass fractions for all constituents are uniform across each TV leaflet and show no variance between any two TV leaflets. The two-way analysis of variance (ANOVA) was performed, and p-values < 0.05 were considered as statistically significant, pointing towards constituent variations.

3. Results

3.1. Biaxial Mechanical Testing Results

J-shape, nonlinear membrane tension-stretch curves were observed for each TV leaflet, in both the circumferential and radial tissue directions (see the left column in Figure 3a–c). Under low tensions (0–2 N/m), there was a relatively linear and compliant response, where minor increases in the applied loading caused substantial increases in the tissue stretches (Figure 3a–c). This low-force regime of the membrane tension-stretch curve contrasted with the stiffer, nearly asymptotic portion of the mechanical behaviors observed under larger applied loads (>10 N/m). For three TV leaflets from the representative heart #1, the tissue stretches changed minimally between the two equibiaxial loading states ($T_C : T_R = 1{:}1$ and 2:2), yielding an increase of 2.09% in the circumferential stretch λ_C and a 4.43% increase in the radial stretch λ_R, with those changes as the average over the TVAL, TVPL, and TVSL specimens (Table 1). Furthermore, under non-equibiaxial loading ($T_C : T_R = 1{:}0.5$ and $T_C : T_R = 0.5{:}1$), the investigated TV leaflets exhibited a larger stretch in the radial direction, regardless of the direction with the dominant load. For example, in the TVAL specimen, a circumferential stretch λ_C of 1.42 was observed, together with a radial stretch λ_R of 1.51 under the circumferentially-dominant loading (i.e., $T_C : T_R = 1{:}0.5$). Interestingly, the TVPL specimen displayed the most compliant mechanical behaviors among all the three TV leaflets in both the tissue's circumferential and radial directions (Table 1 and Figure 3).

Table 1. The quantified tissue stretches of the TVAL, TVPL, and TVSL specimens under various biaxial tension protocols, together with the pixels within the tissue's region of interest (ROI) corresponding to the pSFDI-quantified results of the load-dependent CFA. Values are reported as mean ± SEM (n = 3).

$T_C : T_R$	Circumferential Stretch λ_C			Radial Stretch λ_R			Pixels in the Tissue ROI (pSFDI)		
	TVAL	TVPL	TVSL	TVAL	TVPL	TVSL	TVAL	TVPL	TVSL
PPC	1.00	1.00	1.00	1.00	1.00	1.00	368,508	434,703	377,398
1:1 [†]	1.35 ± 0.03	1.43 ± 0.02	1.39 ± 0.01	1.59 ± 0.06	1.71 ± 0.03	1.77 ± 0.04	536,726	624,611	541,161
0.5:1	1.25 ± 0.02	1.28 ± 0.01	1.24 ± 0.01	1.67 ± 0.05	1.81 ± 0.05	1.87 ± 0.04	550,747	626,641	533,910
1:0.5	1.35 ± 0.03	1.49 ± 0.02	1.43 ± 0.02	1.52 ± 0.01	1.55 ± 0.01	1.61 ± 0.05	505,182	567,840	545,645
2:2	1.40	1.50	1.38	1.75	1.87	1.81	584,990	697,596	625,324

[†] Note that the targeted membrane tension is 25 N/m with respect to the equibiaxial loading protocol, i.e., $T_C : T_R$ = (25 N/m): (25 N/m).

The mean ± standard error of the mean (SEM) values of the tissue stretches from the three porcine hearts ($n = 3$) were determined and reported as follows: (i) greater extensibility in the radial direction under equibiaxial loading (i.e., $T_C : T_R = 1{:}1$, TVAL: $\lambda_C = 1.35 \pm 0.03$, $\lambda_R = 1.59 \pm 0.06$; TVPL: $\lambda_C = 1.43 \pm 0.02$, $\lambda_R = 1.71 \pm 0.03$; TVSL: $\lambda_C = 1.39 \pm 0.01$, $\lambda_R = 1.77 \pm 0.04$); (ii) smaller changes in the circumferential extensibility under circumferentially-dominant loading (i.e., $T_C : T_R = 1{:}0.5$), compared to the radial direction (i.e., TVAL: $\lambda_C = 1.35 \pm 0.03$, $\lambda_R = 1.52 \pm 0.01$; TVPL: $\lambda_C = 1.49 \pm 0.02$, $\lambda_R = 1.55 \pm 0.01$; TVSL: $\lambda_C = 1.43 \pm 0.02$, $\lambda_R = 1.61 \pm 0.05$); (iii) under radially-dominant loading (i.e., $T_C : T_R = 0.5{:}1$), the TVSL displaying the greatest extensibility among the TV leaflets (TVAL: $\lambda_C = 1.25 \pm 0.02$, $\lambda_R = 1.67 \pm 0.05$; TVPL: $\lambda_C = 1.28 \pm 0.01$, $\lambda_R = 1.81 \pm 0.05$; TVSL: $\lambda_C = 1.24 \pm 0.01$, $\lambda_R = 1.87 \pm 0.04$). The greater equibiaxial loading protocol (i.e., $T_C : T_R = 2{:}2$) was only used for TV leaflets extracted from heart #1, based on the observed minimal changes between protocol $T_C : T_R = 2{:}2$ and protocol $T_C : T_R = 1{:}1$.

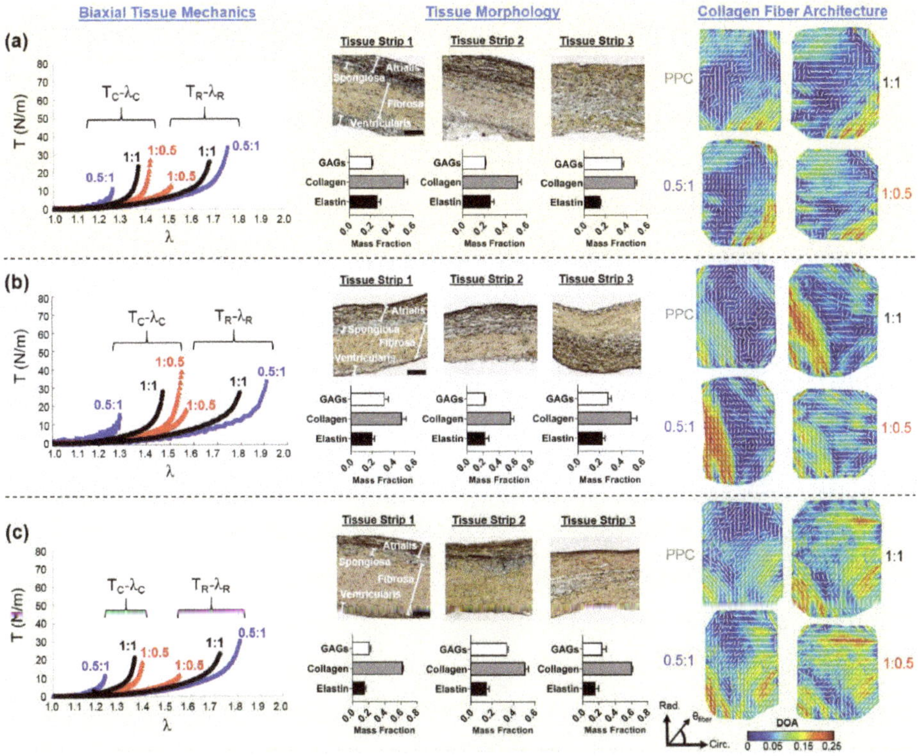

Figure 3. Representative heart #1 quantifications, as follows: (**left**) membrane tension (T) versus tissue stretch (λ) results of biaxial testing experiments (subscripts C and R stand for the circumferential and radial directions, respectively), (**middle**) histology-based evaluations of tissue's morphology and constituents, and (**right**) the pSFDI-quantified collagen fiber architecture for: (**a**) the TVAL, (**b**) the TVPL, and (**c**) the TVSL specimens. In the right column, the dash lines represent the predicted fiber orientation angle and the colormaps denote the DOA.

3.2. Histological Results

The thicknesses and percent mass compositions for each tissue layer of the TV leaflets, as found by histological analysis, are summarized in Table 2. The TVSL was shown to be the thickest leaflet across all three hearts (783.1 ± 62.6 μm), while also containing the largest percent mass composition for the ventricularis layer (12.5%), and the greatest mean total thickness for the fibrosa (363.3 ± 41.1 μm). The TVPL presented the greatest percent mass composition for both the atrialis and the fibrosa layers (A = 33.3%; F = 50.0%) among all leaflets, with the second largest overall mean total thickness (705.4 ± 22.2 μm). Finally, the smallest mean total thickness across all leaflets occurred in the TVAL (614.0 ± 23.9 μm). Quantifications of the TV leaflets' morphological constituents from the representative heart #1 showed that all three TV leaflets are composed primarily of collagen fibers (see the middle column in Figure 3a–c).

First of all, by comparing the different regions within the same TV leaflet tissue (Table 3), significant regional variance in the elastin was noted between Regions 2 and 3 of the TVAL ($p = 0.023$). For the TVAL, the GAGs were also found to vary significantly between Region 1 and Region 3 ($p = 0.002$), and between Regions 2 and 3 ($p = 0.001$). For the TVPL, significant differences were found in the GAG contents between Regions 1 and 2 ($p = 0.025$). In contrast, the TVSL exhibited significant differences in both the collagen ($p = 0.029$) and the GAGs ($p = 0.004$) between Region 1 and Region 2, while only varying in the GAG contents between Regions 2 and 3 ($p = 0.017$).

Secondly, by comparing similar regional tissue strips across all TV leaflets from heart #1 (Table 4), the TVAL and the TVPL deviated significantly only in the GAG contents for Region 1 ($p = 0.029$), whereas the TVAL and the TVSL differed significantly, in both the elastin ($p = 0.018$) and the GAGs ($p = 0.004$) for Region 1, and in the collagen ($p = 0.019$) and the GAGs ($p = 0.004$) for Region 3. The TVSL showed significant differences when compared to TVPL in all the 3 regions, primarily for the collagen content ($p = 0.006$) and the GAG contents ($p = 0.016$) in Region 1, for the GAG contents ($p = 0.007$) in Region 2, and for the collagen content ($p = 0.034$) in Region 3.

Table 2. Histologically quantified thickness and percent composition of the four tissue layers (i.e., atrialis, spongiosa, fibrosa, and ventricularis) for the TVAL, TVPL, and TVSL specimens. Values are reported as mean ± SEM ($n = 3$).

Tissue Layer	Thickness (Respective %)		
	TVAL	**TVPL**	**TVSL**
Atrialis (A)	168.8 ± 18.1 μm (27.5%)	235.1 ± 6.3 μm (33.3%)	230.8 ± 16.3 μm (29.5%)
Spongiosa (S)	77.6 ± 12.4 μm (12.6%)	86.2 ± 11.7 μm (12.2%)	90.8 ± 13.8 μm (11.6%)
Fibrosa (F)	297.6 ± 15.6 μm (48.5%)	289.1 ± 15.4 μm (50.0%)	363.3 ± 41.1 μm (46.4%)
Ventricularis (V)	69.7 ± 4.5 μm (11.4%)	73.8 ± 3.2 μm (10.5%)	98.2 ± 14.9 μm (12.5%)
Total Thickness	614.0 ± 23.9 μm	705.4 ± 22.2 μm	783.1 ± 62.6 μm

Table 3. Statistical comparisons (p-values) of the constituent mass fractions, between any two regions within each of the three representative heart #1 TV leaflets (see Figure 1d).

	Within the TVAL			Within the TVPL			Within the TVSL		
	Collagen	**Elastin**	**GAGs**	**Collagen**	**Elastin**	**GAGs**	**Collagen**	**Elastin**	**GAGs**
Region 1 vs. Region 2	0.593	0.377	0.933	0.266	0.720	0.025	0.029	0.983	0.004
Region 1 vs. Region 3	0.189	0.283	0.002	0.937	0.767	0.449	0.889	0.999	0.809
Region 2 vs. Region 3	0.686	0.023	0.001	0.429	0.997	0.244	0.073	0.989	0.017

Table 4. Statistical comparisons (p-values) of the constituent mass fraction between any two TV leaflets, regarding the three regions for the representative heart #1 (see Figure 1d).

	Within Region 1			Within Region 2			Within Region 3		
	Collagen	Elastin	GAGs	Collagen	Elastin	GAGs	Collagen	Elastin	GAGs
TVAL vs. TVPL	0.166	0.924	0.029	0.769	0.604	0.964	0.963	0.169	0.054
TVAL vs. TVSL	0.253	0.311	0.952	0.981	0.018	0.004	0.019	0.999	0.004
TVPL vs. TVSL	0.006	0.510	0.016	0.659	0.121	0.007	0.034	0.176	0.441

3.3. Load-Dependent Collagen Fiber Architecure

For each TV leaflet, the load-dependent changes in both the DOA and the collagen fiber orientation were quantified throughout the entire tissue's region of interest (see the right column in Figure 3a–c), with the representative specimen's histograms presented in Figure 4, whereas the results of the regional analysis, based on the 3 × 3 grid array, are shown in Figure 5, at the unloaded state, to highlight the TV leaflets' intrinsic characteristics. Furthermore, regional analysis regarding varying biaxial loading protocols can be found in Appendix B.

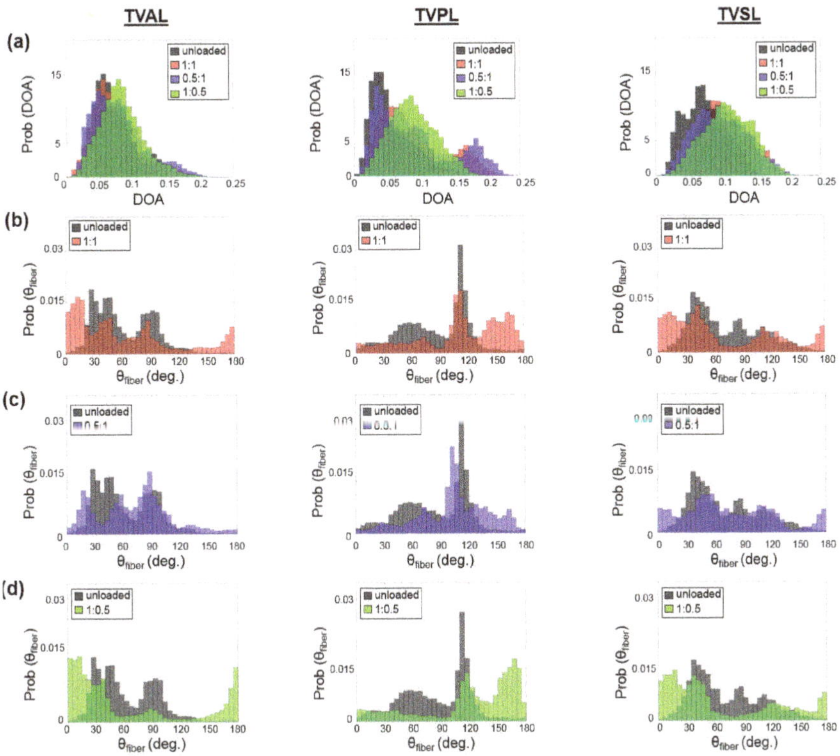

Figure 4. Comparisons of the load-dependent collagen fiber architecture for the entire tissue's ROI described in Table 1. (see the right column in Figure 3a–c): (**a**) the predicted DOA, (**b**) the predicted θ_{fiber} between unloaded and equibiaxial loading ($T_C : T_R = 1{:}1$), (**c**) the predicted θ_{fiber} between unloaded and radially-dominate loading ($T_C : T_R = 0.5{:}1$), and (**d**) the predicted θ_{fiber} between unloaded and circumferentially-dominated loading ($T_C : T_R = 1{:}0.5$).

Four primary trends were observed in the quantification of the collagen fiber architecture of the three representative TV leaflets from heart #1, with an attempt to relate the collagen fiber spatial alignment and reorientation to the direction of applied loading.

- Firstly, for the non-equibiaxial loading protocols, the collagen fiber orientations θ_{fiber} displayed a shift towards the direction of the maximum applied loading (Figure 4). Moreover, collagen fiber networks became more aligned, as evidenced by the increasing DOA within the central regions of the tissues (Figures A2–A4).
- Secondly, the largest changes in the mean collagen fiber orientation θ_{fiber} were observed in the TVPL under equibiaxial and radially-dominant loading conditions (Table 5), with differences of 37.5% ($T_C : T_R = 1{:}1$), 36.0% ($T_C : T_R = 2{:}2$), and 42.2% ($T_C : T_R = 0.5{:}1$), compared to the post-preconditioning (PPC) state, also referred to as the "unloaded" state in the remaining discussion.
- Thirdly, the largest percent change in the quantified DOA, as compared to the PPC state, occurred under the radially-dominant loading ($T_C : T_R = 0.5{:}1$), with a 10.8% increase for the TVAL, a 39.1% increase for the TVPL, and a 47.4% increase for the TVSL, respectively (Table 5). Under circumferentially-dominant loading ($T_C : T_R = 1{:}0.5$), a minimal change of 1.4% in DOA was seen for the TVAL, with a 27.3% and a 32.4% increase for the TVPL and TVSL, respectively (Table 5).
- Fourthly, across all the loading protocols, the TVSL displayed the largest increase in the DOA (47.3%) under the radially-dominant loading, whereas the smallest change was found for the TVAL (1.4%), under equibiaxial and circumferentially-dominant loading (Table 5).

Table 5. Mean, skewness, and standard deviation (SD) for the predicted θ_{fiber} and the DOA of the entire tissue's ROI (see right column in Figure 3a–c) for the TVAL, TVPL, and TVSL specimens from heart #1, considering various biaxial tension loading conditions.

	DOA (TVAL)			**DOA (TVPL)**			**DOA (TVSL)**		
$T_C : T_R$	Mean	Skewness	SD	Mean	Skewness	SD	Mean	Skewness	SD
PPC	0.074	0.873	0.033	0.064	0.708	0.036	0.074	0.314	0.031
1:1	0.075	0.879	0.035	0.089	0.654	0.048	0.104	0.155	0.036
1:0.5	0.073	1.264	0.038	0.088	0.730	0.059	0.098	0.317	0.041
0.5:1	0.082	0.564	0.032	0.089	0.257	0.034	0.109	−0.012	0.036
2:2	0.072	0.930	0.034	0.087	0.700	0.048	0.102	0.142	0.037
	θ_{fiber} (TVAL)			**θ_{fiber} (TVPL)**			**θ_{fiber} (TVSL)**		
$T_C : T_R$	Mean	Skewness	SD	Mean	Skewness	SD	Mean	Skewness	SD
PPC	61.6°	0.588	30.8°	85.9°	−0.377	34.6°	70.9°	0.621	36.3°
1:1	60.6°	0.906	52.3°	112.4°	−0.698	46.3°	67.2°	0.655	53.5°
1:0.5	69.7°	0.348	38.3°	103.3°	−0.462	37.2°	77.1°	0.398	46.9°
0.5:1	61.6°	0.889	63.4°	122.3°	−1.042	49.3°	63.6°	0.799	56.8°
2:2	60.1°	0.928	52.5°	111.1°	−0.665	47.2°	66.8°	0.655	53.6°

Moreover, the quantified load-dependent changes of the CFAs for the three TV leaflets of the representative heart #1, considering various biaxial loading protocols, are further elaborated in the following subsections. Further results for heart #2 and heart #3 can be found in Appendix C.

3.3.1. Changes in the CFA Associated with the Equibiaxial Loading States

The TVAL specimen displayed minimal changes in θ_{fiber} from 61.6° in the PPC state to 60.6° and 60.1° under $T_C : T_R = 1{:}1$ and $T_C : T_R = 2{:}2$, respectively. Similar trends were observed for the spatial alignment of the fibers (i.e., DOA), varying from 0.074 in the PPC state to 0.075 under $T_C : T_R = 1{:}1$

and to 0.072 under $T_C : T_R = 2{:}2$. In contrast, the TVPL showed profound changes in θ_{fiber}, yielding a 30.8% difference from the PPC state (85.9°) to the $T_C : T_R = 1{:}1$ loading protocol (112.4°) and a 29.3% difference when compared to the $T_C : T_R = 2{:}2$ loading state (111.1°). Similarly, the quantified DOA for the PPC state displayed increases of 37.5% and 36.0% compared to $T_C : T_R = 1{:}1$ and $T_C : T_R = 2{:}2$ loading protocols, respectively. For the TVSL, θ_{fiber} varied from the PPC state (70.9°), by 5.5% for the $T_C : T_R = 1{:}1$ state (67.2°) and 5.8% for the $T_C : T_R = 2{:}2$ state (66.8°), whereas the DOA increased by 40.5% and 37.8% under $T_C : T_R = 1{:}1$ and $T_C : T_R = 2{:}2$, respectively, when compared to the PPC state.

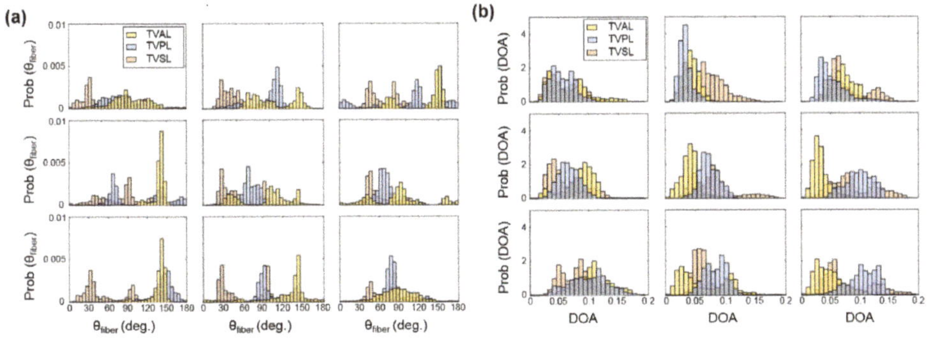

Figure 5. 3 × 3 grid comparisons of (**a**) the predicted θ_{fiber}, and (**b**) the predicted DOA in the unloaded state for the TVAL, TVPL, and TVSL specimens from heart #1.

3.3.2. Changes in the CFA Associated with the Circumferentially-Dominant Loading ($T_C : T_R = 1{:}0.5$)

For all TV leaflets, θ_{fiber} shifted towards the tissues' circumferential direction with an increased DOA under circumferentially-dominant loading (Figure 3a–c). Specifically, the TVAL displayed a 13.1% difference in θ_{fiber} between the PPC state (61.6°) and the circumferentially-loaded state (69.7°), together with a negligible change in the DOA (unloaded = 0.074 versus loaded = 0.073). For the TVPL, a 20.3% change in θ_{fiber} was observed, along with a 27.3% increase in the DOA, whereas the TVSL displayed a shift in θ_{fiber}, from 70.9° (PPC) to 77.1° (circumferentially-dominant loaded) and a greater change in the quantified DOA (32.4%).

3.3.3. Changes in the CFA Associated with the Radially-Dominant Loading ($T_C : T_R = 0.5{:}1$)

When radially-dominant loading was considered, the CFAs of the TV leaflet specimens tended to reorient towards the tissues' radial direction, with greater changes in the quantified DOAs throughout the tissue compared to other loading protocols. Specifically, the TVAL showed a minimal change in θ_{fiber} under radially-dominant loading conditions, but a 10.8% increase in the DOA was observed between the PPC (0.074) and radially-dominant loaded states (0.082). In contrast, the TVPL's collagen fibers reoriented towards the radial direction (122.3°), with a 42.4% change between the PPC and loaded state, while the DOA increased by 39.1%. For the TVSL, the quantified θ_{fiber} varied from 70.9° (PPC) to 63.6° (loaded), together with a 47.4% increase in the quantified DOAs.

4. Discussion

4.1. Mechanics-Related Observations

The nonlinear stress-strain response for each TV leaflet was consistent with the findings from the previous studies [14,17,53]. In the present study, for all three TV leaflets, the radial direction of the tissues displayed greater stretches than that of the circumferential direction, under both equibiaxial and non-equibiaxial tensions. This anisotropic tissue mechanical response has also been observed by Pokutta-Paskaleva et al. (2019) [19] and Mathur et al. (2019) [15], through the similar biaxial testing method, among others [14,54]. Essential behaviors of soft collagenous tissues, such as heart valve leaflets, have been further characterized through various fiber reorientation theories and rheological constitutive models. Lanir et al. (1979, 1983) [55,56] and Fung et al. (1984) [57] have contributed to these models for various fibrous soft tissues, taking into consideration the structure of the tissues' constituents and their mechanical properties, yet, further experimentation is needed to quantify the complex mechanical interactions between constituents that yield this stress-strain response. The anisotropic responses of the TV leaflets, although well documented, have not yet been examined within the context of the pSFDI-based collagen microstructural quantification, as shown in the current pilot study. Supplementing these well-known biomechanical trends within the TV leaflets, together with such novel microstructural quantifications, will ultimately lead to an improved understanding of collagen fiber alignment and reorientation in response to mechanical loading.

4.2. Collagen Fiber Architecture-Related Observations

We found in this pilot study that the CFA of each TV leaflet tissue is related to the directional-dependence of the tissue-level mechanics. As previously discussed, each tissue showed a greater extensibility in the radial direction (Table 1), which may be correlated to the collagen fibers' initial circumferential orientations (see the right column in Figure 3a–c). As increasing tension was applied to the tissue, the asymptotic regime of the membrane tension-stretch curve is associated with the rotation of crimped collagen fibers towards the radial direction. These fibers were then straightened, exhibiting the low-force linear regime of the curve. The collagen fibers' ability to reorient in response to the applied loading was also shown to result in the higher spatial alignment. This increased alignment of the collagen fibers is exemplified under the radially-dominant loading conditions, where the largest changes in DOA values were found in comparison with the PPC (i.e., unloaded) state (Table 5).

Similar trends in the load-dependent changes in the CFAs were also observed under non-equiaxial loading. Specifically, for all three TV leaflets, the collagen fibers were found to shift towards the direction of dominant loading (see the right column in Figure 3a–c), yielding an increase in the quantified DOAs after fiber reorientation. In contrast, when circumferentially-dominant loading was considered, the quantified θ_{fiber} became more closely aligned with the tissues' circumferential directions (Figure 4). These observations in our pilot study are in a good agreement with those findings from the previous studies on the other heart valve leaflets and the TV leaflets with chemical fixation [58,59]. By employing our combined experimental approach, the collagen fiber architecture of the same TV leaflet can be quantified at different equibiaxial and non-equibiaxial loading states, without the use of chemical fixation. This improvement on the previous collagen microstructural observations [37,41] permits further investigations into how the constituent compositions of each TV leaflet tissue contributes to the respective mechanical-microstructure responses under varying loads.

Apart from examining the load-dependent changes among the CFAs for all three TV leaflets, each specimen was also observed in the "unloaded", or PPC state. The ROI for each TV leaflet was divided into 9 sub-regions (Figure 2), and the collagen fiber orientation and DOA of each sub-region was extracted to further investigate the intrinsic characteristics of the TV leaflets. The TVAL and TVSL leaflets each displayed primarily circumferentially-dominant CFAs throughout all sub-regions of the tissue, with some variation in the radially-oriented collagen fibers of the tissues in the belly portion and near the annulus (Figure 5). The TVPL, on the other hand, exhibited a predominantly radially-oriented CFA throughout the entirety of the tissue, with θ_{fiber} closely aligned with 90°. The DOA of the TV leaflets remained consistent throughout each section of the tissue (Figure 5), with no statistically significant changes in spatial alignment between the collagen fibers. This assertion can be connected to the collagen fibers' naturally crimped state, where with greater load, the CFA will straighten and recruit multiple fiber families to compensate for the induced stress on the tissue. Observing the natural CFA and spatial alignment in TV leaflets can be used to further explain the mechanics-microstructure relationship, that has been shown to vary regionally across TV anterior leaflet specimens [36].

4.3. Study Limitations and Future Work

Throughout this pilot study, we encountered difficulty in determining an average thickness for each sample, primarily due to non-uniformity and surface imperfections of the tissues, by using a contact-based caliper measurement technique. This limitation was moderately alleviated by measuring the thickness across the leaflet in three different regions and averaging the observed thicknesses. Another limitation of this pilot study is the tine-based biaxial loading technique used. When observing the collagen fiber microstructure via the pSFDI-based technique, the tines inserted in the tissue were captured in the images associated with each phase shift, producing edge artifacts on the CFA map (e.g., see the right column in Figure 3a–c). An extension of this limitation is the displacement-controlled biaxial testing methods used to emulate the target membrane tensions similar to the force-controlled biaxial mechanical tests. This was done by performing displacement-controlled loading to the specimen size associated with the peak loading stress, but due to the stress relaxation of the tissue when determining the peak specimen size, there was a reduction in the target stress values. In regards to the histological limitations presented in this pilot study, we attempted to excise the 3 tissue strips within the observed ROI, but our calculations did not account for the portions of the tissue outsides of the ROI, which may potentially alter the observed trends. Another limitation of this work is the examination of only three representative porcine hearts ($n = 3$). Definite conclusions and trends cannot be drawn regarding the TV leaflets without including a much larger sample size.

Despite these restrictions, our pilot study provides a first look into the investigations of the mechanical-microstructural relationships within TV leaflets. Subsequent studies may warrant a more in-depth examination of the TV with a larger sample size, to draw statistically supported conclusions. Regarding other potential future extensions, the microstructural-mechanical relationship in the collagenous heart valve leaflet tissue can also be examined under various mechanical tests to quantify behaviors such as the preconditioning effect, the stress-relaxation effect, and the creep effect, informing fiber kinematic models, such as the previous ones developed for the mitral valve and the aortic valve [33,34,60,61]. This combined experimental approach could also be used to investigate other collagenous tissues, utilizing its unique capability to elucidate the microstructural-mechanical relationship and inform high-fidelity constitutive models.

5. Conclusions

In this pilot study, we have presented a novel systematic framework for characterizing collagenous tissues that utilize both the histology-based morphological assessment and the pSFDI modality in conjunction with biaxial mechanical testing, allowing for a direction examination of the interrelationship between tissue mechanics and collagen microstructures in response to mechanical loads. In the case of the TV leaflets, observing these load-dependent microstructural changes will lead to further developments of improved computational models. Such enhanced computational modeling tools could not only aid in a better understanding of TV function as well as its associated diseases, but they could also be used for patient-specific surgical planning and treatment options. Preoperative guidance in assessing the achievability of a successful tricuspid valve repair, thereby avoiding a less desirable valve replacement, would be of great benefit. The application of this experimental approach is readily applicable to other cardiovascular collagenous tissues (e.g., the aortic valve cusps, and the pulmonary valve cusps), to complement the current understanding in the field. The proposed systematic framework also provides new frontiers and understanding within the field of cardiovascular biomechanics, offering a potential to advance the development of novel clinical therapeutics for TV diseases.

Author Contributions: Conceptualization, R.B., K.M.L., A.M., H.M.B., C.-H.L.; Data curation, L.T.H., S.V.J., K.E.K.; Formal analysis, S.V.J., K.E.K., D.W.L., C.J.R., Y.W.; Funding acquisition, H.M.B., C.-H.L.; Project administration, C.-H.L.; Writing—Original draft, L.T.H., S.V.J., K.E.K.; Writing—Review and editing, L.T.H., S.V.J., D.W.L., C.J.R., R.A.T., R.B., K.M.L., A.M., H.M.B., Y.W., C.-H.L. All authors have read and agreed to the published version of the manuscript.

Funding: This research was funded by American Heart Association: 16SDG27760143 and Presbyterian Health Foundation: C5122401.

Acknowledgments: C.-H.L. was in part supported by the start-up funds from the School of Aerospace and Mechanical Engineering, the IBEST-OUHSC Seed Funding, and the funding from the Research Council at the University of Oklahoma (OU).

Conflicts of Interest: The authors of this paper have no financial or personal relationships with other people or organizations that could inappropriately influence (bias) our work.

Nomenclature

Category	Abbreviation	Definition
Anatomy	TV	Tricuspid valve
	TVAL	Tricuspid valve anterior leaflet
	TVPL	Tricuspid valve posterior leaflet
	TVSL	Tricuspid valve septal leaflet
	RV	Right ventricle
	FTR	Functional tricuspid regurgitation
	ECM	Extracellular matrix
Tissue Layers	A	Atrialis
	S	Spongiosa
	F	Fibrosa
	V	Ventricularis
Tissue Layer Constituents	CFA	Collagen fiber architecture
	GAGs	Glycosaminoglycans
	PGs	Proteoglycans
Instrumentation	pSFDI	Polarized spatial frequency domain imaging
	SALS	Small angle light scattering
	SHG	Second harmonic generation
	FOV	Field of view
	ROI	Region of interest
Collagen Fiber Quantification	θ_{fiber}	Collagen fiber orientation
	DOA	Degree of optical anisotropy
Mechanics	C	Tissue's circumferential direction
	R	Tissue's radial direction
	T	Membrane tension
	PPC	Post-preconditioning
	λ	Tissue stretch

Appendix A. Quantification of Collagen Fiber Orientation and Degree of Optical Anisotropy

The pSFDI imaging technique combines the ability of co-polarized imaging to quantify the birefringent fiber structures with the depth-discrimination capabilities of SFDI. Interested readers can refer to more details in [44,45,62]. During the polarized spatial frequency domain imaging procedure, polarization-state images (2560 × 2048 pixels) for each phase shift were obtained, and the acquired 37 images were then smoothed via an in-house MATLAB program using convolution with a 5 × 5 uniform kernel. After smoothing, each image was then combined at each pixel to create the predicted birefringent reflected intensity:

$$I_{out} = \frac{1}{3}\left(I^{0^\circ} + I^{120^\circ} + I^{240^\circ}\right), \tag{A1}$$

where I^{0°, I^{120°, and I^{240° are the pixel-wise intensities at each respective phase shift. The global maximum for each of these intensity functions occurs when the θ_{fiber} is both perpendicular and parallel with the polarizer's transmission axis, $\theta_{polarizer}$ (Figure A1).

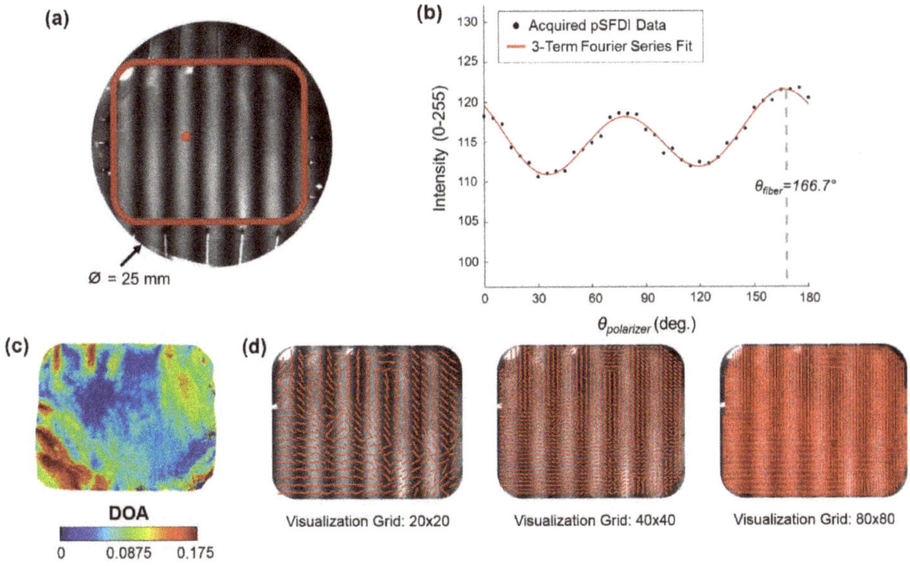

Figure A1. Demonstration of the pSFDI modality to capture the CFA: (**a**) an experimental photo of a representative TV leaflet specimen mounted to the BioTester system, with a defined region of interest as delimited by the red box, and (**b**) the quantified intensity-$\theta_{polarizer}$ relationship of a selected pixel point (see the red dot in (**a**)), together with the fitting of the 3-term Fourier series in Equation (A2) to the acquired data for predicting the collagen fiber orientation angle θ_{fiber}. (**c**) The quantified DOA throughout the ROI, and (**d**) vector plots of the quantified θ_{fiber} with an increasing level of visualization grid.

The quantified birefringent reflected intensity is then approximated by a three-term Fourier cosine series:

$$\frac{I_{out}}{\tau_{sys}} = \alpha_0 + \alpha_2\left[2(\theta_{fiber} - \theta_{polarizer})\right] + \alpha_4\left[4(\theta_{fiber} - \theta_{polarizer})\right], \quad (A2)$$

where α_0, α_2, and α_4 are the three Fourier coefficients and τ_{sys} is a system dependent coefficient that includes non-birefringent intensity modifiers (Figure A1). The optical anisotropy permits a quantitative examination of the local dispersion of the collagen fibers within the specimen, which can be expressed as the degree of optical anisotropy (DOA) (Figure A2):

$$DOA = \frac{\alpha_2 + \alpha_4}{\alpha_0 + \alpha_2 + \alpha_4}. \quad (A3)$$

Please refer to more details about the step-by-step algorithmic procedures in Section 2.3 of [45].

Appendix B. Spatial Heterogeneity of CFAs

To further investigate the spatial heterogeneity of the collagen fiber architecture for each TV leaflet, a regional analysis was performed, by dividing the tissue's ROI into 3 × 3 sub-regions (Figure 2). Variations in the fiber orientation angle were observed across different sub-regions of the tissue, as some sub-regions of the tissue showed a significant shift in the fiber orientation angle in response to the mechanical loads, while other sub-regions remained unchanged (Figures A2–A4).

For example, the TVAL exhibited circumferentially-oriented fibers in the unloaded state, but when radially loaded, the fibers in the center sub-region of the tissue shifted more towards the radial direction compared to the upper portion of the tissue (Figure A2b). The differences in collagen fiber reorientation in response to loading throughout these sub-regions can be connected to the regional variations in biaxial mechanical response previously observed in TV leaflets [36]. The coupling of both the regionally-varied mechanical and microstructural responses can be used to validate affine kinematic theories regarding this relationship. As for the regional variations in the alignment of collagen fibers for each TV leaflet tissue, the belly portion, along with the rough zone (i.e., tissue strip 3, see Figure 1d), showed higher DOA values than the upper portion of the tissue.

To supplement these findings, a histological analysis for each TV leaflet displayed differences in the constituent mass fraction across the tissue domain, further highlighting the potential underlying mechanisms that contribute to collagen fiber reorientation in response to applied loading. The connection between the collagen fiber alignment and the histomorphological compositions has yet to be observed within literature. For example, previous studies are largely limited by optical clearing solutions and a more confined FOV [40,43], and, therefore, regional analyses of the load-dependent CFAs may not be feasible. Our observations, presented in this article, provide unique, spatially-varying collagen fiber architectural information that can be implemented into current rheological models for an accurate representation of the bulk tissue response [55–57].

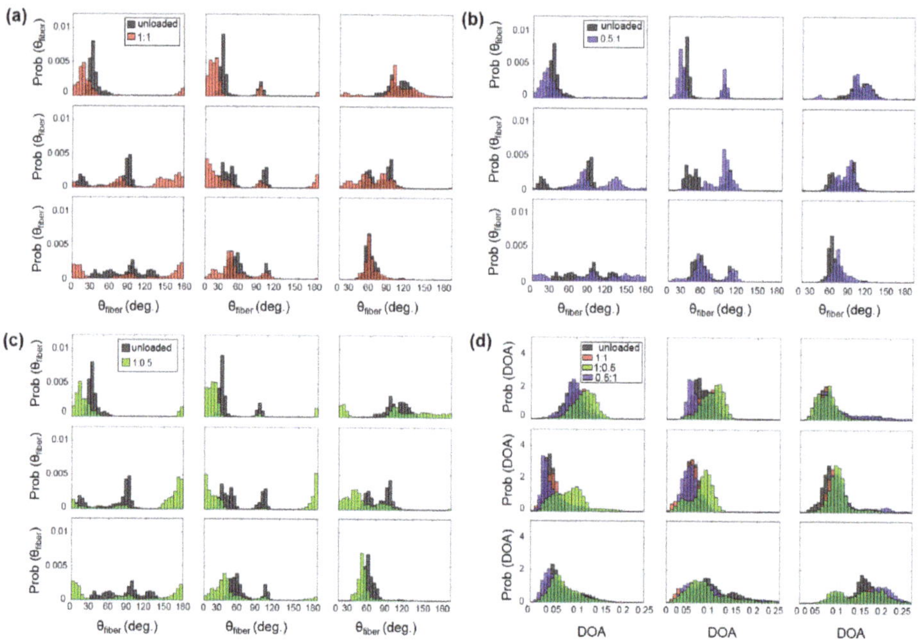

Figure A2. 3 × 3 grid comparisons of the predicted θ_{fiber} : (**a**) between unloaded and equibiaxial loading, (**b**) between unloaded and radially-dominate loading, and (**c**) between unloaded and circumferentially-dominate loading. (**d**) Comparison of the predicted DOA among all the states for the TVAL specimen of heart #1.

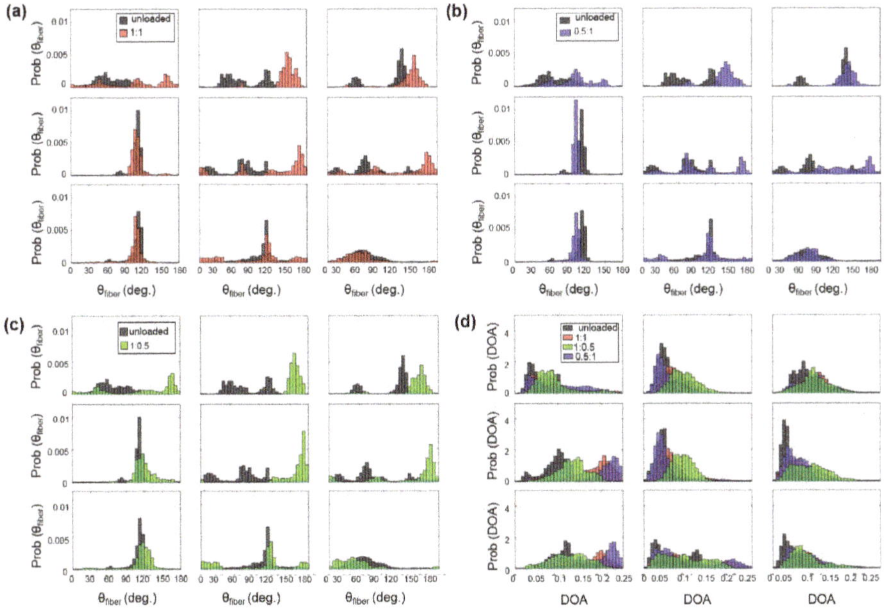

Figure A3. 3 × 3 grid comparisons of the predicted θ_{fiber} : (**a**) between unloaded and equibiaxial loading, (**b**) between unloaded and radially-dominate loading, and (**c**) between unloaded and circumferentially-dominate loading. (**d**) Comparison of the predicted DOA among all the states for the TVPL specimen of heart #1.

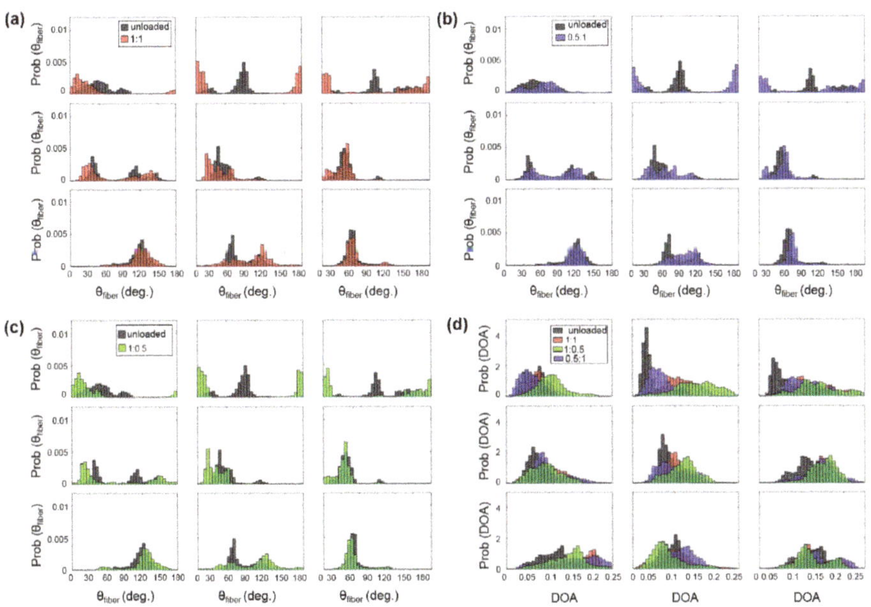

Figure A4. 3 × 3 grid comparisons of the predicted θ_{fiber} : (**a**) between unloaded and equibiaxial loading, (**b**) between unloaded and radially-dominate loading, and (**c**) between unloaded and circumferentially-dominated loading. (**d**) Comparison of the predicted DOA among all the states for the TVSL specimen of heart #1.

Appendix C. Quantified Load-Dependent Changes in the CFA for Heart #2 and Heart #3

Table A1. The percentage changes in mean, skewness, and standard deviation (SD) for the predicted θ_{fiber} and the DOA of the entire tissue's ROI for the TVAL, TVPL, and TVSL specimens of heart #2.

	DOA (TVAL)			**DOA (TVPL)**			**DOA (TVSL)**		
$T_C : T_R$	Mean	Skewness	SD	Mean	Skewness	SD	Mean	Skewness	SD
PPC	0.069	0.791	0.025	0.071	1.377	0.029	0.062	0.410	0.026
1:1	−1.6°	12.4°	9.5°	−5.3°	−10.0°	27.0°	14.3°	−23.1°	12.2°
1:0.5	12.2°	−5.2°	41.1°	6.1°	−8.1°	39.9°	43.5°	−47.6°	13.5°
0.5:1	−3.3°	−18.0°	−0.6°	−7.4°	−12.8°	6.7°	15.0°	−16.4°	19.2°
	θ_{fiber} (TVAL)			**θ_{fiber} (TVPL)**			**θ_{fiber} (TVSL)**		
$T_C : T_R$	Mean	Skewness	SD	Mean	Skewness	SD	Mean	Skewness	SD
PPC	93.4°	−0.365	30.0°	105.3°	−0.891	26.3°	106.0°	−0.462	31.9°
1:1	−1.4°	−18.6°	15.2°	6.5°	−19.4°	47.1°	0.3°	27.9°	54.1°
1:0.5	5.5°	12.5°	−2.1°	14.3°	−16.5°	2.6°	11.5°	54.0°	3.4°
0.5:1	−9.9°	−130.9°	25.6°	−3.5°	−46.5°	78.4°	−13.8°	−94.0°	68.9°

Table A2. The percent changes in mean, skewness, and standard deviation (SD) for the predicted θ_{fiber} and the DOA of the entire tissue's ROI, for the TVAL, TVPL, and TVSL specimens of heart #3.

	DOA (TVAL)			**DOA (TVPL)**			**DOA (TVSL)**		
$T_C : T_R$	Mean	Skewness	SD	Mean	Skewness	SD	Mean	Skewness	SD
PPC	0.072	0.861	0.026	0.065	2.774	0.032	0.082	0.515	0.028
1:1	0.3°	−5.8°	20.4°	25.8°	−82.9°	6.6°	4.2°	5.4°	12.0°
1:0.5	12.9°	−35.8°	22.3°	39.5°	−95.8°	12.2°	12.0°	40.2°	15.9°
0.5:1	−7.1°	13.4°	16.2°	12.8°	−86.9°	−1.5°	1.2°	−38.5°	14.9°
	θ_{fiber} (TVAL)			**θ_{fiber} (TVPL)**			**θ_{fiber} (TVSL)**		
$T_C : T_R$	Mean	Skewness	SD	Mean	Skewness	SD	Mean	Skewness	SD
PPC	113.1°	−0.256	22.7°	110.6	−0.948	34.3°	85.0°	0.639	29.4°
1:1	−2.1°	12.9°	35.5°	−5.9°	−35.4°	−5.4°	−5.9°	21.0°	6.2°
1:0.5	−1.1°	−27.9°	12.1°	−3.0°	14.7°	−22.5°	6.9°	−34.2°	−11.7°
0.5:1	−6.7°	59.8°	60.1°	−13.5°	−70.3°	32.6°	−17.7°	98.9°	18.5°

References

1. Arsalan, M.; Walther, T.; Smith, R.L., II; Grayburn, P.A. Tricuspid regurgitation diagnosis and treatment. *Eur. Heart J.* **2015**, *38*, 634–638. [CrossRef] [PubMed]
2. Anyanwu, A.C.; Adams, D.H. Functional tricuspid regurgitation in mitral valve disease: Epidemiology and prognostic implications. *Semin. Thorac. Cardiovasc. Surg.* **2010**, *22*, 69–75. [CrossRef] [PubMed]
3. Di Mauro, M.; Bezante, G.P.; Di Baldassarre, A.; Clemente, D.; Cardinali, A.; Acitelli, A.; Salerni, S.; Penco, M.; Calafiore, A.M.; Gallina, S. Functional tricuspid regurgitation: An underestimated issue. *Int. J. Cardiol.* **2013**, *168*, 707–715. [CrossRef] [PubMed]
4. Braunwald, N.S.; Ross, J., Jr.; Morrow, A.G. Conservative management of tricuspid regurgitation in patients undergoing mitral valve replacement. *Circulation* **1967**, *35*, I-63–I-69. [CrossRef]
5. Dreyfus, G.D.; Corbi, P.J.; Chan, K.M.J.; Bahrami, T. Secondary tricuspid regurgitation or dilatation: Which should be the criteria for surgical repair? *Ann. Thorac. Surg.* **2005**, *79*, 127–132. [CrossRef]
6. Cyron, C.J.; Humphrey, J.D. Growth and remodeling of load-bearing biological soft tissues. *Meccanica* **2017**, *52*, 645–664. [CrossRef]

7. Driessen, N.J.B.; Cox, M.A.J.; Bouten, C.V.C.; Baaijens, F.P.T. Remodelling of the angular collagen fiber distribution in cardiovascular tissues. *Biomech. Model. Mechanobiol.* **2008**, *7*, 93–103. [CrossRef]
8. Song, J.-M.; Jang, M.-K.; Kim, Y.-J.; Kim, D.-H.; Kang, D.-H.; Song, J.-K. Right ventricular remodeling determines tricuspid valve geometry and the severity of functional tricuspid regurgitation: A real-time 3-dimensional echocardiography study. *Korean Circ. J.* **2010**, *40*, 448–453. [CrossRef] [PubMed]
9. Kramer, K.E.; Ross, C.J.; Laurence, D.W.; Babu, A.R.; Wu, Y.; Towner, R.A.; Mir, A.; Burkhart, H.M.; Holzapfel, G.A.; Lee, C.-H. An investigation of layer-specific tissue biomechanics of porcine atrioventricular heart valve leaflets. *Acta Biomater.* **2019**, *96*, 368–384. [CrossRef]
10. Jett, S.V.; Laurence, D.W.; Kunkel, R.P.; Babu, A.R.; Kramer, K.E.; Baumwart, R.; Towner, R.A.; Wu, Y.; Lee, C.-H. An investigation of the anisotropic mechanical properties and anatomical structure of porcine atrioventricular heart valves. *J. Mech. Behav. Biomed. Mater.* **2018**, *87*, 155–171. [CrossRef]
11. Tseng, H.; Grande-Allen, K.J. Elastic fibers in the aortic valve spongiosa: A fresh perspective on its structure and role in overall tissue function. *Acta Biomater.* **2011**, *7*, 2101–2108. [CrossRef] [PubMed]
12. Hinton, R.B.; Yutzey, K.E. Heart valve structure and function in development and disease. *Annu. Rev. Physiol.* **2011**, *73*, 29–46. [CrossRef] [PubMed]
13. Stella, J.A.; Sacks, M.S. On the biaxial mechanical properties of the layers of the aortic valve leaflet. *J. Biomech. Eng.* **2007**, *129*, 757–766. [CrossRef] [PubMed]
14. Khoiy, K.A.; Amini, R. On the biaxial mechanical response of porcine tricuspid valve leaflets. *J. Biomech. Eng.* **2016**, *138*, 104504. [CrossRef]
15. Mathur, M.; Jazwiec, T.; Meador, W.D.; Malinowski, M.; Goehler, M.; Ferguson, H.; Timek, T.A.; Rausch, M.K. Tricuspid valve leaflet strains in the beating ovine heart. *Biomech. Model. Mechanobiol.* **2019**, *18*, 1351–1361. [CrossRef]
16. Meador, W.D.; Mathur, M.; Rausch, M.K. Tricuspid valve biomechanics: A brief review. In *Advances in Heart Valve Biomechanics*; Springer: Berlin/Heidelberg, Germany, 2018; pp. 105–114.
17. Pham, T.; Sulejmani, F.; Shin, E.; Wang, D.; Sun, W. Quantification and comparison of the mechanical properties of four human cardiac valves. *Acta Biomater.* **2017**, *54*, 345–355. [CrossRef]
18. Lim, K.O. Mechanical properties and ultrastructure of normal human tricuspid valve chordae tendineae. *Jpn. J. Physiol.* **1980**, *30*, 455–464. [CrossRef]
19. Pokutta-Paskaleva, A.; Sulejmani, F.; DelRocini, M.; Sun, W. Comparative mechanical, morphological, and microstructural characterization of porcine mitral and tricuspid leaflets and chordae tendineae. *Acta Biomater.* **2019**, *85*, 241–252. [CrossRef]
20. Ross, C.J.; Laurence, D.W.; Hsu, M.-C.; Baumwart, R.; Zhao, Y.D.; Mir, A.; Burkhart, H.M.; Holzapfel, G.A.; Wu, Y.; Lee, C.-H. Mechanics of Porcine Heart Valves' Strut Chordae Tendineae Investigated as a Leaflet–Chordae–Papillary Muscle Entity. *Ann. Biomed. Eng.* **2020**. [CrossRef]
21. Basu, A.; He, Z. Annulus tension on the tricuspid valve: An in-vitro study. *Cardiovasc. Eng. Technol.* **2016**, *7*, 270–279. [CrossRef]
22. Makaryus, A.N.; Ismail, H.; Makaryus, J.N.; Fan, D. Geometric comparison of the mitral and tricuspid valve annulus: Insights from three dimensional transesophageal echocardiography. *World J. Cardiol.* **2017**, *9*, 757. [CrossRef] [PubMed]
23. Malinowski, M.; Jazwiec, T.; Goehler, M.; Quay, N.; Bush, J.; Jovinge, S.; Rausch, M.K.; Timek, T.A. Sonomicrometry-derived 3-dimensional geometry of the human tricuspid annulus. *J. Thorac. Cardiovasc. Surg.* **2018**, *157*, 1452–1461. [CrossRef] [PubMed]
24. Rausch, M.K.; Malinowski, M.; Wilton, P.; Khaghani, A.; Timek, T.A. Engineering analysis of tricuspid annular dynamics in the beating ovine heart. *Ann. Biomed. Eng.* **2018**, *46*, 443–451. [CrossRef]
25. Rausch, M.K.; Malinowski, M.; Meador, W.D.; Wilton, P.; Khaghani, A.; Timek, T.A. The effect of acute pulmonary hypertension on tricuspid annular height, strain, and curvature in sheep. *Cardiovasc. Eng. Technol.* **2018**, *9*, 365–376. [CrossRef]
26. Ring, L.; Rana, B.S.; Kydd, A.; Boyd, J.; Parker, K.; Rusk, R.A. Dynamics of the tricuspid valve annulus in normal and dilated right hearts: A three-dimensional transoesophageal echocardiography study. *Eur. Heart J. Cardiovasc. Imaging* **2012**, *13*, 756–762. [CrossRef] [PubMed]
27. Dreyfus, G.D.; Martin, R.P.; Chan, K.J.; Dulguerov, F.; Alexandrescu, C. Functional tricuspid regurgitation: A need to revise our understanding. *J. Am. Coll. Cardiol.* **2015**, *65*, 2331–2336. [CrossRef]

28. Hahn, R.T. Imaging of the Tricuspid Valve: Transoesophageal Echocardiography. In *Practical Manual of Tricuspid Valve Diseases*; Springer: Berlin/Heidelberg, Germany, 2018; pp. 117–126.
29. Kim, J.B.; Jung, S.-H.; Choo, S.J.; Chung, C.H.; Lee, J.W. Clinical and echocardiographic outcomes after surgery for severe isolated tricuspid regurgitation. *J. Thorac. Cardiovasc. Surg.* **2013**, *146*, 278–284. [CrossRef] [PubMed]
30. Nishi, H.; Toda, K.; Miyagawa, S.; Yoshikawa, Y.; Fukushima, S.; Kawamura, M.; Yoshioka, D.; Saito, T.; Ueno, T.; Kuratani, T. Tricuspid annular dynamics before and after tricuspid annuloplasty–Three-dimensional transesophageal echocardiography. *Circ. J.* **2015**, *79*, 873–879. [CrossRef]
31. Utsunomiya, H.; Itabashi, Y.; Mihara, H.; Kobayashi, S.; De Robertis, M.A.; Trento, A.; Shiota, T. Usefulness of 3D echocardiographic parameters of tricuspid valve morphology to predict residual tricuspid regurgitation after tricuspid annuloplasty. *Eur. Heart J. Cardiovasc. Imaging* **2017**, *18*, 809–817. [CrossRef]
32. Driessen, N.J.B.; Bouten, C.V.C.; Baaijens, F.P.T. Improved prediction of the collagen fiber architecture in the aortic heart valve. *J. Biomech. Eng.* **2005**, *127*, 329–336. [CrossRef]
33. Liao, J.; Yang, L.; Grashow, J.; Sacks, M.S. The relation between collagen fibril kinematics and mechanical properties in the mitral valve anterior leaflet. *J. Biomech. Eng.* **2007**, *129*, 78–87. [CrossRef]
34. Lee, C.-H.; Zhang, W.; Liao, J.; Carruthers, C.A.; Sacks, J.I.; Sacks, M.S. On the presence of affine fibril and fiber kinematics in the mitral valve anterior leaflet. *Biophys. J.* **2015**, *108*, 2074–2087. [CrossRef] [PubMed]
35. Meador, W.D.; Mathur, M.; Sugerman, G.P.; Jazwiec, T.; Malinowski, M.; Bersi, M.R.; Timek, T.A.; Rausch, M.K. A detailed mechanical and microstructural analysis of ovine tricuspid valve leaflets. *Acta Biomater.* **2020**, *102*, 100–113. [CrossRef] [PubMed]
36. Laurence, D.W.; Ross, C.J.; Jett, S.V.; Johns, C.H.; Echols, A.L.; Baumwart, R.; Towner, R.A.; Liao, J.; Bajona, P.; Wu, Y.; et al. An investigation of regional variations in the biaxial mechanical properties and stress relaxation behaviors of porcine atrioventricular heart valve leaflets. *J. Biomech.* **2019**, *83*, 16–27. [CrossRef] [PubMed]
37. Chen, X.; Nadiarynkh, O.; Plotnikov, S.; Campagnola, P.J. Second harmonic generation microscopy for quantitative analysis of collagen fibrillar structure. *Nat. Protoc.* **2012**, *7*, 654–669. [CrossRef] [PubMed]
38. Schriefl, A.J.; Wolinski, H.; Regitnig, P.; Kohlwein, S.D.; Holzapfel, G.A. An automated approach for three-dimensional quantification of fibrillar structures in optically cleared soft biological tissues. *J. R. Soc. Interface* **2013**, *10*, 20120760. [CrossRef] [PubMed]
39. Zhang, Y.; Akins, M.L.; Murari, K.; Xi, J.; Li, M.-J.; Luby-Phelps, K.; Mahendroo, M.; Li, X. A compact fiber-optic SHG scanning endomicroscope and its application to visualize cervical remodeling during pregnancy. *Proc. Natl. Acad. Sci. USA* **2012**, *109*, 12878–12883. [CrossRef] [PubMed]
40. Gaul, R.T.; Nolan, D.R.; Lally, C. Collagen fibre characterisation in arterial tissue under load using SALS. *J. Mech. Behav. Biomed. Mater.* **2017**, *75*, 359–368. [CrossRef]
41. Sacks, M.S.; Smith, D.B.; Hiester, E.D. A small angle light scattering device for planar connective tissue microstructural analysis. *Ann. Biomed. Eng.* **1997**, *25*, 678–689. [CrossRef]
42. Goth, W.; Lesicko, J.; Sacks, M.S.; Tunnell, J.W. Optical-based analysis of soft tissue structures. *Annu. Rev. Biomed. Eng.* **2016**, *18*, 357–385. [CrossRef]
43. Plotnikov, S.; Juneja, V.; Isaacson, A.B.; Mohler, W.A.; Campagnola, P.J. Optical clearing for improved contrast in second harmonic generation imaging of skeletal muscle. *Biophys. J.* **2006**, *90*, 328–339. [CrossRef] [PubMed]
44. Goth, W.; Potter, S.; Allen, A.C.B.; Zoldan, J.; Sacks, M.S.; Tunnell, J.W. Non-destructive reflectance mapping of collagen fiber alignment in heart valve leaflets. *Ann. Biomed. Eng.* **2019**, *47*, 1250–1264. [CrossRef] [PubMed]
45. Jett, S.V.; Hudson, L.T.; Baumwart, R.; Bohnstedt, B.N.; Mir, A.; Burkhart, H.M.; Holzapfel, G.A.; Wu, Y.; Lee, C.-H. Integration of polarized spatial frequency domain imaging (pSFDI) with a biaxial mechanical testing system for quantification of load-dependent collagen architecture in soft collagenous tissues. *Acta Biomater.* **2020**, *102*, 149–168. [CrossRef]
46. Duginski, G.A.; Ross, C.J.; Laurence, D.W.; Johns, C.H.; Lee, C.-H. An investigation of the effect of freezing storage on the biaxial mechanical properties of excised porcine tricuspid valve anterior leaflets. *J. Mech. Behav. Biomed. Mater.* **2020**, *101*, 103438. [CrossRef] [PubMed]
47. Ross, C.J.; Laurence, D.W.; Wu, Y.; Lee, C.-H. Biaxial mechanical characterizations of atrioventricular heart valves. *J. Vis. Exp.* **2019**, *146*, e59170. [CrossRef] [PubMed]
48. Fung, Y.C. On mathematical models of stress-strain relationship for living soft tissues. *Polym. Mech.* **1975**, *11*, 726–740. [CrossRef]

49. Cuccia, D.J.; Bevilacqua, F.; Durkin, A.J.; Tromberg, B.J. Modulated imaging: Quantitative analysis and tomography of turbid media in the spatial-frequency domain. *Opt. Lett.* **2005**, *30*, 1354–1356. [CrossRef]
50. Cuccia, D.J.; Bevilacqua, F.P.; Durkin, A.J.; Ayers, F.R.; Tromberg, B.J. Quantitation and mapping of tissue optical properties using modulated imaging. *J. Biomed. Opt.* **2009**, *14*, 024012. [CrossRef]
51. Mazhar, A.; Saggese, S.; Pollins, A.C.; Cardwell, N.L.; Nanney, L.B.; Cuccia, D.J. Noncontact imaging of burn depth and extent in a porcine model using spatial frequency domain imaging. *J. Biomed. Opt.* **2014**, *19*, 086019. [CrossRef]
52. Ruifrok, A.C.; Johnston, D.A. Quantification of histochemical staining by color deconvolution. *Anal. Quant. Cytol. Histol.* **2001**, *23*, 291–299.
53. Heyden, S.; Nagler, A.; Bertoglio, C.; Biehler, J.; Gee, M.W.; Wall, W.A.; Ortiz, M. Material modeling of cardiac valve tissue: Experiments, constitutive analysis and numerical investigation. *J. Biomech.* **2015**, *48*, 4287–4296. [CrossRef] [PubMed]
54. Jett, S.V.; Laurence, D.W.; Kunkel, R.P.; Babu, A.R.; Kramer, K.E.; Baumwart, R.; Towner, R.A.; Wu, Y.; Lee, C.-H. Biaxial mechanical data of porcine atrioventricular valve leaflets. *Data Brief* **2018**, *21*, 358–363. [CrossRef] [PubMed]
55. Fung, Y.C. Structure and Stress-Strain Relationship of Soft Tissues. *Am. Zool.* **1984**, *24*, 13–22. [CrossRef]
56. Lanir, Y. A structural theory for the homogeneous biaxial stress-strain relationships in flat collagenous tissues. *J. Biomech.* **1979**, *12*, 423–436. [CrossRef]
57. Lanir, Y. Constitutive equations for fibrous connective tissues. *J. Biomech.* **1983**, *16*, 1–12. [CrossRef]
58. Pant, A.D.; Thomas, V.S.; Black, A.L.; Verba, T.; Lesicko, J.G.; Amini, R. Pressure-induced microstructural changes in porcine tricuspid valve leaflets. *Acta Biomater.* **2018**, *67*, 248–258. [CrossRef]
59. Kunzelman, K.S.; Cochran, R.P. Stress/strain characteristics of porcine mitral valve tissue: Parallel versus perpendicular collagen orientation. *J. Card. Surg.* **1992**, *7*, 71–78. [CrossRef]
60. Stella, J.A.; Liao, J.; Sacks, M.S. Time-dependent biaxial mechanical behavior of the aortic heart valve leaflet. *J. Biomech.* **2007**, *40*, 3169–3177. [CrossRef]
61. Anssari-Benam, A.; Screen, H.R.C.; Bucchi, A. Insights into the micromechanics of stress-relaxation and creep behaviours in the aortic valve. *J. Mech. Behav. Biomed. Mater.* **2019**, *93*, 230–245. [CrossRef]
62. Yang, B.; Lesicko, J.; Sharma, M.; Hill, M.; Sacks, M.S.; Tunnell, J.W. Polarized light spatial frequency domain imaging for non-destructive quantification of soft tissue fibrous structures. *Biomed. Opt. Express* **2015**, *6*, 1520–1533. [CrossRef]

bioengineering

MDPI

Communication

Mechanical Response Changes in Porcine Tricuspid Valve Anterior Leaflet Under Osmotic-Induced Swelling

Samuel D. Salinas, Margaret M. Clark and Rouzbeh Amini *

Department of Biomedical Engineering, The University of Akron, Akron, OH 44325, USA
* Correspondence: ramini@uakron.edu; Tel.: +1-(330)-972-5232

Received: 31 May 2019; Accepted: 13 August 2019; Published: 15 August 2019

Abstract: Since many soft tissues function in an isotonic in-vivo environment, it is expected that physiological osmolarity will be maintained when conducting experiments on these tissues ex-vivo. In this study, we aimed to examine how not adhering to such a practice may alter the mechanical response of the tricuspid valve (TV) anterior leaflet. Tissue specimens were immersed in deionized (DI) water prior to quantification of the stress–strain responses using an in-plane biaxial mechanical testing device. Following a two-hour immersion in DI water, the tissue thickness increased an average of 107.3% in the DI water group compared to only 6.8% in the control group, in which the tissue samples were submerged in an isotonic phosphate buffered saline solution for the same period of time. Tissue strains evaluated at 85 kPa revealed a significant reduction in the radial direction, from 34.8% to 20%, following immersion in DI water. However, no significant change was observed in the control group. Our study demonstrated the impact of a hypo-osmotic environment on the mechanical response of TV anterior leaflet. The imbalance in ions leads to water absorption in the valvular tissue that can alter its mechanical response. As such, in ex-vivo experiments for which the native mechanical response of the valves is important, using an isotonic buffer solution is essential.

Keywords: biaxial mechanical testing; cardiac valves; osmotic swelling

---

1. Introduction

The properties of many soft tissues can best be obtained by conducting experiments in an environment that resembles in-vivo conditions. In the realm of biomechanics, the characterization of soft tissue mechanical properties has traditionally relied on benchtop experiments such as uniaxial or biaxial tensile extension tests [1,2]. For these experiments, both for the purpose of tissue storage and during the course of the experiments, tissue samples are generally immersed in a buffer solution (e.g., phosphate buffered saline (PBS)). By virtue of its non-toxicity to cells and its pH buffering capability, PBS is widely used in biological studies [3].

The regulation of osmolarity is also of great importance in maintaining cell and tissue viability. Researchers in previous studies have quantified the effects of hypo- and hyper-osmolarity on soft tissues [4–8]. The low concentration of ions in a hypo-osmotic solution ultimately leads to tissue swelling over time. The subsequent changes in morphology of the tissue and the potential for damage to its constituents could lead to alteration of the mechanical responses [9,10]. Moreover, Lanir et al. have shown the effects of swelling and their correlation with residual stresses, as demonstrated in the left ventricle and aortic tissues of murine models [5,10,11]. Despite such strong evidence of mechanical dependence on normal osmolarity, deionized (DI) water has been employed in some studies in lieu of isotonic solutions [12,13]. In one particular in-vitro study conducted on heart valves, the assumption of no potential difference between DI water and PBS as it pertains to the mechanical responses of the tissues was adopted [12]. Notwithstanding the importance of the findings of these studies, isotonic solutions have been generally used in similar ex-vivo valvular studies to prevent changes in the mechanical responses of heart valves [14–17]. Since no previous experiments have been conducted to specifically show the effects of hypotonicity on the mechanical response of cardiac valves, we performed experiments on porcine tricuspid valve (TV) anterior leaflets in order to guide future research in heart valve biomechanics.

The TV, which is located on the pulmonary side of the heart, is composed of three leaflets: anterior, posterior and septal leaflets. The study of the biomechanics of this valve, albeit nascent in comparison to the study of mitral valve biomechanics, has seen an emergence in interest [16,18–26]. The TV is characterized by having a larger orifice than the mitral valve as well as having thinner leaflets [27]. The extracellular matrix (ECM) in both heart valves is comprised primarily of collagen, elastin, and proteoglycans; and hence, it plays a role in the mechanical response of the valve tissue [28–30]. In this brief study, we hypothesize that the swelling effect due to the exposure of TV anterior leaflets to DI water will alter their mechanical response.

2. Materials and Methods

2.1. Specimen Preparation

Porcine hearts ($n = 14$) were acquired from a local slaughterhouse (3-D Meats, Dalton, OH, USA) and transported in chilled PBS back to our laboratory. Consistent with our previous methodology [31], upon isolating the TV apparatus, we identified and excised the anterior leaflet. The tissue was later trimmed to a smaller square size (approximately 11 mm × 11 mm) using a custom-made tissue phantom while ensuring that the axes of the tissue phantom coincided with the radial and circumferential anatomical directions of the tissue samples as described previously [31]. The radial direction was defined perpendicular to the TV annulus and the circumferential direction was defined as the direction perpendicular to the aforementioned radial direction [32].

Prior to mounting the anterior leaflet specimen on a custom-built biaxial tensile machine [31,33], the thickness of each specimen was measured using a thickness gauge. Five readings were taken from each sample with the average value used in our calculations. Next, four glass submillimeter markers were attached on the surface of the leaflet for optical tracking of tissue deformation. Suture lines were attached around the edges of the tissue. The dimensions of the trimmed tissue enclosed by the suture lines were 7.6 mm × 7.6 mm.

2.2. Biaxial Testing Protocol

The maximum right ventricular pressure for a normal person is defined as 30 mmHg [34,35]. From our previous study of the porcine TV, we found that the average thickness of the anterior leaflet was 313 μm [31]. In our prior work, we also employed Laplace's law to arrive at an estimated stress value for the leaflets. Given the above parameters, the maximum target stress used in this study was calculated to be 127 kPa. A total of five loading protocols, listed in Table 1, were employed.

Table 1. Loading protocol for radial and circumferential directions.

Protocol	Radial (kPa)	Circum. (kPa)
1	127	127
2	95.25	127
3	127	95.25
4	63.5	127
5	127	63.5

Each protocol consisted of ten loading/unloading cycles. Only data from the tenth cycle was used in our analysis; the first nine were used for pre-conditioning purposes. The bath was filled with room temperature (21 °C) PBS and the specimen was loaded on the biaxial actuators. A tare load of 0.5 grams was used throughout the biaxial testing of the anterior leaflets. It is important to note that the load applied to achieve the desired stress was dependent upon the sample thickness. Hence, each sample had unique loads applied to it.

2.3. Tissue Swelling Application

Following all five loading protocols, tissue samples were unmounted and placed in DI water. In previous in-house tests (data not shown), we had observed that the maximum swelling of porcine TV leaflets occurred after they were immersed in DI water for two hours. As such, we used a two-hour submergence period for all samples in this study. Following the two-hour soaking period, the specimen thickness was measured again. Before being remounted on the biaxial testing machine, the samples were retrimmed to the 11 mm × 11 mm specimen size mentioned above. All protocols from Table 1 were then repeated. Similarly, our control group ($n = 14$), which used a subset of specimens soaked in PBS for 2 hours, underwent the same testing procedure as specimens that were soaked in DI water. Because the control samples did not change in size, no specimen retrimming was necessary.

2.4. Data Processing

Data collected from biaxial testing was analyzed using an internally developed program in MATLAB (MathWorks, Nantick, MA, USA). Positional data obtained from tracking the surface-mounted glass fiducial markers allowed for the calculation of the deformation gradient tensor, F, as described previously [31,36].

The Green–Lagrangian strain tensor, E, was then calculated:

$$E = \frac{1}{2}\left(F^T F - I\right) \tag{1}$$

where I is the identity matrix.

The load applied on the specimen allowed for the calculation of the first Piola-Kirchoff stress:

$$P_{rr} = \frac{F_r}{A},\ P_{cc} = \frac{F_c}{A} \tag{2}$$

where F is the force applied by the actuators and A is the cross-sectional area, which is defined as the product of the length (7.6 mm) and thickness of the sample. The double subscripts rr and cc designate the radial and circumferential normal stresses, respectively. Likewise, the single subscripts r and c refer to the applied force in the radial and circumferential directions, respectively.

2.5. Statistical Analysis

To determine the effect of soaking in DI water on the mechanical response of the leaflet, a Student's paired t-test was used. The null hypothesis for this analysis was that the average strain at an equibiaxial load of 85 kPa following DI water exposure was equivalent to the average strain prior to DI water submersion. A value of $p \leq 0.05$ was considered as significant for this test. Only the equibiaxial data were used in the current analysis; data from other protocols may be used for future analyses, if needed.

3. Results

Following two hours of soaking in DI water, the tissue exhibited less compliance in the radial direction as compared to the circumferential direction, as can be noticed from the results in Figure 1. Visual examination of the equibiaxial data (Figure 1) shows that the mechanical response in the control group exhibited a slight increase in radial strain. The additional protocols from Table 1 display similar behaviors, primarily concerning the radial compliance. Relative to the DI water group, the change in stress–strain response, albeit different, was not as aberrant in the control group. Subsequent protocols in the control group also showed a slight compliance in the radial direction similar to what is displayed in Figure 1. The stress–strain responses for the additional protocols, from Table 1, are provided in the Supplementary Materials.

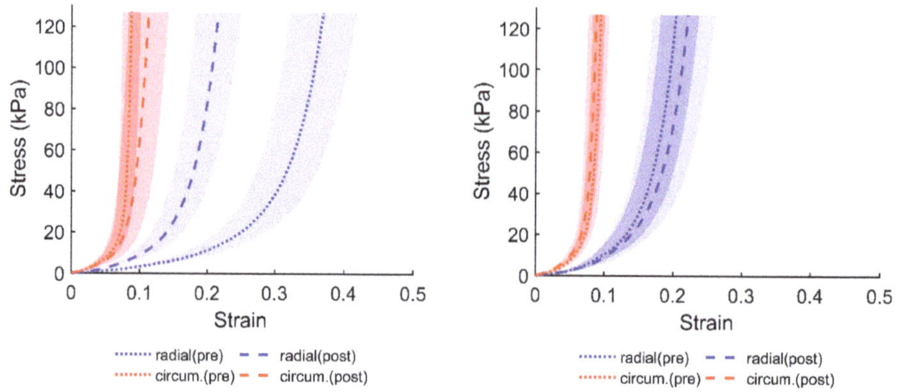

(a) (b)

Figure 1. Average equibiaxial response of the anterior leaflet (i.e., Protocol 1 in Table 1) from: (**a**) DI water group; (**b**) control group (soaked in PBS). Shaded regions represent standard error.

The effect that DI water had on the leaflet shape and color was obvious upon visual inspection, since the tissue absorbed water during the soaking process. Following the two-hour soaking in DI water, the tissue thickness increased by 107.3% versus 6.8% in the control group that was soaked in PBS (Table 2). A Student's paired t-test comparing pre- and post-treatment thickness across both DI and Control groups revealed that while both groups were significantly different, the p-value for the DI treatment group (10^{-5}) was much smaller than that of the control ($p = 0.002$).

Table 2. Average thickness of anterior leaflets in DI water group and control group.

Heart	DI-Pre	DI-Post	Control-Pre	Control-Post
1	424	924	269	299
2	391	627	237	325
3	314	662	342	393
4	259	490	312	350
5	246	467	287	284
6	317	548	233	259
7	254	482	287	284
8	332	416	223	226
9	337	614	246	251
10	299	599	223	228
11	322	609	348	378
12	350	548	343	396
13	224	1061	335	365
14	315	1046	269	279
Average	313	650	289	308
Std. Dev.	53	201	44	57
p-value	<0.001		0.002	

Note: All thickness values reported in this table are in microns.

At a mean normal ventricular pressure of 25 mmHg [37], the physiological strain using Laplace's law (as mentioned above) was approximated to be 85 kPa. As shown in Figure 2, the treatment group exhibited a significant ($p = 0.0026$) radial reduction (34.8% to 20%) following exposure to DI water. The change in circumferential strain was neither visually detectable (8.4% to 10.4%) nor statistically different ($p = 0.5176$). Although the control group, in Figure 2, showed a minimal change in radial strain (19% to 20.6%) and circumferential strain (9% to 8.3%), the strains were not found to be significantly different in either the radial ($p = 0.426$) or the circumferential ($p = 0.546$) direction.

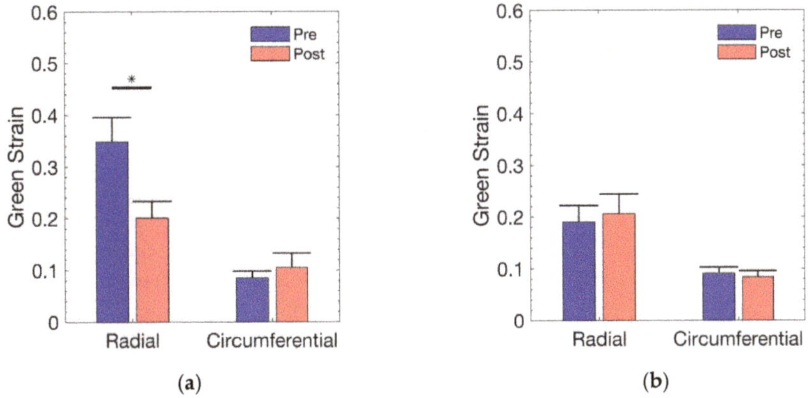

Figure 2. Average equibiaxial strain across all samples at an estimated physiological stress level of 85 kPa in: (**a**) DI water group; (**b**) control group. Error bars represent standard error.

4. Discussion

The aim of this study was to determine the validity of the use of DI water, in lieu of isotonic solutions, as a viable medium for the handling of heart valve tissues during experiments that rely on obtaining native mechanical responses of the tissue. PBS, as a commonly used isotonic solution, does present various benefits for experiments involving biological specimens. Primarily, the isotonic properties of PBS ensure cell and tissue viability by approximating a physiological environment that is appropriate for data collection in research. Our study has shown that for equibiaxial tests, the response in the radial direction is greatly affected by the hypo-osmotic environment introduced by DI water. Although the changes in the circumferential direction seem to be much smaller, due to the tensorial nature of strain, any changes in one component of the strain (in this case the normal strain in the radial direction) is indicative of a completely different state of deformation in the tissue.

The stiffer tissue response in the radial direction (as shown in Figure 1) may be indicative of a restrictive ECM environment as a result of tissue swelling and the stretching of fibers [11]. The submersion of soft tissues in solutions with different osmotic conditions is known to affect tissue morphology by changing the intrafibrillar water [7]. This change was clearly seen in the TV samples following their submersion in DI water. In fact, not only was the specimen thickness affected (Table 2), but the overall dimensions of the tissue were also altered. Interestingly, statistical analysis performed on the thickness in both groups revealed that both were significantly different, albeit the DI treatment group had a much smaller p-value. While a significant thickness difference was not expected in the control group, such differences did not lead to significant changes in the mechanical responses of the tissue.

It has been shown that ECM collagen fibers are generally undulated in the unloaded state and, with increases in deformation leading to uncrimping of the fibers, they become stiffer [38]. The TV anterior leaflet, however, has a complex microstructure that affects its tissue-level mechanical response [39]. As such, understanding the overall effects of swelling-induced uncrimping and their influence on the mechanical properties of the TV leaflets will require further investigation.

In a previous study, Pierce et al. examined the effects of exposure to hypotonic solutions on the mechanical responses of the mitral valve annulus as a pilot study [12]. They used an indentation method and compared the measured forces for the same level of indentation depth in samples submerged in DI water as compared to those immersed in an isotonic saline solution. Even with a small sample size, the mean force recorded for the control group was smaller than that of the hypotonic group, indicating the potential stiffening of the tissue. Although no statistically significant difference was found, it is worth noting that a small sample size was used, and the standard deviations were relatively large. While caution should be taken in comparing biaxial testing and indentation and also in extending the findings for one type of cardiac valve to another, we believe that with an increased sample size, the investigations of Pierce et al. would likely corroborate the findings of our study.

Our study was not without limitations. Although the statistical analysis performed on the control group did not indicate a significant difference in response due to immersion in an isotonic PBS solution, one should recognize that time also plays a degenerative role in tissue integrity. Our current method was not able to isolate such potential effects on the tissue mechanical responses. It is possible that with longer immersion in an isotonic PBS solution at room temperature (21 °C), tissue degeneration leads to alteration of the mechanical responses of the tissue. In addition, the focus of this study was on the role of isotonicity on the biomechanical responses of the tricuspid valves, especially as it pertains to ex-vivo setups [14,15]. However, in functioning valves, the viscosity of the flow could significantly affect the flow profile and subsequently alter the deformation of the valve leaflets. As shown by Biswas et al., even viscosity-matched media may not biomechanically function in a manner similar to native blood, and such limitations should always be considered in the interpretation of the results of ex-vivo studies [40]. Another limiting factor in interpretation of our data is the existence of a significant difference between the radial responses at estimated physiological stress in the DI group prior to DI water exposure and in those of the control group (p-value = 0.0174, Student's paired t-test). Since no known errors existed in our methodology, we attributed such differences to possible variations in the porcine hearts obtained from the local slaughterhouse. Despite such differences, in our opinion, the important outcome of this study, i.e., exposure to DI water significantly changes the mechanical responses of the tricuspid valve leaflets, is still valid as it relies on the statistical comparison within each of the two groups rather than a comparison between the two groups. Lastly, tissue swelling encompasses not only physical changes in specimen thickness but also volume. Since the biaxial tensile machine is able to track the fiducial marker configuration, any changes in tissue can be tracked. However, following DI water application and tissue swelling, the physical changes in the specimen required that the tissue be retrimmed to the 11 mm × 11 mm size discussed in Section 2.1. This measure required that a new set of markers be attached, thereby losing the original marker configuration that would allow for the quantification of the induced physical changes using the marker tracking tool in our biaxial testing device.

5. Conclusions

The effect of DI water on the anterior TV leaflet yielded a mechanical response that was significantly different from the response of valves that were immersed in an isotonic PBS solution. Such outcomes further support the importance of using an isotonic solution when conducting experiments that require mimicking the in-vivo mechanical response of cardiac valve tissues.

Supplementary Materials: The following are available online at http://www.mdpi.com/2306-5354/6/3/70/s1, Figure S1: Protocol 2 for DI water group, Figure S2: Protocol 3 for DI water group, Figure S3: Protocol 4 for DI water group, Figure S4: Protocol 5 for DI water group, Figure S5: Protocol 2 for control group, Figure S6: Protocol 3 for control group, Figure S7: Protocol 4 for control group, Figure S8: Protocol 5 for control group.

Author Contributions: All authors were involved in preliminary literature survey; M.C. performed the experiments; S.S. was involved in data analysis; R.A. contributed to conceptualization of the research idea, funding acquisition, and supervision of the study. All authors contributed to preparation of this manuscript.

Funding: This research was funded in part by a grant from the American Heart Association (Grant No. 16SDG27770088) and National Science Foundation (CAREER Award No. 1846715).

Acknowledgments: The thoughtful comments from Sheila Pearson and Francis Loth at The University of Akron are acknowledged.

Conflicts of Interest: The authors declare no conflict of interest.

References

1. Witzenburg, C.M.; Dhume, R.Y.; Shah, S.B.; Korenczuk, C.E.; Wagner, H.P.; Alford, P.W.; Barocas, V.H. Failure of the porcine ascending aorta: Multidirectional experiments and a unifying microstructural model. *J. Biomech. Eng.* **2017**, *139*, 031005. [CrossRef] [PubMed]
2. Witzenburg, C.; Raghupathy, R.; Kren, S.M.; Taylor, D.A.; Barocas, V.H. Mechanical changes in the rat right ventricle with decellularization. *J. Biomech.* **2012**, *45*, 842–849. [CrossRef] [PubMed]
3. Uquillas, J.A.; Kishore, V.; Akkus, O. Effects of phosphate-buffered saline concentration and incubation time on the mechanical and structural properties of electrochemically aligned collagen threads. *Biomed. Mater.* **2011**, *6*, 035008. [CrossRef] [PubMed]
4. Alkhouli, N.; Bell, J.; Tham, J.C.; Winlove, C.P.; Liversedge, N.; Welbourn, R.; Green, E.; Knight, B.; Mansfield, J.; Kos, K.; et al. The mechanical properties of human adipose tissues and their relationships to the structure and composition of the extracellular matrix. *Am. J. Physiol. Metab.* **2013**, *305*, E1427–E1435. [CrossRef] [PubMed]
5. Lanir, Y.; Hayam, G.; Abovsky, M.; Zlotnick, A.Y.; Uretzky, G.; Nevo, E.; Ben-Haim, S.A. Effect of myocardial swelling on residual strain in the left ventricle of the rat. *Am. J. Physiol.-Heart Circ. Physiol.* **1996**, *270*, H1736–H1743. [CrossRef] [PubMed]
6. Azeloglu, E.U.; Albro, M.B.; Thimmappa, V.A.; Ateshian, G.A.; Costa, K.D. Heterogeneous transmural proteoglycan distribution provides a mechanism for regulating residual stresses in the aorta. *Am. J. Physiol. Circ. Physiol.* **2007**, *294*, H1197–H1205. [CrossRef] [PubMed]
7. Lanir, Y. Osmotic swelling and residual stress in cardiovascular tissues. *J. Biomech.* **2012**, *45*, 780–789. [CrossRef] [PubMed]
8. Powell, T.A.; Amini, R.; Oltean, A.; Barnett, V.A.; Dorfman, K.D.; Segal, Y.; Barocas, V.H. Elasticity of the Porcine Lens Capsule as Measured by Osmotic Swelling. *J. Biomech. Eng.* **2010**, *132*, 091008. [CrossRef]
9. Lai, V.K.; Nedrelow, D.S.; Lake, S.P.; Kim, B.; Weiss, E.M.; Tranquillo, R.T.; Barocas, V.H. Swelling of collagen-hyaluronic acid co-gels: An in vitro residual stress model. *Ann. Biomed. Eng.* **2016**, *44*, 2984–2993. [CrossRef]
10. Guo, X.; Lanir, Y.; Kassab, G.S. Effect of osmolarity on the zero-stress state and mechanical properties of aorta. *Am. J. Physiol. Circ. Physiol.* **2007**, *293*, H2328–H2334. [CrossRef]
11. Lanir, Y. Mechanisms of residual stress in soft tissues. *J. Biomech. Eng.* **2009**, *131*, 044506. [CrossRef] [PubMed]
12. Pierce, E.L.; Sadri, V.; Ncho, B.; Kohli, K.; Shah, S.; Yoganathan, A.P. Novel in vitro test systems and insights for transcatheter mitral valve design, part I: Paravalvular leakage. *Ann. Biomed. Eng.* **2019**, *47*, 381–391. [CrossRef] [PubMed]
13. Rambod, E.; Beizai, M.; Shusser, M.; Gharib, M. A physical model describing the mechanism for formation of gaseous microbubbles in patients with mechanical heart valves. *ASAIO J.* **2008**, *45*, 133. [CrossRef]
14. Leopaldi, A.M.; Vismara, R.; Lemma, M.; Valerio, L.; Cervo, M.; Mangini, A.; Contino, M.; Redaelli, A.; Antona, C.; Fiore, G.B. In vitro hemodynamics and valve imaging in passive beating hearts. *J. Biomech.* **2012**, *45*, 1133–1139. [CrossRef] [PubMed]
15. Amini Khoiy, K.; Biswas, D.; Decker, T.N.; Asgarian, K.T.; Loth, F.; Amini, R. Surface strains of porcine tricuspid valve septal leaflets measured in ex vivo beating hearts. *J. Biomech. Eng.* **2016**, *138*, 111006. [CrossRef]
16. Amini Khoiy, K.; Asgarian, K.T.; Loth, F.; Amini, R. Dilation of tricuspid valve annulus immediately after rupture of chordae tendineae in ex-vivo porcine hearts. *PLoS ONE* **2018**, *13*, e0206744. [CrossRef] [PubMed]
17. Vismara, R.; Gelpi, G.; Prabhu, S.; Romitelli, P.; Troxler, L.G.; Mangini, A.; Romagnoni, C.; Contino, M.; Van Hoven, D.T.; Lucherini, F.; et al. Transcatheter edge-to-edge treatment of functional tricuspid regurgitation in an ex vivo pulsatile heart model. *J. Am. Coll. Cardiol.* **2016**, *68*, 1024–1033. [CrossRef] [PubMed]
18. Kong, F.; Pham, T.; Martin, C.; McKay, R.; Primiano, C.; Hashim, S.; Kodali, S.; Sun, W. Finite element analysis of tricuspid valve deformation from multi-slice computed tomography images. *Ann. Biomed. Eng.* **2018**, *46*, 1112–1127. [CrossRef]

19. Pham, T.; Sulejmani, F.; Shin, E.; Wang, D.; Sun, W. Quantification and comparison of the mechanical properties of four human cardiac valves. *Acta Biomater.* **2017**, *54*, 345–355. [CrossRef]
20. Pokutta-Paskaleva, A.; Sulejmani, F.; DelRocini, M.; Sun, W. Comparative mechanical, morphological, and microstructural characterization of porcine mitral and tricuspid leaflets and chordae tendineae. *Acta Biomater.* **2019**, *85*, 241–252. [CrossRef]
21. Jett, S.; Laurence, D.; Kunkel, R.; Babu, A.R.; Kramer, K.; Baumwart, R.; Towner, R.; Wu, Y.; Lee, C.-H. Biaxial mechanical data of porcine atrioventricular valve leaflets. *Data Brief* **2018**, *21*, 358–363. [CrossRef] [PubMed]
22. Laurence, D.; Ross, C.; Jett, S.; Johns, C.; Echols, A.; Baumwart, R.; Towner, R.; Liao, J.; Bajona, P.; Wu, Y.; et al. An investigation of regional variations in the biaxial mechanical properties and stress relaxation behaviors of porcine atrioventricular heart valve leaflets. *J. Biomech.* **2019**, *83*, 16–27. [CrossRef] [PubMed]
23. Rausch, M.K.; Malinowski, M.; Wilton, P.; Khaghani, A.; Timek, T.A. Engineering analysis of tricuspid annular dynamics in the beating ovine heart. *Ann. Biomed. Eng.* **2018**, *46*, 443–451. [CrossRef] [PubMed]
24. Malinowski, M.; Jazwiec, T.; Goehler, M.; Quay, N.; Bush, J.; Jovinge, S.; Rausch, M.K.; Timek, T.A. Sonomicrometry-derived 3-dimensional geometry of the human tricuspid annulus. *J. Thorac. Cardiovasc. Surg.* **2019**, *157*, 1452–1461.e1. [CrossRef] [PubMed]
25. Meador, W.D.; Mathur, M.; Rausch, M.K. Tricuspid valve biomechanics: A brief review. In *Advances in Heart Valve Biomechanics*; Springer: Cham, Switzerland, 2018; pp. 105–114.
26. Pant, A.D.; Thomas, V.S.; Black, A.L.; Verba, T.; Lesicko, J.G.; Amini, R. Pressure-induced microstructural changes in porcine tricuspid valve leaflets. *Acta Biomater.* **2018**, *67*, 248–258. [CrossRef] [PubMed]
27. Misfeld, M.; Sievers, H.-H. Heart valve macro- and microstructure. *Philos. Trans. R. Soc. B Biol. Sci.* **2007**, *362*, 1421–1436. [CrossRef] [PubMed]
28. Combs, M.D.; Yutzey, K.E. Heart valve development. *Circ. Res.* **2009**, *105*, 408–421. [CrossRef] [PubMed]
29. Hilton, R.B.; Yutzey, K.E. Heart valve structures and function in development and disease. *Annu. Rev. Physiol.* **2011**, *73*, 29–46.
30. Schoen, F.J. Evolving concepts of cardiac valve dynamics. *Circulation* **2008**, *118*, 1864–1880. [CrossRef]
31. Amini Khoiy, K.; Amini, R. On the biaxial mechanical response of porcine tricuspid valve leaflets. *J. Biomech. Eng.* **2016**, *138*, 104504. [CrossRef] [PubMed]
32. Rezakhaniha, R.; Fonck, E.; Genoud, C.; Stergiopulos, N. Role of elastin anisotropy in structural strain energy functions of arterial tissue. *Biomech. Model. Mechanobiol.* **2011**, *10*, 599–611. [CrossRef] [PubMed]
33. Amini Khoiy, K.; Abdulhai, S.; Glenn, I.C.; Ponsky, T.A.; Amini, R. Anisotropic and nonlinear biaxial mechanical response of porcine small bowel mesentery. *J. Mech. Behav. Biomed. Mater.* **2018**, *78*, 154–163. [CrossRef] [PubMed]
34. Friedman, B.J.; Lozner, E.C.; Curfman, G.D.; Herzberg, D.; Rolett, E.L. Characterization of the human right ventricular pressure-volume relation: Effect of dobutamine and right coronary artery stenosis. *J. Am. Coll. Cardiol.* **1984**, *4*, 999–1005. [CrossRef]
35. Seward, J.B.; Tajik, A.J.; Fyfe, D.A.; Hagler, D.J.; Currie, P.J.; Chan, K.-L.; Nishimura, R.A.; Reeder, G.S.; Mair, D.D. Continuous wave doppler determination of right ventricular pressure: A simultaneous Doppler-catheterization study in 127 patients. *J. Am. Coll. Cardiol.* **2010**, *6*, 750–756.
36. Lake, S.P.; Barocas, V.H. Mechanical and structural contribution of non-fibrillar matrix in uniaxial tension: A collagen-agarose co-gel Model. *Ann. Biomed. Eng.* **2011**, *39*, 1891–1903. [CrossRef]
37. Rubin, L.J. Primary pulmonary hypertension. *N. Engl. J. Med.* **1997**, *336*, 111–117. [CrossRef] [PubMed]
38. Jan, N.-J.; Sigal, I.A. Collagen fiber recruitment: A microstructural basis for the nonlinear response of the posterior pole of the eye to increases in intraocular pressure. *Acta Biomater.* **2018**, *72*, 295–305. [CrossRef]
39. Thomas, V.S.; Lai, V.K.; Amini, R. A Computational multi-scale approach to investigate mechanically-induced changes in tricuspid valve anterior leaflet microstructure. *Acta Biomater.* **2019**, *94*, 524–535. [CrossRef]
40. Biswas, D.; Casey, D.M.; Crowder, D.C.; Steinman, D.A.; Yun, Y.H.; Loth, F. Characterization of transition to turbulence for blood in a straight pipe under steady flow conditions. *J. Biomech. Eng.* **2016**, *138*, 071001. [CrossRef]

bioengineering

MDPI

Review

Mechanics and Microstructure of the Atrioventricular Heart Valve Chordae Tendineae: A Review

Colton J. Ross [1], Junnan Zheng [2], Liang Ma [2], Yi Wu [1] and Chung-Hao Lee [1,3,*]

[1] Biomechanics and Biomaterials Design Laboratory, School of Aerospace and Mechanical Engineering, The University of Oklahoma, Norman, OK 73019, USA; cjross@ou.edu (C.J.R.); yiwu@ou.edu (Y.W.)
[2] Department of Cardiovascular Surgery, The First Affiliated Hospital of Zhejiang University, Hangzhou 310058, China; zhengjunnan@zju.edu.cn (J.Z.); ml1402@zju.edu.cn (L.M.)
[3] Institute for Biomedical Engineering, Science and Technology (IBEST), The University of Oklahoma, Norman, OK 73019, USA
* Correspondence: ch.lee@ou.edu; Tel.: +1-405-325-4842

Received: 10 February 2020; Accepted: 10 March 2020; Published: 12 March 2020

Abstract: The atrioventricular heart valves (AHVs) are responsible for directing unidirectional blood flow through the heart by properly opening and closing the valve leaflets, which are supported in their function by the chordae tendineae and the papillary muscles. Specifically, the chordae tendineae are critical to distributing forces during systolic closure from the leaflets to the papillary muscles, preventing leaflet prolapse and consequent regurgitation. Current therapies for chordae failure have issues of disease recurrence or suboptimal treatment outcomes. To improve those therapies, researchers have sought to better understand the mechanics and microstructure of the chordae tendineae of the AHVs. The intricate structures of the chordae tendineae have become of increasing interest in recent literature, and there are several key findings that have not been comprehensively summarized in one review. Therefore, in this review paper, we will provide a summary of the current state of biomechanical and microstructural characterizations of the chordae tendineae, and also discuss perspectives for future studies that will aid in a better understanding of the tissue mechanics–microstructure linking of the AHVs' chordae tendineae, and thereby improve the therapeutics for heart valve diseases caused by chordae failures.

Keywords: the mitral valve; the tricuspid valve; collagen fiber architecture; glycosaminoglycan; uniaxial mechanical testing; in-vitro flow loops; polarized spatial frequency domain imaging

---

1. Introduction

The atrioventricular heart valves (AHVs) regulate the unidirectional flow of blood through the heart chambers by the cyclic opening and closing of soft tissue leaflets. These leaflets are supported in their functions by the chordae tendineae, which attach to the papillary muscles. The chordae structures help to distribute load to the papillary muscles during systolic closure and prevent leaflet flail into the atria. In the case of chordae failure, such as elongation, rupture, thickening, retraction, or calcification, the result is a backflow of blood from the ventricle into the atria during the valve's systolic closure–known as AHV regurgitation [1,2]. Uncontrolled severe AHV regurgitation can ultimately lead to left or right heart failure and yield poor outcome including potential years of life lost.

Chordae rupture affects the overall atrioventricular heart valve behaviors and can be a primary cause of valve regurgitation [3–5]. Khoiy et al. (2018) used an in vitro valve apparatus to observe changes in the tricuspid valve closure after induced chordae rupture, and found that ruptured chordae caused up to an 8.8% dilation of the annulus and consequent regurgitation [6]. In another in vitro study for the mitral valve, it was found that a significantly lower ventricular pressure was required to cause leaflet prolapse caused by chordae rupture [7]. Chordae failure plays a major role in valve regurgitation, however current surgical treatments have their pros and cons.

The major surgical treatments for chordae failure include shortening, transposition, and replacement. Chordal shortening involves making an incision in the papillary muscles and using a suture to pull a portion of the elongated chordae into the incised area [8]. The shortening method is undesirable, however, as it has a significantly lower freedom from recurrent mitral regurgitation as compared to chordal transposition (74 ± 9% and 96 ± 2%, respectively) [9]. Further, chordal shortening was also found to be inferior to chordal replacement, with the shortened chordae being more vulnerable to post-operation chordae rupture [10]. Chordal transposition, on the other hand, is performed by either attaching a portion of leaflet with chordae attachment from a non-prolapsing leaflet to a prolapsing leaflet, or by resection of the prolapsing leaflet's chordal attachments. This technique also has drawbacks. For example, there is an approximate 13% recurrence rate of mild regurgitation in the long-term after mitral valve repair [11]. Chordal replacement is the current standard for chordae repair, in which the ruptured chordae is replaced with a synthetic material such as expanded polytetrafluoroethylene (ePTFE) sutures [10,12]. Despite the prevalence of the procedure, there are still issues such as elongation of the synthetic chordae, rupture of the native chordae, calcification, or recurrent prolapse potentially caused by an elastic modulus higher than that of the native chordae [13–16]. Various treatment methods for failed chordae tendineae can be improved and a better understanding of AHV diseases can be facilitated by studying the microstructure and mechanics of the native chordae tendineae. A more comprehensive understanding of the chordae tendineae will be especially beneficial to the refinement of AHV computational models and tissue engineering of chordae tendineae.

In this review paper, we will summarize the recent works in the microstructural and mechanical characterizations of the AHV chordae tendineae, as such a thorough review article is very limited in previous literature. In addition, we will also give some commentary remarks regarding the state of the art of research and perspective for future studies that would be beneficial to further the chordae biomechanics research field.

2. Overview on the Anatomy and Morphology of the AHV Chordae Tendineae

Chordae attaching to the leaflets are considered as true chordae, while those attaching elsewhere, such as the ventricular wall, are considered false chordae [17]. For this review, the focus will be on the true chordae, and any discussions henceforth are to be considered as referring to true chordae tendineae. Generally, chordae originate from the papillary muscle as a singular strand, which then either remains straight or branches into fan shapes prior to a leaflet insertion [17]. The chordae tendineae can be more specifically classified based on their respective insertion region to the leaflet: *basal chordae* attaching to the leaflet base; *marginal chordae* attaching to the leaflet free edge; and the notably-thicker *strut chordae* anchoring at the central belly region of the AHV leaflet (Figure 1a). Some differences exist between the chordae subsets. First, the strut chordae are the most critical subset for bearing load and are of special focus in the previous literature [18–20]. Secondly, in the tricuspid valve (TV), researchers have reported the marginal and basal chordae to be of a similar thickness. Thirdly, thicknesses of the chordae can also vary based on their leaflet of attachment, such as the mitral valve (MV) anterior leaflet chordae being thicker than the MV posterior leaflet chordae [21].

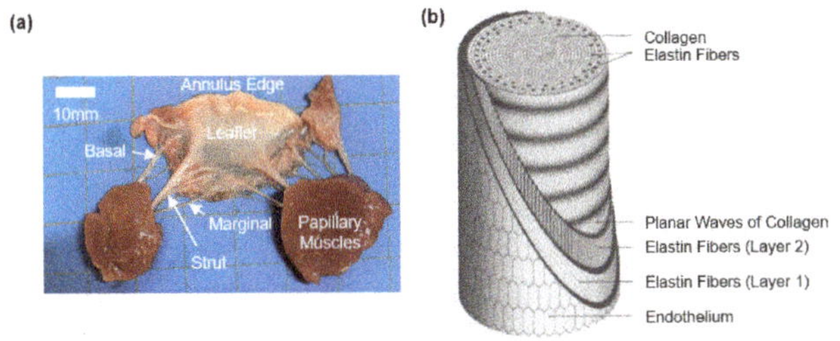

Figure 1. (**a**) An excised porcine mitral valve anterior leaflet with basal, strut, and marginal chordae tendineae. (**b**) Schematic of the microstructural components of the chordae tendineae (image from Millington-Sanders et al. [22] with permission from Wiley Global).

The number of chordae in the human, porcine, and ovine atrioventricular valve specimens have been analyzed in previous studies. Lam et al. and Silver et al. studied the human MV and TV, respectively, and reported an average of 25 chordae per valve [18,19]. Interestingly, De Almeida et al. noted that fetal hearts had fewer connective tissues (CT) and that there were proportionally fewer "muscular" (thicker, muscular texture) chordae as compared to the adult heart (reflected in both the MV and the TV) [23]. In comparison, we found in our previous study an average of 30 chordae in porcine MV and 35 chordae in the porcine TV, and an average of 15 and 24 chordae in ovine MV and TV, respectively [24]. These differences in the chordae anatomical features should be considered when discussing findings of porcine or ovine chordae studies and their translations to human heart anatomy.

It is worth noting that the different subsets of chordae (marginal, basal, and strut) are present in human, porcine, and ovine hearts but that the quantity can vary. However, to the authors' knowledge, there is no study on quantifying the respective number of each chordae subset within the human, porcine, or ovine AHVs. Such important anatomical investigations are the first step to gaining insight to the chordae tendineae morphology, and further studies would be beneficial towards developing AHV computational models.

3. Chordae Tendineae Microstructure

3.1. Microstructures of Human AHV Chordae Tendineae

Regarding the chordae microstructure, a previous study by Fenoglio et al. showed that the chordae tendineae of the human mitral valve are composed of a core of collagen fibers surrounded by an elastin sheath [25]. Lim and Boughner (1977) took a closer look at the human MV chordae microstructure and described two forms of collagen: (i) a mostly straight, dense, collagen fiber core and (ii) widely spaced collagen fibers that wrap around the straight collagen fiber core with some angle of alignment against the primary axis [26]. In a later study, Lim's research group used scanning electron microscopy (SEM) and transmission electron microscopy (TEM) to analyze the human TV chordae and found that the TV chordae were similar to the MV chordae, but that the TV chordae had a greater collagen fiber density and a smaller fibral diameter, owing to a lower force load experienced by the TV [27]. In later years, Millington-Sanders et al. investigated the microstructure of human MV chordae via SEM and light microscopy, and found an intricate layered structure (from outermost to innermost, Figures 1b, 2 and 3): (i) a layer of endothelial cells, (ii) an elastin sheath with fibers oriented at inclined angles with respect to the longitudinal axis, (iii) a longitudinally oriented elastin sheath, (iv) undulated collagen fibers aligned circumferentially, and (v) a core of straight collagen fibers with sparsely dispersed longitudinal elastic fibers [22].

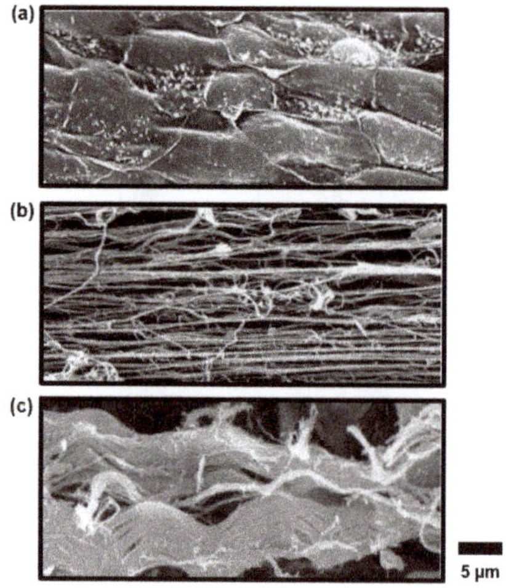

Figure 2. Scanning electron microscope (SEM) images of human mitral valve (MV) chordae tendineae, demonstrating (**a**) a layer of endothelial cells, (**b**) elastin fibers, and (**c**) undulated collagen fibers (image from Millington-Sanders et al. [22] with permission from Wiley Global).

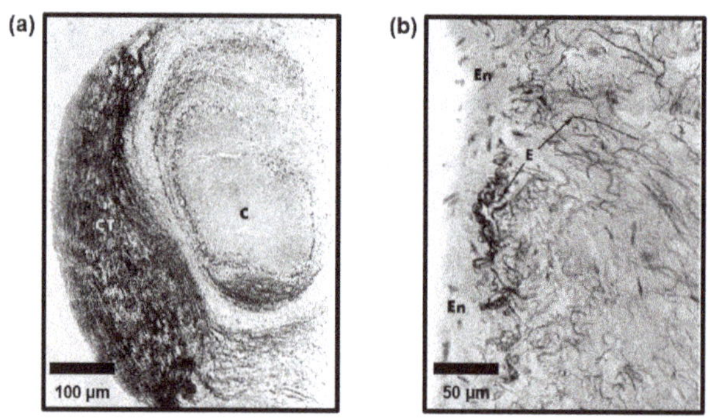

Figure 3. (**a**) A micrograph of a cross section of the chordae tendineae showing the collagen core (C) and the connective tissues (CT). (**b**) Within the connective tissue, a disorganized network of elastin fibers (E) and endothelium (En) exists (images from Millington-Sanders et al. [22] with permission from Wiley Global).

3.2. *Effects of Disease on Human AHV Chordae Tendineae Microstructure*

Other research groups have studied the effects of disease on the chordae microstructures. Grande-Allen et al. [28] studied myxomatous human mitral valve chords of the posterior leaflet, with their key findings summarized as follows: (i) water contents of chordae was higher in valves with unileaflet and bileaflet prolapse relative to normal chordae; (ii) myxomatous valve chordae had significantly higher glycosaminoglycan (GAG) contents than the normal (healthy) chordae; and (iii) the relative increases in the GAG contents of myxomatous valve chordae were higher than those observed in the valve leaflets, suggesting that the disease might have a greater influence on the chordae biochemistry. Other researchers have also sought to understand the effects of a floppy mitral valve on the chordae. Several primary findings include: (i) floppy mitral valve chordae corresponded to greater collagen alterations and acid mucopolysaccharide accumulations (i.e., a proteoglycan that can contribute to calcification in severe cases) [29–31]; (ii) the increased proteoglycan contents could play a role in the degradation or defective formation of elastin and collagen; and (iii) the floppy mitral valve chordae had a disrupted collagen fiber core and surface fibrosis [32–34]. Lis et al. [33] also investigated the rheumatic (inflamed) chordae and found thickened collagen cores and minimal surface fibrosis.

3.3. Comparisons of the Chordae Microstructures Between Different Species

Due to the challenges in obtaining human chordae, porcine chordae have been widely used as a comparative model owing to their similarities to the human heart. However, differences have been found [35,36]. To elaborate, Ritchie et al. (2005, 2006) histologically examined porcine MV chordae, and they did not observe the distinct elastin layer as found by Millington-Sanders et al. for the human MV chordae [22,37]. Ritchie et al., however, did observe blood vessels in the chordae, supporting earlier findings by Duran and Gunning in fetal calf hearts [38]. The translation to human chordae structures should not be assumed, but the finding is interesting, as it describes another subtle role of the chordae tendineae—a structure through which nutrients can be provided to the valve leaflets. Liao et al. (2009) also noted a difference between the porcine and human MV chordae, finding the collagen fibers to be a 3D, wavy structure, as opposed to the planar, undulated collagen fibers observed in the human chordae [39]. Comprehensive studies on the human chordae microstructure using more modernized techniques would be valuable for confirmation of the similarities between the various species.

3.4. Comparisons of the Microstructures Between Chordae Subsets

The three subsets of chordae tendineae have varied microstructures. In a study of porcine MV chordae, it was found that the marginal chordae have a larger fiber density and a smaller fibral diameter than the basal or strut chords [40]. Liao and Vesely (2004) analyzed the GAG contents of the chordae and found GAG concentrations in a decreasing order: marginal, basal, and strut [41]. They hypothesized that the observed varied GAG concentrations among chordae subsets may factor into the differences in the mechanical properties and the structural functions, such as GAG-mediated fibril-to-fibril linkage.

3.5. Microstructures of the Chordae Insertion Regions

Another integral part of the chordae's microstructure is the chordae-leaflet insertion region, where the highly aligned collagen fibers of the chordae transition into the more complex collagen fiber architecture of the leaflets. Chen et al. (2014) studied the strut chordae-leaflet insertions of porcine mitral valves and found that collagen fibers in the leaflet closer to the annulus were more circumferentially aligned, and that those collagen fibers became more radially aligned and uniform approaching the leaflet-chordae transition [42]. The chordae-papillary muscle insertion region has also been examined, and it was found that both human and porcine chordae exhibited a smooth, continuous endocardium endothelium between the chordae and papillary muscles [43]. However, differences were observed in the collagen fiber connection to the muscle, with human hearts containing an organized cross-network while porcine hearts were more random in architecture.

3.6. Microstructures of the Artificial Chordae

Artificial chordae tendineae, generally made of expanded polytetrafluoroethylene, are one of the primary treatment options for failed chordae [44]. The ePTFE sutures are composed of a high molecular weight compound of carbon and fluorine, and made porous with a surface charge that reduces thrombogenicity [16]. When implanted in the body, the artificial chordae will, over time, become fully encapsulated with dense fibrous tissue, covered by a layer of endothelial cells, while maintaining normal mechanical function [45]. However, due to the porous nature of the suture, the artificial chordae can become calcified, leading to eventual rupture [16,46]. Future research efforts may be devoted to refining the microstructure of the artificial chordae to reduce the chances of disease-based failure.

4. Tissue Mechanics of the Chordae Tendineae

In addition to gaining an understanding of the chordae's microstructure and morphology, research efforts have been made to investigate the mechanical properties of the chordae tendineae. Studies of the chordae biomechanics include: (i) uniaxial tensile testing; (ii) stress-relaxation testing; (iii) chordae-leaflet insertion region testing; and (iv) in vitro flow loop testing (Figure 4).

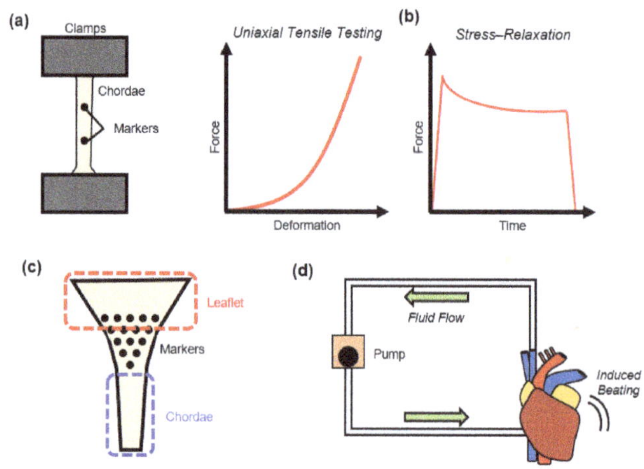

Figure 4. Mechanical characterizations of the atrioventricular heart valve (AHV) chordae tendineae through: (**a**) uniaxial mechanical testing, (**b**) stress-relaxation testing, (**c**) chordae-leaflet insertion region deformation tracking, and (**d**) an in vitro heart simulating flow loop.

4.1. *Uniaxial Mechanical Testing of the Chordae Tendineae*

The most prevalent method for investigating the chordae mechanics in previous literature is through uniaxial tensile testing. Generally, in these studies chordae were fully separated from their valve attachments and placed into a hydraulic uniaxial tensile testing machine (Figure 4a). Then, the chordae specimens were preconditioned, followed by either cyclic force loading and unloading to and from a target load, or loading until tissue rupture.

4.1.1. Uniaxial Tensile Characterizations of Human Chordae Tendineae

One of the earliest documented studies using this method is from Lim and Boughner (1975). They applied monotonically increased uniaxial loading to human MV chordae until rupture. The three primary findings from their chordae characterization study were: (i) chordae were less extensible at an increasing strain rate, (ii) larger cross-sectional areas corresponded to a lower extensibility, and (iii) chordae ruptured at a strain of 21.4% and a stress of 3.1×10^8 dyne/cm^2 [47]. The rupture stress was consistent with a previous study, whereas the rupture strain varied (21.4% vs. 40%) [48]. In a subsequent study by Lim and Boughner (1976), human MV chordae were treated with enzymes to remove the elastin sheath prior to uniaxial mechanical testing. It was found that the removal of the elastin sheath did not significantly affect the elastic response of the chordae tissue. However the inter-specimen variability may overlap with the variation of their quantified mechanical results, as different specimens were used for control and treatment groups [49]. Future investigations using a unified testing scheme would be beneficial, where a specimen is tested as a control specimen, treated to remove the elastin sheath, and tested again for a direct comparison of the changes in the tissue mechanics due to elastin depletion. Lim (1980) also conducted mechanical testing for the TV chordae and found that the TV chordae were less extensible than their MV counterparts [27]. A potential shortcoming of these studies, however, was that machine cross-head displacements were used for quantifying tissue stretch, as opposed to calculating tissue strains using fiducial markers. Other researchers have also characterized human MV chordae using the more accurate fiducial marker approach, such as Zuo et al., for investigating the age-dependent changes in the chordae mechanical properties [21]. They observed stiffer and less extensible chordae as compared to earlier human chordae studies. Differences in the tissue mechanics could be possibly attributed to the use of the marker-based quantification of tissue strains, or potential variations in the patient age. Clark (1973) also studied the effects of freezing on human chordae tensile characteristics and observed that freezer storage resulted in a stiffer mechanical response [50].

4.1.2. Effects of Disease on the Tensile Characteristics of Human Chordae Tendineae

Other researchers have studied the effects of disease on the human chordae tendineae's uniaxial tensile behaviors. Barber et al. studied chordae from myxomatous valves and observed that diseased chordae had significantly lower moduli (40.4 ± 10.2 vs. 132 ± 15 MPa) and failure stresses (6.0 ± 0.6 MPa vs. 25.7 ± 1.8 MPa) than healthy chordae, but the extensibility and failure strain were similar [51]. Lim et al. also analyzed myxomatous chordae of the tricuspid valve and found higher extensibilities, lower rupture stresses, and similar rupture strains in diseased specimens compared to healthy specimens [52]. Differences in the myxomatous chordae behaviors between the AHVs could be attributed to differences in the chordae morphology, however more detailed studies are necessary. To better understand the effects of calcification, Casado et al. analyzed calcified marginal chordae by means of quasi-static tensile testing and observed that the diseased chordae were three to seven times more compliant than normal chordae [53].

4.1.3. Mechanical Characterizations of Porcine Chordae Tendineae

There has also been extensive investigations of chordae mechanics through the study of porcine chordae tendineae. For example, Ritchie et al. (2006) uniaxially tested porcine chordae with their primary finding being that quantifying stretch using machine cross-head displacements corresponded to different extensibilities compared to using graphite markers [54]. Pokutta-Paskaleva et al. characterized porcine MV and TV chordae using the marker-based approach and observed that: (i) the strut chordae were stiffer than the marginal and basal chordae; (ii) the basal chordae had greater extensibilities than the marginal chordae; (iii) the MV chordae were stiffer than their TV counterparts; and (iv) the chordae attaching to the TV septal leaflet were more extensible than the chordae attaching to the other two TV leaflets (Figure 5) [55]. Sedransk et al. characterized the rupture of the porcine MV chordae using a tensile testing device and found that the marginal chordae ruptured at 68% less load and 28% less strain than the basal chordae [56]. They also found that the chordae from the MV posterior leaflet ruptured at 43% less load and 22% less strain than ones from the MV anterior leaflet.

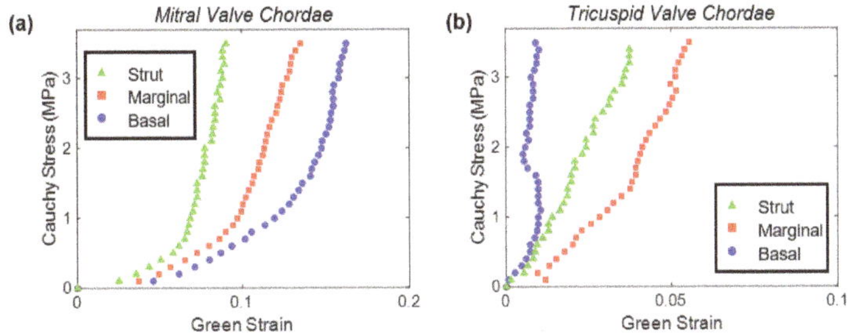

Figure 5. Results from the uniaxial mechanical testing of the strut, marginal, and basal chordae tendineae from: (**a**) the mitral valve anterior leaflet, and (**b**) the tricuspid valve anterior leaflet. (Plots reformatted from the data reported in Pokutta-Paskaleva et al. [55]).

Unique mechanical–morphological findings from other studies on the chordae include: (i) decellularized and glutaraldehyde cross-linked chordae (a method for replacing human chordae with porcine chordae) had a longer fatigue life and a lower creep rate than the native chordae [57], (ii) fatigue-induced micro-cracks in the collagen structures can cause increased creep behaviors [58], and (iii) glutaraldehyde-fixed chordae have decreased storage moduli compared to native structures [59].

It is worthwhile to note that while there are many studies characterizing MV chordae, the information regarding TV chordae mechanics is limited, and future investigations are warranted. In our lab, we made a recent contribution to the understanding of MV and TV strut chordae mechanics under a novel uniaxial tensile testing scheme [60]. In this study, we performed mechanical characterizations of the chordae tendineae where the attachment regions were preserved, resulting in a leaflet-chordae-papillary muscle entity with full planar deformation of the insertion regions. From this unique experimental setting, we observed different chordae mechanics than those reported in previous literature, which may be due to distributions of stress and deformations from the chordae to the leaflet and papillary muscle structures. A similar preliminary study (n = 1 strut chordae for each valve) was recently performed, but in this study the authors bisected the entity such that only a leaflet-chordae segment and a chordae-papillary muscle segment were tested [61]. Reported results from the uniaxial mechanical testing of chordae tendineae are summarized in Table 1.

Table 1. Comparisons of the reported uniaxial mechanical testing results (tissue stretch versus Cauchy stress) of the chordae tendineae of the mitral valve anterior leaflet (MVAL) and tricuspid valve anterior leaflet (TVAL).

		MVAL Strut		**TVAL Strut**	
Study	Species	Tissue Stretch λ (−)	Cauchy Stress (MPa)	Tissue Stretch λ (−)	Cauchy Stress (MPa)
Pokutta-Paskaleva et al. (2019) [55]	porcine (n = not provided)	1.09	3.5	1.04	3.5
Ritchie et al. (2006) [54]	porcine (n = not provided)	1.05	0.89 to 1.18	–	–
Liao and Vesely (2003) [40]	porcine (n = 16)	1.16 ± 0.03 (mean ± SD)	0.75 ± 0.15 (mean ± SD)	–	–
Zuo et al. (2016) [21]	ovine (n = 18)	1.07 ± 0.08 (mean ± SD)	24 (mean ± SD)	–	–
Ross et al. (2020) [60]	porcine (n = 12)	1.03 ± 0.01 (mean ± SEM)	1.59 ± 0.16 (mean ± SEM)	1.02 ± 0.01 (mean ± SEM)	2.71 ± 0.10 (mean ± SEM)
		MVAL Marginal		**TVAL Marginal**	
Pokutta-Paskaleva et al. (2019) [55]	porcine (n = not provided)	1.13	3.5	1.05	3.5
Kunzelman and Cochran (1990) [62]	porcine (n = 31)	1.09	1.96 ± 0.20 (mean ± SEM)	–	–
Liao and Vesely (2003) [40]	porcine (n = 16)	1.04 ± 0.01 (mean ± SD)	5.22 ± 3.30 (mean ± SD)	–	–
		MVAL Basal		**TVAL Basal**	
Pokutta-Paskaleva et al. (2019) [55]	porcine (n = not provided)	1.15	3.5	1.01	3.5
Kunzelman and Cochran (1990) [62]	porcine (n = 29)	1.12	1.57 ± 0.05 (mean ± SEM)	–	–
Liao and Vesely (2003) [40]	porcine (n = 20)	1.08 ± 0.03 (mean ± SD)	2.41 ± 0.81 (mean ± SD)	–	–

4.2. *Stress-Relaxation Testing of the Chordae Tendineae*

To supplement the information on the chordae uniaxial tensile characteristics, Liao and Vesely (2004) conducted stress-relaxation testing of porcine MV chordae [41]. In the testing, they displaced tissues to the deformation associated with an initial 150 g load and observed the stress-relaxation behaviors over 100 seconds (Figure 4b). They observed strut chordae to have the fastest and greatest relaxation behavior (49.1 ± 5.4%), followed by the basal chordae (42.4 ± 8.3%), and then the marginal chordae (33.2 ± 4.7%). They proposed that the differences could be attributed to differences in the GAG contents, with chordae subsets containing fewer GAGs corresponding to greater relaxation behaviors. In other words, the marginal chordae had the largest GAG concentrations, and the smallest relaxation behaviors, whereas the strut chordae had the greatest relaxation behavior, but the lowest GAG contents. In previous studies of the valve leaflets, in contrast, fewer GAG contents corresponded to less stress relaxation [63,64]. The differences in the trends between the relaxation behaviors and the GAG contents may be due to the GAGs serving different roles in the mechanical behaviors for the leaflets or the chordae. Future studies are warranted to better elucidate the GAG contributions to mechanical behaviors in the chordae, such as through stress-relaxation testing of the tissues before and after enzymatic removal of the GAGs. Additionally, stress-relaxation studies could be performed on the TV chordae to better understand the differences in chordae mechanics for each AHV.

4.3. Load-Dependent Collagen Fiber Architecture of the Chordae-Leaflet Insertion Region

There are few studies focused on the mechanics of the chordae-leaflet insertion area. Padala et al. performed such investigations on porcine MV strut chordae using an in vitro flow loop by tracking an array of markers across the surface of the insertion region (Figure 4c) [65]. From this, they found that the edges of the insertion region stretched more than the center, which may be attributed to the chordae-leaflet transition region microstructure. Chen et al. (2004) also examined the porcine MV strut chordae region using a biaxial testing system [42]. In their testing they fixed the leaflet on three edges via suture hooks, and on the fourth edge attached the chordae to a string. They found that along the insertion region the radial extensibility decreased, and the stiffness increased. Another recent preliminary study was published to analyze the leaflet-chordae and papillary muscle-chordae insertion areas using X-ray diffraction [61]. Their preliminary finding was that the leaflet and papillary muscle insertions have a higher molecular strain than the rest of the chordae, suggesting those areas are more rupture vulnerable. Limitations of that study include: a relatively small sample size (n = 1 strut chordae for each valve), a very limited field of view, and the bisection of the chordae.

Extensive studies on the chordae-leaflet insertion have not yet to be performed for TV chordae tissues. In our lab, we recently completed a pilot study by using the polarized spatial frequency domain imaging (pSFDI) modality [66,67] to analyze the load-dependent collagen fiber orientations of the strut chordae insertions of porcine MV and TV anterior leaflets (MVAL and TVAL). Briefly, in the study we loaded leaflet-chordae-papillary muscle entities to physiologically representative loading (MVAL: 1.4 N and TVAL: 1.2 N), and pSFDI was performed to quantify the collagen fiber architecture at various states of the force-deformation curve (Figure 6).

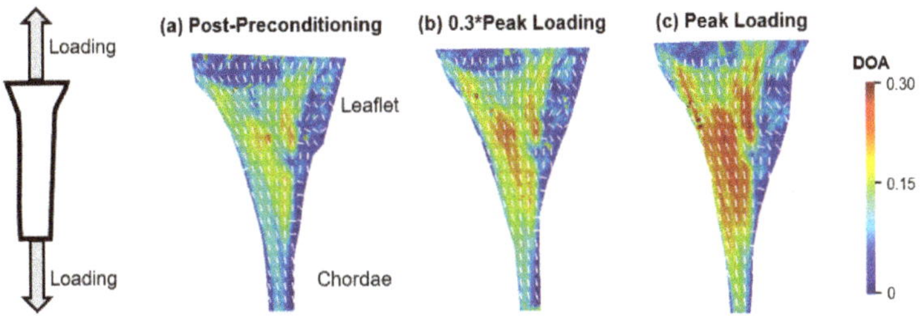

Figure 6. Quantifications of the load-dependent collagen fiber architecture of a representative porcine mitral valve strut chordae-leaflet insertion region using the polarized spatial frequency domain imaging (pSFDI) modality: (**a**) at the post-preconditioning configuration, (**b**) at intermediate loading (0.3 * peak loading), and (**c**) at the peaking loading. White, dashed lines denote the quantified collagen fiber orientation angles, and the colormaps show the degree of optical anisotropy (DOA)—a metric related to the degree of alignment for the underlying collage fiber networks (a warmer color denoting a better aligned collagen fiber network).

4.4. In Vitro Flow Loop Testing of Chordae Tendineae

The use of in vitro flow loops is another experimental way to quantify the mechanics of the chordae tendineae (Figure 4d). Ritchie et al. (2006) used the Georgia Tech left heart simulator in conjunction with a marker-tracking approach to analyze the mechanical responses of porcine MV strut chordae [54]. They found that the chordae experienced a strain rate of 75.3 ± 3.43% during systolic closure and a strain rate of −54.8 ± −56.6% during diastolic opening. Furthermore, there was a constant plateau of the chordae strain between 3.75% and 4.29% during valve closure, indicating a minimal creeping response. Padala et al. used the same in vitro flow loop to quantify the tissue mechanics of the MV strut chordae-leaflet insertion (see Section 4.3) [65]. To the best of our knowledge, no studies have been conducted using in vitro flow loop methods for TV chordae—such future studies would be beneficial to make the connection between the chordae's mechanical behaviors and the overall function of the tricuspid valve.

4.5. Mechanics of Artificial Chordae

The ePTFE artificial chordae used in chordae replacement have been analyzed for their mechanical properties, and how they compare to native chordae mechanics. In general, it has been found that ePTFE chordae do not have similar mechanical properties to native chordae. In particular, Caimmi et al. [68] found that ePTFE chordae were of higher compliance than their native counterparts, and that the stiffness of ePTFE chordae increases with the length of the implant. These findings would be beneficial to the refinement of therapeutics to better emulate native structures.

5. Closing Remarks and Future Prospects

There are several studies on MV strut chordae, but there is limited information pertaining to other subsets of MV chordae, and even more limited information on TV chordae. It would be worthwhile to investigate these under-represented structures through either in vitro flow loop or extensive uniaxial mechanical testing procedures, including tensile tests and stress-relaxation. For example, it would be interesting to understand how the GAG contents of the chordae affect the stress-relaxation behaviors through enzymatic procedures (similar to a previous study from our lab on AHV leaflets [64]). In addition, the stress-relaxation behavior could be analyzed in connection to the amount of the mechanical force acting on the tissues [65]. It would also be useful to better understand the contributions of the elastin sheath to the chordae's overall mechanical behaviors through testing the same tissues before and after sheath removal. Other useful information could be found through new microstructural quantification technologies, such as polarized spatial frequency domain imaging, allowing for an understanding of the load-dependent collagen fiber architecture, especially at the insertion regions [66,67,69].

Findings from these studies are all critical to the future development and refinement of computational models of AHVs. Some of these models employ one set of material properties for all chordae, not considering the differences in the tissue mechanics and the structure of various chordae subsets (i.e., basal, marginal, and strut) [70–72]. There are, however, some computational models which consider these subsets, providing richer and more realistic predictions of heart valve biomechanical function [73–75]. To further improve these models, Khalighi et al. developed a topological and geometric mapping technique for the chordae tendineae of the MV, including branching, probable leaflet insertions, and cross-sectional areas [76]. A similar mapping has been realized for TV chordae as well [77]. Later, Khalighi et al. also proposed a simplified model of MV chordae that is functionally equivalent and significantly reduces computational complexity [78]. Simulation tools such as these are critical to therapeutic refinement and patient-specific surgical planning [74,79–81], as can be found in the simulation of MV annuloplasty and its effects on the forces experienced by the chordae tendineae [82]. Moving forward, future studies could include modeling of the microstructure of the chordae tendineae (Figures 1–3), and developing computational models of the under-investigated

TV chordae [83]. Another study would be incorporating the properties of diseased chordae/valve leaflets to predict the effects of diseased conditions, such as valve calcification, on the hemodynamics and homeostasis of the atrioventricular heart valves. Computational models incorporating the chordae tendineae mechanics and microstructure information will be critical for applications such as the development of microstructurally informed constitutive models, models with collagen fiber recruitment and reorientation predictions, and the growth and remodeling framework.

In summary, there have been foundational strides towards a better understanding of the chordae tendineae of the AHVs. Specifically, the morphology and microstructure of the chordae have been well-defined for both human and porcine chordae, for each of the chordal subsets, and for the MV and the TV. However, there are discrepancies between different studies that require further and more systematic investigations. Currently, there are no standard protocols for investigating chordae mechanics or microstructure. Through a review of previous research efforts, future efforts may be better guided for more detail and greater consistency. Moreover, the tissue mechanics of porcine MV strut chordae have been well characterized, but future studies are warranted regarding the mechanics of TV chordae, as well as the linking of the tissue mechanics and microstructures for human chordae. Furthermore, there are limited investigations regarding the chordae-leaflet insertion region other than our preliminary results for porcine MV strut chordae. Through more comprehensive investigations such as those mentioned above, efforts towards improved therapies and treatment outcomes can be better informed.

Author Contributions: Writing–original draft, C.J.R. and C.-H.L; writing–review and editing, J.Z., L.M., Y.W., and C.-H.L; funding acquisition, C.-H.L; project administration, C.-H.L. All authors have read and agreed to the published version of the manuscript.

Funding: Support from the American Heart Association Scientist Development Grant (SDG) Award (16SDG27760143) and the Presbyterian Health Foundation Team Science Grants (C5122401) is gratefully acknowledged. CHL was in part supported by the institutional start-up funds from the School of Aerospace and Mechanical Engineering (AME), the IBEST-OUHSC Seed Funding for Interdisciplinary Research, and research funding through the Faculty Investment Program from the Research Council at the University of Oklahoma.

Conflicts of Interest: The authors declare no conflict of interest.

References

1. Waller, B.F.; Howard, J.; Fess, S. Pathology of mitral valve stenosis and pure mitral regurgitation—Part I. *Clin. Cardiol.* **1994**, *17*, 330–336. [CrossRef] [PubMed]
2. Waller, B.F.; Howard, J.; Fess, S. Pathology of tricuspid valve stenosis and pure tricuspid regurgitation—Part I. *Clin. Cardiol.* **1995**, *18*, 97–102. [CrossRef] [PubMed]
3. Sanders, C.A.; Austen, W.G.; Harthorne, J.W.; Dinsmore, R.E.; Scannell, J.G. Diagnosis and surgical treatment of mitral regurgitation secondary to ruptured chordae tendineae. *N. Engl. J. Med.* **1967**, *276*, 943–949. [CrossRef] [PubMed]
4. Orszulak, T.A.; Schaff, H.V.; Danielson, G.K.; Piehler, J.M.; Pluth, J.R.; Frye, R.L.; McGoon, D.C.; Elveback, L.R. Mitral regurgitation due to ruptured chordae tendineae: Early and late results of valve repair. *J. Thorac. Cardiovasc. Surg.* **1985**, *89*, 491–498. [CrossRef]
5. Jeresaty, R.M.; Edwards, J.E.; Chawla, S.K. Mitral valve prolapse and ruptured chordae tendineae. *Am. J. Cardiol.* **1985**, *55*, 138–142. [CrossRef]
6. Khoiy, K.A.; Asgarian, K.T.; Loth, F.; Amini, R. Dilation of tricuspid valve annulus immediately after rupture of chordae tendineae in ex-vivo porcine hearts. *PLoS ONE* **2018**, *13*, e0206744. [CrossRef]
7. Espino, D.M.; Hukins, D.W.; Shepherd, D.E.; Buchan, K.G. Mitral valve repair: An in-vitro comparison of the effect of surgical repair on the pressure required to cause mitral valve regurgitation. *J. Heart. Valve Dis.* **2006**, *15*, 375.
8. Carpentier, A.; Adams, D.H.; Filsoufi, F. *Carpentier's Reconstructive Valve Surgery From Valve Analysis to Valve Reconstruction*; Saunders Elsevier Health Sciences: Philadelphia, PA, USA, 2011.

9. Smedira, N.G.; Selman, R.; Cosgrove, D.M.; McCarthy, P.M.; Lytle, B.W.; Taylor, P.C.; Apperson-Hansen, C.; Stewart, R.W.; Loop, F.D. Repair of anterior leaflet prolapse: Chordal transfer is superior to chordal shortening. *J. Thorac. Cardiovasc. Surg.* **1996**, *112*, 287–292. [CrossRef]
10. Phillips, M.R.; Daly, R.C.; Schaff, H.V.; Dearani, J.A.; Mullany, C.J.; Orszulak, T.A. Repair of anterior leaflet mitral valve prolapse: Chordal replacement versus chordal shortening. *Ann. Thorac. Surg.* **2000**, *69*, 25–29. [CrossRef]
11. Sousa, U.M.; Grare, P.; Jebara, V.; Fuzelier, J.F.; Portoghese, M.; Acar, C.; Relland, J.; Mihaileanu, S.; Fabiani, J.N.; Carpentier, A. Transposition of chordae in mitral valve repair. Mid-term results. *Circulation* **1993**, *88*, II35–II38.
12. Revuelta, J.M.; Garcia-Rinaldi, R.; Gaite, L.; Val, F.; Garijo, F. Generation of chordae tendineae with polytetrafluoroethylene stents: Results of mitral valve chordal replacement in sheep. *J. Thorac. Cardiovasc. Surg.* **1989**, *97*, 98–103. [CrossRef]
13. Tabata, M.; Kasegawa, H.; Fukui, T.; Shimizu, A.; Sato, Y.; Takanashi, S. Long-term outcomes of artificial chordal replacement with tourniquet technique in mitral valve repair: A single-center experience of 700 cases. *J. Thorac. Cardiovasc. Surg.* **2014**, *148*, 2033–2038.e2031. [CrossRef] [PubMed]
14. Cochran, R.P.; Kuzelman, K.S. Comparison of viscoelastic properties of suture versus porcine mitral valve chordae tendineae. *J. Card. Surg.* **1991**, *6*, 508–513. [CrossRef] [PubMed]
15. Colli, A.; Manzan, E.; Rucinskas, K.; Janusauskas, V.; Zucchetta, F.; Zakarkaitė, D.; Aidietis, A.; Gerosa, G. Acute safety and efficacy of the NeoChord procedure. *Interact. Cardiovasc. Thorac. Surg.* **2015**, *20*, 575–581. [CrossRef]
16. Butany, J.; Collins, M.J.; David, T.E. Ruptured synthetic expanded polytetrafluoroethylene chordae tendinae. *Cardiovasc. Pathol.* **2004**, *13*, 182–184. [CrossRef]
17. Gunnal, S.A.; Wabale, R.N.; Farooqui, M.S. Morphological study of chordae tendinae in human cadaveric hearts. *Heart Views* **2015**, *16*, 1–12. [CrossRef]
18. Lam, J.H.C.; Ranganathan, N.; Wigle, E.D.; Silver, M.D. Morphology of the human mitral valve: I. Chordae tendineae: A new classification. *Circulation* **1970**, *41*, 449–458. [CrossRef]
19. Silver, M.D.; Lam, J.H.C.; Ranganathan, N.; Wigle, E.D. Morphology of the human tricuspid valve. *Circulation* **1971**, *43*, 333–348. [CrossRef]
20. Lomholt, M.; Nielsen, S.L.; Hansen, S.B.; Andersen, N.T.; Hasenkam, J.M. Differential tension between secondary and primary mitral chordae in an acute in-vivo porcine model. *J. Heart. Valve Dis.* **2002**, *11*, 337–345.
21. Zuo, K.; Pham, T.; Li, K.; Martin, C.; He, Z.; Sun, W. Characterization of biomechanical properties of aged human and ovine mitral valve chordae tendineae. *J. Mech. Behav. Biomed. Mater.* **2016**, *62*, 607–618. [CrossRef]
22. Millington-Sanders, C.; Meir, A.; Lawrence, L.; Stolinski, C. Structure of chordae tendineae in the left ventricle of the human heart. *J. Anat.* **1998**, *192*, 573–581. [CrossRef] [PubMed]
23. De Almeida, M.T.B.; Aragao, I.C.S.A.; Aragao, F.M.S.A.; Reis, F.P.; Aragao, J.A. Morphological study on mitral valve chordae tendineae in the hearts of human fetuses. *Int. J. Anat. Var.* **2019**, *12*, 17–20.
24. Jett, S.V.; Laurence, D.W.; Kunkel, R.P.; Babu, A.R.; Kramer, K.E.; Baumwart, R.; Towner, R.A.; Wu, Y.; Lee, C.-H. An investigation of the anisotropic mechanical properties and anatomical structure of porcine atrioventricular heart valves. *J. Mech. Behav. Biomed. Mater.* **2018**, *87*, 155–171. [CrossRef] [PubMed]
25. Fenoglio, J.J., Jr.; Pham, T.D.; Wit, A.L.; Bassett, A.L.; Wagner, B.M. Canine mitral complex: Ultrastructure and electromechanical properties. *Circul. Res.* **1972**, *31*, 417–430. [CrossRef] [PubMed]
26. Lim, K.O.; Boughner, D.R. Scanning electron microscopical study of human mitral valve chordae tendineae. *Arch. Pathol. Lab. Med.* **1977**, *101*, 236–238.
27. Lim, K.O. Mechanical properties and ultrastructure of normal human tricuspid valve chordae tendineae. *Jpn. J. Physiol.* **1980**, *30*, 455–464. [CrossRef]
28. Grande-Allen, K.J.; Griffin, B.P.; Ratliff, N.B.; Cosgrove, D.M.; Vesely, I. Glycosaminoglycan profiles of myxomatous mitral leaflets and chordae parallel the severity of mechanical alterations. *J. Am. Coll. Cardiol.* **2003**, *42*, 271–277. [CrossRef]
29. Baker, P.B.; Bansal, G.; Boudoulas, H.; Kolibash, A.J.; Kilman, J.; Wooley, C.F. Floppy mitral valve chordae tendineae: Histopathologic alterations. *Hum. Pathol.* **1988**, *19*, 507–512. [CrossRef]

30. Hollander, W. Unified concept on the role of acid mucopolysaccharides and connective tissue proteins in the accumulation of lipids, lipoproteins, and calcium in the atherosclerotic plaque. *Exp. Mol. Pathol.* **1976**, *25*, 106–120. [CrossRef]
31. Fornes, P.; Heudes, D.; Fuzellier, J.-F.; Tixier, D.; Bruneval, P.; Carpentier, A. Correlation between clinical and histologic patterns of degenerative mitral valve insufficiency: A histomorphometric study of 130 excised segments. *Cardiovasc. Pathol.* **1999**, *8*, 81–92. [CrossRef]
32. Tamura, K.; Fukuda, Y.; Ishizaki, M.; Masuda, Y.; Yamanaka, N.; Ferrans, V.J. Abnormalities in elastic fibers and other connective-tissue components of floppy mitral valve. *Am. Heart J.* **1995**, *129*, 1149–1158. [CrossRef]
33. Lis, Y.; Burleigh, M.C.; Parker, D.J.; Child, A.H.; Hogg, J.; Davies, M.J. Biochemical characterization of individual normal, floppy and rheumatic human mitral valves. *Biochem. J.* **1987**, *244*, 597–603. [CrossRef] [PubMed]
34. Icardo, J.M.; Colvee, E.; Revuelta, J.M. Structural analysis of chordae tendineae in degenerative disease of the mitral valve. *Int. J. Cardiol.* **2013**, *167*, 1603–1609. [CrossRef]
35. Thein, E.; Hammer, C. Physiologic barriers to xenotransplantation. *Curr. Opin. Organ. Transplant.* **2004**, *9*, 186–189. [CrossRef]
36. Crick, S.J.; Sheppard, M.N.; Ho, S.Y.; Gebstein, L.; Anderson, R.H. Anatomy of the pig heart: Comparisons with normal human cardiac structure. *J. Anat.* **1998**, *193*, 105–119. [CrossRef] [PubMed]
37. Ritchie, J.; Warnock, J.N.; Yoganathan, A.P. Structural characterization of the chordae tendineae in native porcine mitral valves. *Ann. Thorac. Surg.* **2005**, *80*, 189–197. [CrossRef]
38. Duran, C.M.G.; Gunning, A.J. The vascularization of the heart valves: A comparative study. *Cardiovasc. Res.* **1968**, *2*, 290–296. [CrossRef]
39. Liao, J.; Priddy, L.B.; Wang, B.; Chen, J.; Vesely, I. Ultrastructure of porcine mitral valve chordae tendineae. *J. Heart. Valve Dis.* **2009**, *18*, 292.
40. Liao, J.; Vesely, I. A structural basis for the size-related mechanical properties of mitral valve chordae tendineae. *J. Biomech.* **2003**, *36*, 1125–1133. [CrossRef]
41. Liao, J.; Vesely, I. Relationship between collagen fibrils, glycosaminoglycans, and stress relaxation in mitral valve chordae tendineae. *Ann. Biomed. Eng.* **2004**, *32*, 977–983. [CrossRef]
42. Chen, L.; Yin, F.C.P.; May-Newman, K. The structure and mechanical properties of the mitral valve leaflet-strut chordae transition zone. *J. Biomech. Eng.* **2004**, *126*, 244–251. [CrossRef] [PubMed]
43. Gusukuma, L.W.; Prates, J.C.; Smith, R.L.; Gusukuma, W.L.; Prates, J.C.; Smith, R.L. Chordae tendineae architecture in the papillary muscle insertion. *Int. J. Morphol.* **2004**, *22*, 267–272. [CrossRef]
44. Salvador, L.; Mirone, S.; Bianchini, R.; Regesta, T.; Patelli, F.; Minniti, G.; Masat, M.; Cavarretta, E.; Valfrè, C. A 20-year experience with mitral valve repair with artificial chordae in 608 patients. *J. Thorac. Cardiovasc. Surg.* **2008**, *135*, 1280–1287.e1281. [CrossRef] [PubMed]
45. Minatoya, K.; Kobayashi, J.; Sasako, Y.; Ishibashi-Ueda, H.; Yutani, C.; Kitamura, S. Long-term pathological changes of expanded polytetrafluoroethylene (ePTFE) suture in the human heart. *J. Heart. Valve Dis.* **2001**, *10*, 139–142.
46. Coutinho, G.F.; Carvalho, L.; Antunes, M.J. Acute mitral regurgitation due to ruptured ePTFE neo-chordae. *J. Heart Valve Dis.* **2007**, *16*, 278–281. [PubMed]
47. Lim, K.O.; Boughner, D.R. Mechanical properties of human mitral valve chordae tendineae: Variation with size and strain rate. *Can. J. Physiol. Pharmacol.* **1975**, *53*, 330–339. [CrossRef]
48. Yamada, H.; Evans, F.G. *Strength of Biological Materials*; Williams & Wilkins: Baltimore, MD, USA, 1970.
49. Lim, K.O.; Boughner, D.R. Morphology and relationship to extensibility curves of human mitral valve chordae tendineae. *Circul. Res.* **1976**, *39*, 580–585. [CrossRef]
50. Clark, R.E. Stress-strain characteristics of fresh and frozen human aortic and mitral leaflets and chordae tendineae. Implications for clinical use. *J. Thorac. Cardiovasc. Surg.* **1973**, *66*, 202–208. [CrossRef]
51. Barber, J.E.; Ratliff, N.B.; Cosgrove, D.M., 3rd; Griffin, B.P.; Vesely, I. Myxomatous mitral valve chordae. I: Mechanical properties. *J. Heart. Valve Dis.* **2001**, *10*, 320–324.
52. Lim, K.O.; Boughner, D.R.; Perkins, D.G. Ultrastructure and mechanical properties of chordae tendineae from a myxomatous tricuspid valve. *Jpn. Heart J.* **1983**, *24*, 539–548. [CrossRef]
53. Casado, J.A.; Diego, S.; Ferreño, D.; Ruiz, E.; Carrascal, I.; Méndez, D.; Revuelta, J.M.; Pontón, A.; Icardo, J.M.; Gutiérrez-Solana, F. Determination of the mechanical properties of normal and calcified human mitral chordae tendineae. *J. Mech. Behav. Biomed. Mater.* **2012**, *13*, 1–13. [CrossRef] [PubMed]

54. Ritchie, J.; Jimenez, J.; He, Z.; Sacks, M.S.; Yoganathan, A.P. The material properties of the native porcine mitral valve chordae tendineae: An *in vitro* investigation. *J. Biomech.* **2006**, *39*, 1129–1135. [CrossRef] [PubMed]
55. Pokutta-Paskaleva, A.; Sulejmani, F.; DelRocini, M.; Sun, W. Comparative mechanical, morphological, and microstructural characterization of porcine mitral and tricuspid leaflets and chordae tendineae. *Acta Biomater.* **2019**, *85*, 241–252. [CrossRef] [PubMed]
56. Sedransk, K.L.; Grande-Allen, K.J.; Vesely, I. Failure mechanics of mitral valve chordae tendineae. *J. Heart. Valve Dis.* **2002**, *11*, 644–650. [PubMed]
57. Gunning, G.M.; Murphy, B.P. The effects of decellularization and cross-linking techniques on the fatigue life and calcification of mitral valve chordae tendineae. *J. Mech. Behav. Biomed. Mater.* **2016**, *57*, 321–333. [CrossRef] [PubMed]
58. Gunning, G.M.; Murphy, B.P. Characterisation of the fatigue life, dynamic creep and modes of damage accumulation within mitral valve chordae tendineae. *Acta Biomater.* **2015**, *24*, 193–200. [CrossRef]
59. Constable, M.; Burton, H.E.; Lawless, B.M.; Gramigna, V.; Buchan, K.G.; Espino, D.M. Effect of glutaraldehyde based cross-linking on the viscoelasticity of mitral valve basal chordae tendineae. *Biomed. Eng.* **2018**, *17*, 93. [CrossRef]
60. Ross, C.J.; Laurence, D.W.; Hsu, M.-C.; Baumwart, R.; Zhao, D.Y.; Mir, A.; Burkhart, H.M.; Holzapfel, G.A.; Wu, Y.; Lee, C.-H. Mechanics of porcine heart valves' strut chordae tendineae investigated as a leaflet-chordae-papillary muscle entity. *Ann. Biomed. Eng.* **2020**. [CrossRef]
61. Madhurapantula, R.S.; Krell, G.; Morfin, B.; Roy, R.; Lister, K.; Orgel, J.P. Advanced methodology and preliminary measurements of molecular and mechanical properties of heart valves under dynamic strain. *Int. J. Mol. Sci.* **2020**, *21*, 763. [CrossRef]
62. Kunzelman, K.S.; Cochran, K.P. Mechanical properties of basal and marginal mitral valve chordae tendineae. *ASAIO J.* **1990**, *36*, M405–M407.
63. Ross, C.J.; Laurence, D.W.; Richardson, J.; Babu, A.R.; Evans, L.E.; Beyer, E.G.; Wu, Y.; Towner, R.A.; Fung, K.-M.; Mir, A.; et al. An investigation of the glycosaminoglycan contribution to biaxial mechanical behaviors of porcine atrioventricular heart valve leaflets. *J. R. Soc. Interface* **2019**, *16*, 20190069. [CrossRef] [PubMed]
64. Krishnamurthy, V.K.; Grande-Allen, K.J. The role of proteoglycans and glycosaminoglycans in heart valve biomechanics. In *Advances in Heart Valve Biomechanics;* Springer: Cham, Switzerland, 2018; pp. 59–79.
65. Padala, M.; Sacks, M.S.; Liou, S.W.; Balachandran, K.; He, Z.; Yoganathan, A.P. Mechanics of the mitral valve strut chordae insertion region. *J. Biomech. Eng.* **2010**, *132*, 081004. [CrossRef] [PubMed]
66. Jett, S.V.; Hudson, L.T.; Baumwart, R.; Bohnstedt, B.N.; Mir, A.; Burkhart, H.M.; Holzapfel, G.A.; Wu, Y.; Lee, C.-H. Integration of polarized spatial frequency domain imaging (pSFDI) with a biaxial mechanical testing system for quantification of load-dependent collagen architecture in soft collagenous tissues. *Acta Biomater.* **2020**, *102*, 149–168. [CrossRef] [PubMed]
67. Goth, W.; Potter, S.; Allen, A.C.B.; Zoldan, J.; Sacks, M.S.; Tunnell, J.W. Non-destructive reflectance mapping of collagen fiber alignment in heart valve leaflets. *Ann. Biomed. Eng.* **2019**, *47*, 1250–1264. [CrossRef] [PubMed]
68. Caimmi, P.P.; Sabbatini, M.; Fusaro, L.; Borrone, A.; Cannas, M. A study of the mechanical properties of ePTFE suture used as artificial mitral chordae. *J. Card. Surg.* **2016**, *31*, 498–502. [CrossRef]
69. Goth, W.; Yang, B.; Lesicko, J.; Allen, A.; Sacks, M.S.; Tunnell, J.W. Polarized spatial frequency domain imaging of heart valve fiber structure. In Proceedings of SPIE, Optical Elastography and Tissue Biomechanics III, San Francisco, CA, USA, 24 June 2016; Volume 9710, p. 971019.
70. Arts, T.; Meerbaum, S.; Reneman, R.; Corday, E. Stresses in the closed mitral valve: A model study. *J. Biomech.* **1983**, *16*, 539–547. [CrossRef]
71. McQueen, D.M.; Peskin, C.S.; Yellin, E.L. Fluid dynamics of the mitral valve: Physiological aspects of a mathematical model. *Am. J. Physiol. Heart Circ. Physiol.* **1982**, *242*, H1095–H1110. [CrossRef]
72. Lim, K.H.; Yeo, J.H.; Duran, C. Three-dimensional asymmetrical modeling of the mitral valve: A finite element study with dynamic boundaries. *J. Heart. Valve Dis.* **2005**, *14*, 386–392.
73. Wang, Q.; Sun, W. Finite element modeling of mitral valve dynamic deformation using patient-specific multi-slices computed tomography scans. *Ann. Biomed. Eng.* **2013**, *41*, 142–153. [CrossRef]

74. Kunzelman, K.; Reimink, M.S.; Verrier, E.D.; Cochran, R.P. Replacement of mitral valve posterior chordae tendineae with expanded polytetrafluoroethylene suture: A finite element study. *J. Card. Surg.* **1996**, *11*, 136–145. [CrossRef]
75. Wenk, J.F.; Ratcliffe, M.B.; Guccione, J.M. Finite element modeling of mitral leaflet tissue using a layered shell approximation. *Med. Biol Eng. Comput* **2012**, *50*, 1071–1079. [CrossRef] [PubMed]
76. Khalighi, A.H.; Drach, A.; Bloodworth, C.H.; Pierce, E.L.; Yoganathan, A.P.; Gorman, R.C.; Gorman, J.H., III; Sacks, M.S. Mitral valve chordae tendineae: Topological and geometrical characterization. *Ann. Biomed. Eng.* **2017**, *45*, 378–393. [CrossRef] [PubMed]
77. Meador, W.D.; Mathur, M.; Sugerman, G.P.; Jazwiec, T.; Malinowski, M.; Bersi, M.R.; Timek, T.A.; Rausch, M.K. A detailed mechanical and microstructural analysis of ovine tricuspid valve leaflets. *Acta Biomater.* **2020**, *102*, 100–113. [CrossRef] [PubMed]
78. Khalighi, A.H.; Rego, B.V.; Drach, A.; Gorman, R.C.; Gorman, J.H., III; Sacks, M.S. Development of a functionally equivalent model of the mitral valve chordae tendineae through topology optimization. *Ann. Biomed. Eng.* **2019**, *47*, 60–74. [CrossRef]
79. Rim, Y.; Laing, S.T.; McPherson, D.D.; Kim, H. Mitral valve repair using ePTFE sutures for ruptured mitral chordae tendineae: A computational simulation study. *Ann. Biomed. Eng.* **2014**, *42*, 139–148. [CrossRef]
80. Aggarwal, A.; Aguilar, V.S.; Lee, C.-H.; Ferrari, G.; Gorman, J.H., III; Gorman, R.C.; Sacks, M.S. Patient-specific modeling of heart valves: From image to simulation. In Proceedings of the International Conference on Functional Imaging and Modeling of the Heart, London, UK, 20–22 June 2013; Sébastien, O., Rueckert, D., Smith, N., Eds.; Springer: Berlin/Heidelberg, Germany, 2013; pp. 141–149.
81. Ionasec, R.I. *Patient-Specific Modeling and Quantification of the Heart Valves from Multimodal Cardiac Images*; Technische Universität München: Munich, Germany, 2010.
82. Siefert, A.W.; Rabbah, J.-P.M.; Pierce, E.L.; Kunzelman, K.S.; Yoganathan, A.P. Quantitative evaluation of annuloplasty on mitral valve chordae tendineae forces to supplement surgical planning model development. *Cardiovasc. Eng. Technol.* **2014**, *5*, 35–43. [CrossRef]
83. Stevanella, M.; Votta, E.; Lemma, M.; Antona, C.; Redaelli, A. Finite element modelling of the tricuspid valve: A preliminary study. *Med. Eng. Phys.* **2010**, *32*, 1213–1223. [CrossRef]

bioengineering

MDPI

Article

Elastin-Dependent Aortic Heart Valve Leaflet Curvature Changes During Cyclic Flexure

Melake D. Tesfamariam, Asad M. Mirza, Daniel Chaparro, Ahmed Z. Ali, Rachel Montalvan, Ilyas Saytashev, Brittany A. Gonzalez, Amanda Barreto, Jessica Ramella-Roman, Joshua D. Hutcheson and Sharan Ramaswamy *

Department of Biomedical Engineering, Florida International University, Miami, FL 33174, USA; mtesf003@fiu.edu (M.D.T.); amirz013@fiu.edu (A.M.M.); dchap015@fiu.edu (D.C.); aali056@fiu.edu (A.Z.A.); rmont065@fiu.edu (R.M.); isaytash@fiu.edu (I.S.); bgonz049@fiu.edu (B.A.G.); abarr247@fiu.edu (A.B.); jramella@fiu.edu (J.R.-R.); jhutches@fiu.edu (J.D.H.)

* Correspondence: sramaswa@fiu.edu; Tel.: +1-(305)-348-2532

Received: 9 March 2019; Accepted: 5 May 2019; Published: 7 May 2019

Abstract: The progression of calcific aortic valve disease (CAVD) is characterized by extracellular matrix (ECM) remodeling, leading to structural abnormalities and improper valve function. The focus of the present study was to relate aortic valve leaflet axial curvature changes as a function of elastin degradation, which has been associated with CAVD. Circumferential rectangular strips (L × W = 10 × 2.5 mm) of normal and elastin-degraded (via enzymatic digestion) porcine AV leaflets were subjected to cyclic flexure (1 Hz). A significant increase in mean curvature ($p < 0.05$) was found in elastin-degraded leaflet specimens in comparison to un-degraded controls at both the semi-constrained (50% of maximum flexed state during specimen bending and straightening events) and fully-constrained (maximally-flexed) states. This significance did not occur in all three flexed configurations when measurements were performed using either minimum or maximum curvature. Moreover, the mean curvature increase in the elastin-degraded leaflets was most pronounced at the instance of maximum flexure, compared to un-degraded controls. We conclude that the mean axial curvature metric can detect distinct spatial changes in aortic valve ECM arising from the loss in bulk content and/or structure of elastin, particularly when there is a high degree of tissue bending. Therefore, the instance of maximum leaflet flexure during the cardiac cycle could be targeted for mean curvature measurements and serve as a potential biomarker for elastin degradation in early CAVD remodeling.

Keywords: aortic valve; calcification; elastin degradation; leaflet; curvature; biomarker; early detection

1. Introduction

Calcific aortic valve disease (CAVD) is characterized by pathological remodeling of the aortic valve leaflets, fibrotic tissue formation, and calcified mineral deposition [1]. CAVD is the most frequent condition that necessitates surgical valve replacement [2,3]. In developed nations, an alarming rate of incidence of CAVD is projected, from 2.5 million cases in 2000 to approximately 4.5 million in 2030 [4]. Worldwide significance is also projected to reach an epidemic level with an estimated third of the global population aged 65 exhibiting at least early clinical signs of CAVD [5]. Risk factors for CAVD include: Older age, genetics, use of tobacco products, rheumatic fever and high blood pressure [6]. Untreated CAVD leads to valve malfunction typically via narrowing or stenosis of the aortic valve, which substantially augments the workload of the left ventricle. This leads to heart failure with current estimates of about 17,000 resulting deaths/year in the United States [7]. Surgery or transcatheter artificial valve replacement procedures are the only viable intervention currently available. Many patients cannot undergo surgery either due to high mortality risks or adverse responses to required anticoagulants when a mechanical valve is employed. Bioprosthetic replacement valves made of animal tissues do not require blood thinners, but typically last only between 8 to 15 years [8]. More effective management of CAVD is needed, including the development of pharmaceutical interventions. Potential non-invasive therapeutic targets have been identified [1], but clinical utility of these treatment options will require early diagnosis prior to irreversible gross valvular remodeling.

Previous investigations have indicated that there may be a correlation between elastin degradation and calcification in several cardiovascular disorders [1,9,10]. For example, Hinton et al. [9] have shown that the onset of CAVD is associated with aortic valve extracellular matrix (ECM) remodeling events, specifically, elastin degradation [4]. The degradation is caused by elastolytic enzymes including matrix metallopeptidases and cathepsins in the pathological ECM remodeling of the valves either via induction of fibrosis or by accelerated degradation of elastin fibers [4]. Elastin, which is predominant on the ventricularis layer proximal to the left ventricle, enhances full aortic valve closure via stretching during the end-diastolic phase of the cardiac cycle during which time the leaflets are subjected to transmembrane pressure stresses. Elastin also permits sufficient bending and re-coil of the aortic valve leaflets during the systolic and early diastolic phases respectively. This ensures that minimal energy losses are expended for the heart to pump blood to the systemic circulation, and that instantaneously after, the leaflets have efficiently transitioned to the coaptation state.

In the current study, we sought to determine if a change in aortic valve leaflet tissue structure due to elastin degradation could be detected and subsequently quantified by curvature measurements. Beyond standard echocardiography for routine diagnosis of valve diseases, when greater spatial and/or temporal resolution is needed, such as for example for interventional and surgical planning purposes, transoesophageal echocardiography or multi-detector computed tomography are used in the acquisition of 3-dimensional heart valve geometry, with each modality having its own merits. For example, transoesophageal echocardiography is able to provide detailed spatial information on the distribution of calcification on the aortic valve and its surrounding vasculature. On the other hand, multi-detector computed tomography can provide detection of highly-tortuous geometries, with high spatial resolution in the order of 500 μm [11] to capture 3-D native aortic valve geometries with high-fidelity. Collectively, while the selection of imaging modalities for a given patient will be specific to the end-objectives, quantification of leaflet curvature in a fairly straightforward manner does currently appear to be feasible. Aortic valve leaflet curvature can be easily measured from imaging modalities such as echocardiography or Computer-Aided Tomography (CT)-derived images. Therefore, detection of changes in curvature based-metrics of the aortic valve leaflet due to abnormal elastin remodeling or loss could serve as a potential biomarker for early detection of CAVD.

2. Materials and Methods

2.1. Tissue Sample Preparation

Following the protocol by Butcher et al. [12] for porcine valvular tissue isolation, young porcine hearts (~5 months old) were harvested from a local abattoir (Mary's Ranch, Miami, FL, USA) from which the aortic valves were resected (n = 3 valves). The valves, along with the aortic root, were placed in cold 1X phosphate buffered saline solution (PBS) and transported to the laboratory. From each aortic valve, the three leaflets were excised from the aortic root and individually placed in a protease inhibitor (PI) solution prepared by dissolving a PI tablet (Sigma-Aldrich, St. Louis, MO, USA) in 50 mL PBS, so as to decelerate naturally-occurring degradation activity. Tissue sectioning began with cutting rectangular strips (10 mm +/− 0.1 mm × 2.5 mm +/− 0.5 mm) from the belly region of the valve with the long axis oriented in the circumferential direction. The remaining surrounding tissue was thoroughly minced, mixed, and subsequently divided into 5 equal parts. Storage conditions for, all tissues were maintained at 4 °C while immersed in the protease inhibitor solution.

Elastin Degradation

Each rectangular tissue strip that was subjected to cyclic flexure (Section 2.2) and curvature assessment (Section 2.2) subsequently underwent elastin degradation, that is, each sample served as its own control. Elastin tissue degradation was performed with the use of an elastase powder (Bio Basic Inc., Toronto, ON, Canada) dissociated in de-ionized (DI) water [13]. In brief, 0.5 mg/mL of elastase solution was prepared in DI water by gently shaking the solution to maintain dissociation of the powder. Elastase degradation of each strip and corresponding specimen of minced tissue was performed for 2 h at 37 °C with 5% CO_2 and 95% humidity. A control group consisted of un-degraded leaflet tissue strips in PI solution, that is, zero hour or non-exposure to the aqueous elastase enzyme.

2.2. *Cyclic Flexure Experiments*

Aortic valve leaflet uniaxial flexure experiments and subsequent curvature measurements were conducted using a mechanical testing instrument available in our laboratory (Electroforce 2300 with Wintest 7 software control, TA instruments, New Castle, DE, USA). Just prior to each experiment, the tissues were washed with 1X PBS and immersed in cold protease inhibitor solution. The tissue samples were measured in length, width and thickness before testing. Each specimen was then carefully affixed between the two grips length-wise (Figure 1). Initially, a caliper was used to allow for a known set distance during the recorded footage, in this case the distance between the left and right side of the grips (Figure 1). The parameters used for the experiment were as follows: Frequency of 1 Hz, time of 30 seconds, valve effective length between the grips was 11 mm and bending distance was from 2.8 units to –5 units (negative displacement) for bending and from –5 units to 2.8 units for stretching (positive displacement).

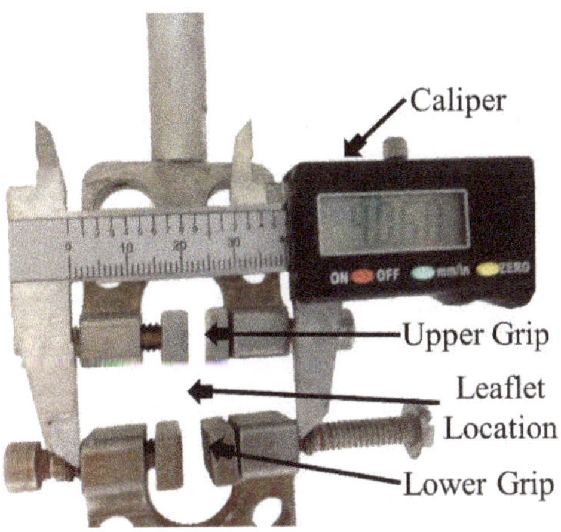

Figure 1. Leaflet strip set-up in mechanical test instrument (Electroforce 2300 with Wintest 7 software control, TA instruments) to facilitate specimen cyclic flexure (bending and straightening events).

A simple camera system (Model #: 8/A1864, Apple Inc., Cupertino, CA, USA) was mounted tangential to the edge of the tissue specimens to record the bending and straightening events during cyclic flexure. The camera was positioned to precisely detect the specimen bending process along the edge of the sample (i.e., a 1-dimensional (1-D) axial edge measurement). Ten cycles of pre-conditioning were first carried out before data collection. Subsequently, each specimen was video-recorded during an entire cyclic flexure event, that is, from the straight specimen configuration, to the initiation of bending until the two sample grips were 0.5 cm apart and subsequently until the specimens returned to its original straight position. The video tracking was conducted on recorded frames using an in-house script based on the controlling parameters defined above (MATLAB, Mathworks, Natick, MA, USA).

The total number of frames corresponding to the total recording time were distributed over the time and position for curvature calculations. The camera was set to record at 240 frames per second (FPS) totaling approximately 7,200 frames per video.

Curvature Assessment

Image frames collected from the cyclic flexure experiments were imported into an in-house script (MATLAB) to calculate the localized mean/minimum (min) and maximum (max) curvature (k_{mean}, k_{min}, k_{max}) by fitting a line along the coordinates of the edges of the sample (Figure 2). For each frame the coordinates of the points along the edges were fitted to n^{th} degree polynomial as per the equation given below:

$$f(x) = a_0x^n + a_1x^{n-1} + \ldots + a_n \quad (1)$$

The curvature, k of the sample at each time point was then computed from the curvature equation [14]:

$$k = \frac{f''}{\left(1 + (f\prime)^2\right)^{3/2}} \quad (2)$$

where $Y\prime$ and Y'' refers to the first and second order derivatives of the polynomial equation, respectively. Since the curvature can vary widely along the edge of the sample the polynomial was fitted piecewise along the edge, resulting in a curvature value for every point (Figure 2C(i,ii,iii)). The minimum, maximum, and mean curvatures were then determined for each frame (Figure 2C(iv)). Validation of the computed curvature was previously conducted by comparing the in-house curvature code with three different circles of known radii. In all three cases, the computed curvature was found to be within 5% of the actual curvature of the circles.

The k_{mean}, k_{min} and k_{max} metrics were computed at 5 temporal point during the flexure cycle. Specifically, these temporal points were as follows: Unconstrained or fully straight specimen positon prior to the bending event (UNC_b), 50% constrained during bending or 50% of the maximum specimen flexure state during bending (50%-C_b), the fully constrained or maximum specimen flexure state possible (FC), 50% constrained during straightening or 50% of the fully straightened specimen state during the straightening event (50%-C_s) and finally, the fully straight or unconstrained specimen at the instance of complete straightening (UNC_s). A total of 30 cycles of cyclic flexure were performed for each tissue specimen. The first 10 were excluded for pre-conditioning purposes. From the remaining 20 cycles, 5 were randomly chosen and their respective curvature was calculated at each of the bending phases and then averaged together. All 5 cycles were within 5% of each other for the minimum, maximum, and mean curvatures.

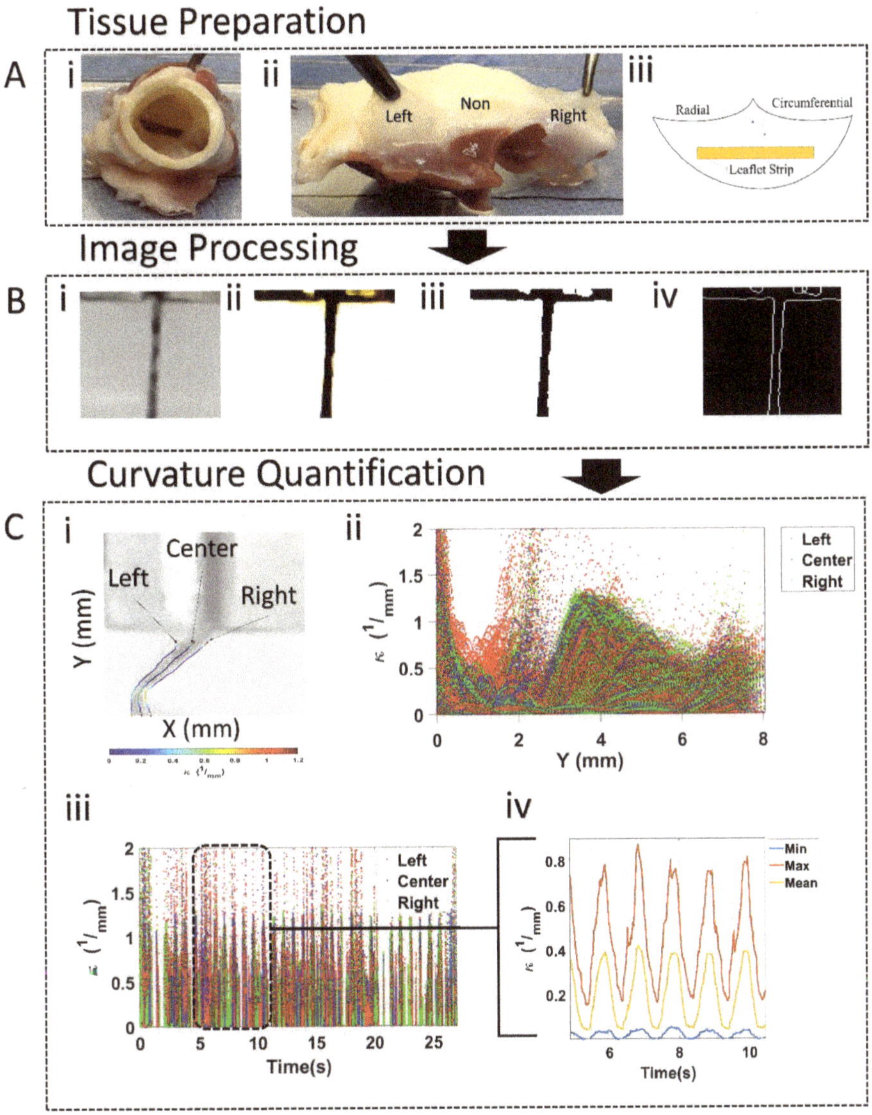

Figure 2. Workflow (**A**) Tissue preparation: i) Porcine aortic valve, ii) split aortic valve revealing the three leaflets, iii) schematic of circumferentially oriented rectangular strips cut from the belly region of porcine aortic valve leaflets. The remaining surrounding tissue was minced and used for the degradation and quantification of elastin in this study. (**B**) Image processing: i) Cropped image according to region of interest, ii) increase background to foreground contrast, iii) threshold image, iv) edge detection. (**C**) Curvature quantification: i) Left/center/right curvatures for a frame of a video, ii) temporal curvature for a given leaflet, iii) spatial curvature for a given leaflet, iv) minimum, Maximum, and mean curvature along center profile for the 5–10 range.

2.3. Elastin Assay

A commercially available Elastin assay kit (Fastin kit, Accurate Chemical Scientific, Westbury, NY, USA) was used to extract, solubilize and quantify elastin present in aortic valve leaflet tissues following the manufacturer's protocol. The extraction of insoluble elastin from tissue samples was done by hot oxalis acid digestion; because tissue samples were very small, one extraction was necessary to digest all the sample present. Absorbance measurements were performed utilizing a spectrophotometer (Synergy HTX Multi-Mode reader, BioTek Instruments, Inc., Winooski, VT, USA). The elastin concentration was calculated for each of the un-degraded and elastin-degraded groups, using a calibration curve to convert measured absorbance values into actual concentrations. Finally, the elastin content of each sample was normalized with respect to its corresponding wet weight.

2.4. Elastin Structure

Unstained two-photon imaging using a home-built laser scanning microscopy system with broadband femtosecond excitation laser (Element 600, Femtolasers, Austria) was performed to evaluate elastin structure of porcine aortic valve leaflets (n = 3 leaflets), both before and after being subjected to a two-hour elastase degradation. The protocol for leaflet elastin degradation was identical to that utilized for degrading elastin in the tissue strips prior to cyclic flexure testing (Section 2.1). Microscopy was first carried out by scanning a laser beam with a pair of galvanometer mirrors (Thorlabs, Newton, NJ, USA) and directing it into a 20×/1 NA water immersion objective (Olympus, Japan), via a dichroic mirror (655spxr, Chroma, Bellows Falls, VT, USA) to separate two-photon signals from a reflected fundamental wavelength. Two photomultiplying tube photodetectors (Hamamatsu, Japan) with respective bandpass filters (400 nm central wavelength/40 nm bandwidth and 480 nm central wavelength/40 nm bandwidth) simultaneously recorded second harmonic generation and two-photon excitation fluorescence (TPEF) signals. Scanning and acquisition was controlled using a custom-written software (MATLAB) and a data acquisition board (National Instruments, Austin, TX, USA). Each image consisted of an average of ten 1000 × 1000-pixel frames acquired at 0.5 frames per second. The second harmonic generation (SHG)-signal was derived from the collagen fibers [15] in the leaflet tissues while TPEF-derived images consisted of signal intensities from elastin protein and cellular autofluorescence [13,16]. Images of un-degraded leaflets with a rich TPEF-signal but which were characterized by an absence of a fibrous structures were interpreted to be mainly autofluorescence and subsequently discarded. On the other hand, images that contained a fibrous network with TPEF-rich signal intensities were regarded as being elastin fibers and were thereby saved for comparison with images derived from the same leaflets after elastin degradation, at the corresponding spatial location. Prior to final evaluation of elastin structure, minor image processing was performed (ImageJ, NIH, Bethesda, MD, USA). Specifically, a 2-pixel mean radius filter was applied to each image, with subsequent square root of the intensity taken, and finally, brightness and contrast adjusted for the final visualization.

2.5. Histological Staining

The non-coronary leaflets from two porcine aortic heart valves were isolated. One leaflet was immersed in 0.5 mg/ml elastase solution for 2 h while the other leaflet was immediately fixed in 10% formalin. After two hours of elastase degradation, the degraded non-coronary leaflet was also immersed in 10% formalin. The leaflets were then rinsed with PBS and embedded in optimal cutting temperature (OCT). The leaflets were subsequently stored in −80 °C overnight. The next day, the embedded samples were cut depth-wise to a thickness of 10 μm/slice and mounted on glass slides (TruBond 380, Newcomer Supply, Middleton, WI, USA), which were allowed to dry at room temperature. Finally, the tissues were stained (Russel–Movat Pentachrome; American MasterTech Scientific, Item #: KTRMP, Lodi, CA, USA) to identify elastin and collagen distributions with the tissues. Each sample was then imaged and processed (ImageJ, NIH-Image, Bethesda, MD, USA) with auto-adjustment of the brightness and contrast.

2.6. Statistical Analysis

Curvature values were presented as the mean ± standard error of the mean (SEM). Significance in curvature measurements was determined via t-tests (GraphPad Software, San Diego, CA, USA) comparing the un-degraded controls to samples exposed to 2 h of enzymatic-solution at each of the 5 temporal-bending states evaluated, that is, UNC_b, 50%-C_b, FC, 50%C_s and UNC_s. Sample size (n) for each of the 5 time-points/group was as follows: n = 9 leaflet strips/time-point for the un-degraded, control group and n = 6 leaflet strips/time-point for group that was subjected to the 2-hour elastase enzymatic degradation. A *p*-value of less than 0.05 ($p < 0.05$) between the two groups was considered to be statistically significant.

3. Results

3.1. Leaflet Shape Changes

Instantaneous image capture showed the changes in curvature at five levels (or temporal positions) of bending and straightening (Table 1). Visible differences between corresponding time-points were observed between the 2 h elastin-degraded and un-degraded groups.

Table 1. Bending/straightening behavior of samples with and without 2 hours (2 h) of elastin degradation.

Group	UNC_b	50%-C_b	FC	50%-C_s	UNC_s
Control 0 h					
2 h					

Curvature Computation

Mean, min, and max curvature for control and elastin-degraded (2 h of elastase degradation) porcine aortic valve leaflet specimens (between 6 and 9 strips/group) were computed at five different temporal positions during the cyclic flexure event (Tables 2–4). Mean curvature (k_{mean}) between the two groups was found to be significantly higher for the elastin degraded tissue ($p < 0.05$) at the 50%-C_b, FC and 50%-C_s flexure states. In the case of the k_{min} metric, only the 50%-C_b exhibited statistical significance between the groups, again with higher curvature in the degraded group. The magnitude of k_{min} values were relatively small compared to k_{mean} and k_{max}. Finally, no significance ($p > 0.05$) between the groups was observed for k_{max}.

Table 2. Mean ± standard error of the mean (SEM) of bending of samples of mean curvature, k_{mean} (1/mm) at five levels of bending for the un-degraded and elastin-degraded porcine aortic valve leaflet specimens after 2 h of elastase exposure (n = 6–9 samples/group). UNC_b, 50%-C_b, stands for unconstrained and semi-constrained during leaflet strip bending, while UNC_s and 50%-C_s refer to these events during specimen straightening; FC stands for the leaflet strip at the fully constrained or fully-flexed state.

Group	Amount of Bending					n
Elastin Degradation	UNC_b	50%-C_b *	FC *	50%-C_s *	UNC_s	
0 h (Control)	0.07 ± 0.04	0.13 ± 0.04	0.19 ± 0.04	0.14 ± 0.04	0.09 ± 0.06	9
2 h	0.11 ± 0.05	0.28 ± 0.01	0.40 ± 0.03	0.29 ± 0.01	0.14 ± 0.01	6
p-Value	$p > 0.05$	$p < 0.05$	$p < 0.05$	$p < 0.05$	$p > 0.05$	

* Indicates significance when compared to its corresponding control value.

Table 3. Mean ± SEM of bending of samples of min curvature, k_{min} (1/mm) at five levels of bending for the un-degraded and elastin-degraded porcine aortic valve leaflet specimens after 2 h of elastase exposure (n = 6–9 samples/group). UNC_b, 50%-C_b, stands for unconstrained and semi-constrained during leaflet strip bending, while UNC_s and 50%-C_s refer to these events during specimen straightening; FC stands for the leaflet strips at the fully constrained or fully-flexed state.

Group	Amount of Bending					n
Elastin Degradation	UNC_b	50%-C_b *	FC	50%-C_s	UNC_s	
0 h (Control)	0.0026 ± 0.0006	0.0055 ± 0.0017	0.0083 ± 0.0058	0.0071 ± 0.0656	0.0067 ± 0.0059	9
2 h	0.0091 ± 0.0099	0.0614 ± 0.0214	0.0606 ± 0.0334	0.0475 ± 0.0313	0.0148 ± 0.0001	6
p-Value	$p > 0.05$	$p < 0.05$	$p > 0.05$	$p > 0.05$	$p > 0.05$	

* Indicates significance when compared to its corresponding control value.

Table 4. Mean ± SEM of bending of samples of max curvature, k_{max} (1/mm) at five levels of bending for the un-degraded and elastin-degraded porcine aortic valve leaflet specimens after 2 h of elastase exposure (n = 6–9 samples/group). UNC_b, 50%-C_b, stands for unconstrained and semi-constrained during leaflet strip bending, while UNC_s and 50%-C_s refer to these events during specimen straightening; FC stands for the leaflet strips at the fully constrained or fully-flexed state.

Group	Amount of Bending					n
Elastin Degradation	UNC_b	50%-C_b	FC	50%-UC_s	UNC_s	
0 h (Control)	0.25 ± 0.12	0.58 ± 0.22	0.88 ± 0.40	0.58 ± 0.20	0.27 ± 0.17	9
2 h	0.37 ± 0.25	0.73 ± 0.08	1.20 ± 0.23	0.83 ± 0.04	0.44 ± 0.04	6
p-Value	$p > 0.05$	$p > 0.05$	$p > 0.05$	$p > 0.05$	$p > 0.05$	

3.2. *Leaflet Elastin Loss*

The normalized elastin content (with respect to specimen wet weight) indicated that the elastin concentration in the elastin-degraded leaflet samples after 2 h of elastase exposure was 2.3-fold lower than the un-degraded samples (n = 3 specimens/degradation-group; Table 5).

Table 5. Mean ± SEM elastin concentration in un-degraded and elastin-degraded porcine aortic valve leaflet specimens after 2 h of elastase exposure. (n = 3 samples/group).

Group	Elastin Degradation	
Elastin (μg/mg)	0 h (Control) 0.161 ± 0.07	2 h 0.069 ± 0.016

3.3. *Elastin Curvature Comparison*

Elastin concentration was plotted against k_{mean}, k_{min} and k_{max} for each temporal position of bending phase (Figure 3A–C respectively) for the two groups assessed in this study (0 Hours of elastase exposure (un-degraded controls) and 2 h of elastase exposure (elastase-degraded leaflet samples). A general trend of increase in curvature between the groups was observed with more elastin-degraded leaflet samples. Interestingly, the change in k_{mean} and k_{max} as a function of elastin loss was most pronounced when the leaflet tissue strips were at the maximum flexed state, that is, FC (Figure 3A,B).

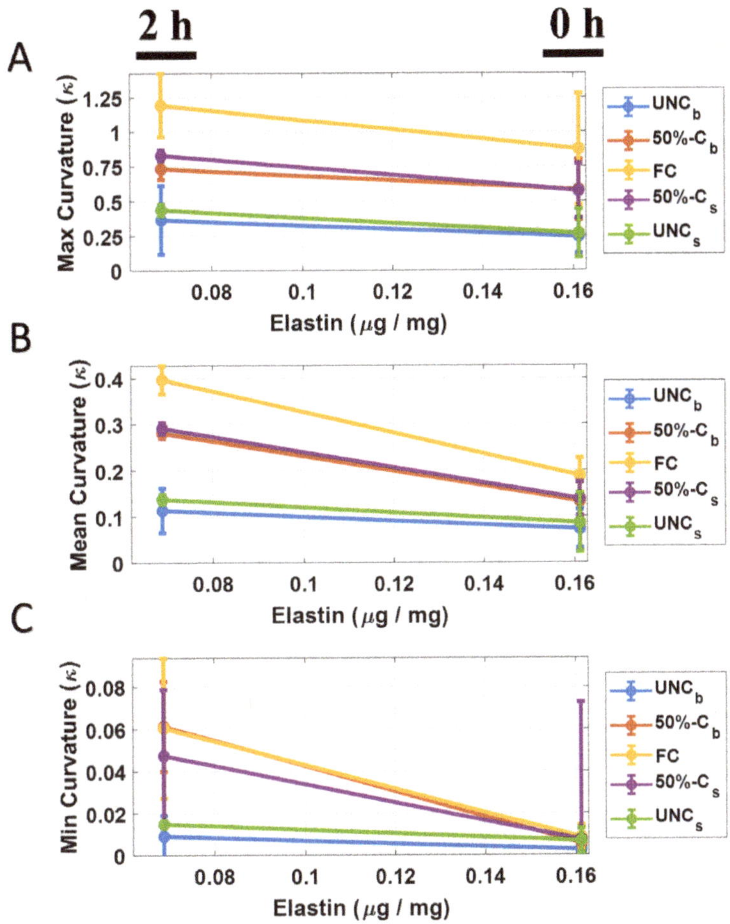

Figure 3. Elastin and curvature comparisons. (**A**) Maximum curvature with mean ± SEM plotted against elastin for 0 h and 2 h degradation time points for all bending phases. (**B**) Mean curvature with mean ± SEM plotted against elastin for 0 h and 2 h degradation time points for all bending phases. (**C**) Minimum curvature with mean ± SEM plotted against elastin for 0 h and 2 h degradation time points for all bending phases.

3.4. Elastin Structure

Un-degraded porcine aortic valve leaflets exhibited characteristic elastin network structure, with elastin fibers running in the radial direction (Figure 4A,C). On the other hand, after the leaflets were degraded with elastase, elastin was still present in the tissues but with a considerable loss in content observed (Figure 4B,D). In addition, there was clear disruption to elastin structure. Specifically, bulk aggregation of elastin was evident and was found to be randomly interspersed in the tissues. In addition, there was a complete loss in elastin fiber orientation and network.

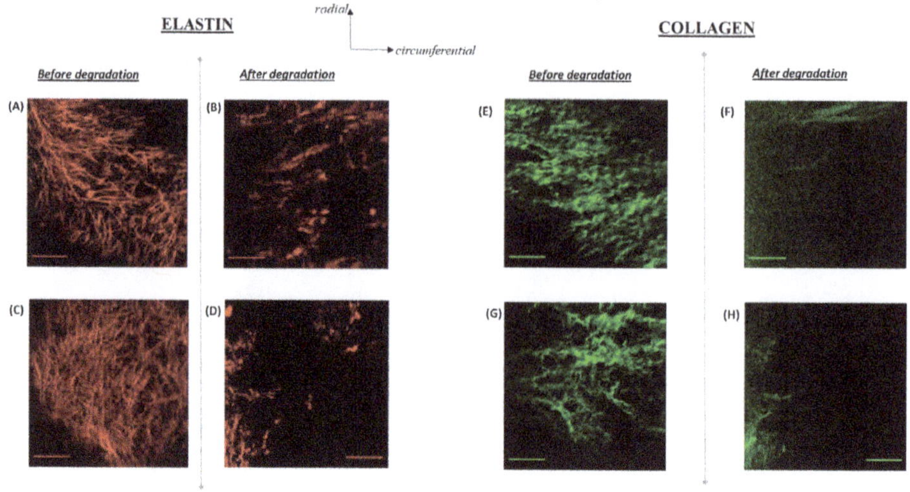

Figure 4. (**A–D**) Elastin structure in aortic valve leaflets before (**A**,**C**) and after (**B**,**D**) elastase-degradation. Images (**A**) and (**B**) were obtained from the non-coronary cusp whereas images (**C**) and (**D**) were from the left coronary cusp. Image sections were taken at a depth of 36 μm from the nearest leaflet surface. Scale bar in all images is 50 μm. Elastin fibers (**A**,**C**) exhibited a strong affinity for alignment in the radial direction. Enzymatic-degradation clearly led to loss of elastin content, orientation and network in the leaflets (**B**,**D**). (**E–H**) Collagen distribution at the same spatial location in which Elastin structure was imaged (**A–D**). Collagen fibers (**E**,**G**) were primarily oriented in circumferential direction. Degradation with elastase enzyme was not specific to elastin as clearly, loss of collagen content, orientation and network in the leaflets also occurred (**F**,**H**).

3.5. Elastin Distribution

Histological evidence (Figure 5) revealed that an abundance of collagen was still visible within the leaflet after elastase-degradation, even though structure was severely disrupted (Figure 4). On the other hand, the elastin component of the non-coronary leaflet's ECM was substantially lost, to the extent that the ventricularis layer was no longer visible (Figure 5).

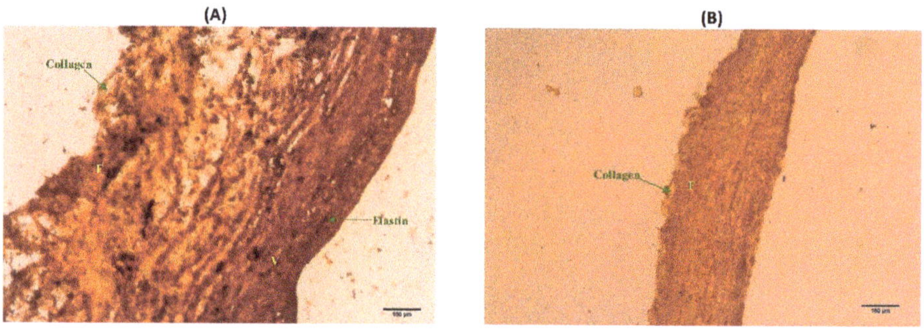

Figure 5. Movat's staining of: (**A**) Normal native porcine aortic valve non-coronary leaflet and (**B**) elastase-degraded (0.5 mg/mL for 2 h) native porcine aortic valve non-coronary leaflet. The extracellular matrix (ECM) components stain as brownish-black for Elastin and yellow-orange for Collagen. "F" indicates the fibrosa side and "V" is the ventricularis side of the leaflet. Note that the ventricularis side of the leaflet was completely lost after elastase degradation (B).

4. Discussion and Conclusions

Matrix components such elastin and collagen are altered during early onset of CAVD [4,15]. Specifically, previous investigations have also described how elastolytic enzymes such as MMPs and cathepsins are involved in ECM remodeling by enhancing degradation of elastic fibers [1]. In particular, elastin degradation in aortic heart valves can serve as an initiating event in CAVD [16]. In the current work, we investigated a simple and rapid assessment for changes in aortic valve leaflet tissue curvature as a function of tissue elastin content and structure. The motivation for identifying such a diagnostic arises from the fact that while drug targets have been identified for the treatment of CAVD [17], early detection of the disease is still sorely needed. Aiming to understand these changes, we developed a methodology to quantify leaflet curvature change as a function of elastin loss and simultaneously, to identify alterations in the elastin structure when degraded. Monitoring leaflet curvature could thereby serve as a biomarker for early diagnosis of CAVD.

Axial curvature (k_{mean}, k_{min} and k_{max}) was found to generally be increased in aortic leaflet tissue strips with degraded elastin. However, the k_{min} metric had very low magnitudes, thereby causing insufficient resolution for it to facilitate accurate distinctions between un-degraded and elastin-degraded specimens (Figure 3). Between the other two curvature metrics investigated (k_{mean} and k_{max}), only k_{mean} exhibited significantly higher curvature states ($p < 0.05$) at multiple flexure configurations when the tissue specimens were elastin-degraded (Table 2). Specifically, these states occurred at 50%-C_b, FC and 50%-C_s. This important finding suggests that elastin loss due to abnormal longitudinal remodeling in the aortic valve can be detected during sufficient levels of bending in the leaflet as its shape configuration changes during the cardiac cycle. From a clinical diagnostic standpoint, the instance of maximum aortic valve leaflet flexure that occurs during the cardiac cycle can be targeted for medical image acquisition via echocardiography and/or CT. The axial 1-D mean curvature can be easily computed and monitored for changes with aging. Rather than focusing on the entire tri-leaflet geometry, a specific leaflet of the aortic valve could additionally be targeted, such as the non-coronary cusp for example, which is prone to calcification [18–20]. Moreover, abnormal aortic valve elastin remodeling events leading to loss in leaflet elastin content is likely to be accompanied by a loss in the radially-oriented elastin fiber network (Figure 4). Therefore, it is important to note that an increase in the axial 1-D mean leaflet curvature measurement in elastin-degraded leaflets is due to some combination of losses in both bulk elastin content and its structure.

The present findings suggest that the elastase degradation protocol is not specific to elastin-alone and not surprisingly, given the physical interconnectedness of leaflet elastin-to-collagen fibers, led to disruption of the collagen structure as well (Figure 4). However, subsequent histological observations (Figure 5) confirmed that elastase-degradation led to the complete loss of elastin distribution within the leaflet whereas collagen was still very much present within the leaflet ECM. This result provides us with confidence that the curvature changes that initiate from the loss of elastin in the aortic valve, while collagen still remains, can be detectable, particularly at high leaflet flexure states.

In conclusion, we have demonstrated that axial mean curvature of flexed, rectangular, circumferentially-aligned aortic heart valve leaflet strips (k_{mean} at 50%-C_b, FC and 50%-C_s configurations) increases significantly ($p < 0.05$) when enzymatically-degraded after 2 h of exposure to elastase. This finding suggests that elastin remodeling events that occur as a precursor to CAVD could be monitored via echocardiography and/or by CT as a potential biomarker for early disease detection.

5. Limitations

The study presented here represents preliminary findings and as such there were several limitations to this work. As can be observed (Figure 4), in addition to elastin, collagen structure was also lost during degradation of tissues with elastase. This is not entirely surprising since elastin fibers are under pre-tensile stress and serve to impose an intrinsic compressive stress on collagen fibers [21]. As elastin degrades, one would expect the collagen orientation to be disrupted (Figure 4). In our case, as stated, some collagen loss was observed (though much less compared to elastin (Figure 5)) in addition to severe alterations to the remaining collagen structure. Moreover, verification tests (unpublished observations) confirmed that elastase solution consisting of 0.5 mg/ml of elastase dissolved in DI water caused thickening of the leaflet tissue specimens. The thickening was even more pronounced when DI water-alone was used in a 2-hour mock degradation of un-degraded control leaflet tissue strips (data not shown). This finding suggests that the elastase solution utilized would not be able to maintain the same level of tissue thickening with varying levels of elastase concentration. Therefore, this thickening effect, coupled with the reported alterations on collagen content and structure, revealed that our investigation was limited by the lack of specificity to elastin degradation alone, which was not permissible via the 2-hour exposure of the leaflet tissues to 0.5 mg/ml aqueous elastase solution. Instead, three known properties of the leaflet tissues were modified after the enzymatic exposure: (i) Elastin (ii) collagen and (iii) the tissue thickness. Indeed, subsequent in vitro investigations will need to focus on adopting additional experimental controls that account for these three parameters, in order to establish a more conclusive interpretation that associates controlled loss of aortic valve elastin with its mean curvature.

Another major limitation is that the curvature analyses herein focused on strips that were circumferentially dissected from the belly region of the porcine aortic leaflet which does not adequately represent valve leaflet geometry and associated in and out of plane deformations (2D and 3D curvature assessments) that occur during the cardiac cycle. Indeed, the commissures and more broadly the spatial locations where the leaflets are attached to the aorta are prone to high flexural states [22] and therefore serve as important valve anatomical targets for curvature computation. While these experiments are not straightforward, for in-plane curvature assessment, bi-axial cyclic flexure experiments would need to be considered to evaluate the resulting rectangular leaflet tissue strip curvature as a function of both the circumferential and radial directions. On the other hand, 3-dimensional curvature maps depicting the in vivo situation would still need to be further quantified via imaging the aortic valve leaflets while mounted in a flow loop that mimics the systemic hemodynamic conditions. These approaches represent important milestones that need to be considered as pre-cursors to eventual longitudinal in vivo curvature assessment of the aortic valve leaflets. In addition, valve tissue stiffness has been shown to be a function of the strain rate and frequency [23–26], specifically that a faster rate of deformation will yield a higher tissue stiffness. In the present study, leaflet strip curvature was assessed while the frequency was held constant at a physiologically-relevant magnitude of 1 Hz. It must be noted nonetheless that in the context of the in vivo situation, that changes in the heart rate could lead to different curvature responses in leaflets and hence, this effect has to be incorporated, for example, via a normalization process of curvature with respect to the heart rate, for more objective quantification of curvature. Furthermore, while we demonstrated that the flexed rather the straight configurations were more sensitive to differences between un-degraded and elastin-degraded specimens, without identifying specific phases in the cardiac cycle in which pronounced bending states occur, a direct application of the current findings to the clinic will not be possible.

Another major limitation is that we only assessed elastin degradation after 2 h of elastase-exposure. The reason for this is that the utilized elastase solution (0.5 mg/mL) did not facilitate controlled and progressive loss of elastin in the tissues as a function of degradation times (data not shown). Therefore, a precise co-relation of k_{mean} with elastin content in the aortic valve ECM, over subtle, moderate and excessive loss of elastin (e.g., such as that previously described by Roach et al. [13]) would be necessary to determine a unique and precise correlation between k_{mean} and leaflet elastin loss.

Finally, although the relative changes in curvature assessment that was performed for each porcine tissue strip (all acquired from pigs at 5 months of age) before and after a severe tissue degradation protocol were observed, absolute differences in elastin content can vary considerably depending on the donor and even from the specific anatomical leaflet chosen (non-coronary, left coronary and right coronary cusps). This study did not attempt to sort tissue strips according to the donor source or anatomical cusp location. Note however, that proteolytic elastin degradation has been shown to initiate CAVD in mouse models [18], and fragmented elastin has been observed at sites of calcification in human aortic valve leaflets [1]. The degree of elastin fragmentation, however, has not been quantified over the course of CAVD progression. Temporal studies are difficult to perform in human patients due to the inability to biopsy aortic valve tissue. Hence, the current study provides important benchmarks of aortic valve leaflet curvature in tissues with intact and severely degraded elastin. Future studies will vary the protease incubation time and utilize the techniques developed to determine the resolution of this curvature-based analysis in identifying biomechanical changes resulting from a range of elastin degradation. Porcine aortic valve leaflets are structurally similar to human specimens, and we expect that the resolution limits identified in these studies would be relevant for utilization in a clinical setting. The absolute measures of curvature that may detect the onset and progression of CAVD, however, would require extensive validation for human patients using relevant imaging modalities.

Author Contributions: M.D.T. conducted specimen preparations, leaflet cyclic flexure testing and wrote parts of the manuscript. A.M.M. performed leaflet cyclic flexure testing and curvature measurements. D.C. performed curvature measurements. A.Z.A. conducted sample preparation and assisted in cyclic flexure testing. R.M. assayed elastin content in specimens. I.S. conducted SHG and TPEF imaging. B.A.G. assisted with cyclic flexure testing and conducted the histology. A.B. assisted with sample preparation and cyclic flexure testing. J.R.-R. provided guidance on SHG/TPEF imaging. J.D.H. assisted with manuscript writing. S.R. planned the study and wrote parts of the manuscript.

Funding: This study was supported by a Coulter SEED grant, from the Department of Biomedical Engineering, College of Engineering and Computing, Florida International University. In addition, we gratefully acknowledge the support of the Herbert Wertheim Foundation and STROBE: A National Science Foundation Science & Technology Center under Grant No. DMR 1548924.

Conflicts of Interest: The authors declare no conflict of interest.

References

1. Perrotta, I.; Camastra, C.; Filice, G.; Di Mizio, G.; Colosimo, F.; Ricci, P.; Tripepi, S.; Amorosi, A.; Triumbari, F.; Donato, G.; et al. New evidence for a critical role of elastin in calcification of native heart valves: Immunohistochemical and ultrastructural study with literature review. *Histopathology* **2011**, *59*, 504–513. [CrossRef]
2. Passik, C.S.; Ackermann, D.M.; Pluth, J.R.; Edwards, W.D. Temporal changes in the causes of aortic stenosis: A surgical pathologic study of 646 cases. *Mayo Clin. Proc.* **1987**, *62*, 119–123. [CrossRef]
3. Rutkovskiy, A.; Malashicheva, A.; Sullivan, G.; Bogdanova, M.; Kostareva, A.; Stensløkken, K.; Fiane, A.; Vaage, J. Valve interstitial cells: The key to understanding the pathophysiology of heart valve calcification. *J. Am. Heart Assoc.* **2017**, *6*, 6. [CrossRef]
4. Yutzey, K.E.; Demer, L.L.; Body, S.C.; Huggins, G.S.; Towler, D.A.; Giachelli, C.M.; Hofmann-Bowman, M.A.; Mortlock, D.P.; Rogers, M.B.; Sadeghi, M.M.; et al. Calcific aortic valve disease: A consensus summary from the Alliance of Investigators on Calcific Aortic Valve Disease. *Arterioscler. Thromb. Vasc. Biol.* **2014**, *34*, 2387–2393. [CrossRef]
5. Engelmayr, G.C.; Sales, V.L.; Mayer, J.E.; Sacks, M.S. Cyclic flexure and laminar flow synergistically accelerate mesenchymal stem cell-mediated engineered tissue formation: Implications for engineered heart valve tissues. *Biomaterials* **2006**, *27*, 6083–6095. [CrossRef] [PubMed]
6. Zeng, Y.; Sun, R.; Li, X.; Liu, M.; Chen, S.; Zhang, P. Pathophysiology of valvular heart disease. *Exp. Ther. Med.* **2016**, *11*, 1184–1188. [CrossRef]
7. van der Ven, C.F.; Wu, P.J.; Tibbitt, M.W.; van Mil, A.; Sluijter, J.P.; Langer, R.; Aikawa, E. In vitro 3D model and miRNA drug delivery to target calcific aortic valve disease. *Clin. Sci. (Lond.)* **2017**, *131*, 181–195. [CrossRef] [PubMed]
8. Barbarash, L.; Rutkovskaya, N.; Barbarash, O.; Odarenko, Y.; Stasev, A.; Uchasova, E. Prosthetic heart valve selection in women of childbearing age with acquired heart disease: A case report. *J. Med. Case Rep.* **2016**, *10*, 51. [CrossRef] [PubMed]
9. Hinton, R.B.; Lincoln, J.; Deutsch, G.H.; Osinska, H.; Manning, P.B.; Benson, D.W.; Yutzey, K.E. Extracellular matrix remodeling and organization in developing and diseased aortic valves. *Circ. Res.* **2006**, *98*, 1431–1438. [CrossRef]

10. Hinton, R.B.; Adelman-Brown, J.; Witt, S.; Krishnamurthy, V.K.; Osinska, H.; Sakthivel, B.; James, J.F.; Li, D.Y.; Narmoneva, D.A.; Mecham, R.P.; et al. Elastin Haploinsufficiency Results in Progressive Aortic Valve Malformation and Latent Valve Disease in a Mouse Model Novelty and Significance. *Circ. Res.* **2010**, *107*, 549–557. [CrossRef] [PubMed]
11. Bleakley, C.; Monaghan, M.J. The Pivotal Role of Imaging in TAVR Procedures. *Curr. Cardiol. Rep.* **2018**, *12*, 9. [CrossRef]
12. Gould, R.A.; Jonathan, T.B. Isolation of valvular endothelial cells. *J. Vis. Exp.* **2010**, *46*, e2158. [CrossRef]
13. Roach, M.R.; Alan, C.B. The reason for the shape of the distensibility curves of arteries. *Can. J. Biochem. Physiol.* **1957**, *35*, 681–690. [CrossRef]
14. Salinas, M.; Rath, S.; Villegas, A.; Unnikrishnan, V.; Ramaswamy, S. Relative effects of fluid oscillations and nutrient transport in the in vitro growth of valvular tissues. *Cardiov. Eng. Technol.* **2016**, *7*, 170–181. [CrossRef]
15. Chen, X.; Nadiarynkh, O.; Plotnikov, S.; Campagnola, P. Second Harmonic Generation Microscopy for Quantitative Analysis of Collagen Fibrillar Structure. *Nat. Protoc.* **2012**, *7*, 654–669. [CrossRef] [PubMed]
16. König, K.; Schenke-Layland, K.; Riemann, I.; Stock, U. Multiphoton autofluorescence imaging of intratissue elastic fibers. *Biomaterials* **2005**, *5*, 495–500. [CrossRef]
17. Hinton, R.B.; Katherine, E.Y. Heart valve structure and function in development and disease. *Ann. Rev. Physiol.* **2011**, *73*, 29–46. [CrossRef]
18. Aikawa, E.; Aikawa, M.; Libby, P.; Figueiredo, J.-L.; Rusanescu, G.; Iwamoto, Y.; Fukuda, D.; Kohler, R.H.; Shi, G.-P.; Jaffer, F.A.; et al. Arterial and aortic valve calcification abolished by elastolytic cathepsin S deficiency in chronic renal disease. *Circulation* **2009**, *119*, 1785–1794. [CrossRef] [PubMed]
19. Hutcheson, J.D.; Aikawa, E.; Merryman, W.D. Potential drug targets for calcific aortic valve disease. *Nat. Rev. Cardiol.* **2014**, *11*. [CrossRef]
20. Yap, C.H.; Saikrishnan, N.; Tamilselvan, G.; Yoganathan, A.P. Experimental measurement of dynamic fluid shear stress on the aortic surface of the aortic valve leaflet. *Biomech. Model. Mechanobiol.* **2012**, *11*, 171–182. [CrossRef]
21. Chow, M.J.; Turcotte, R.; Lin, C.P.; Zhang, Y. Arterial Extracellular Matrix: A Mechanobiological Study of The Contributions and Interactions of Elastin and Collagen. *Biophys. J.* **2014**, *106*, 2684–2692. [CrossRef] [PubMed]
22. Mirnajafi, A.; Raymer, J.M.; McClure, L.R.; Sacks, M.S. The flexural rigidity of the aortic valve leaflet in the commissural region. *J. Biomech.* **2006**, *39*, 2966–2973. [CrossRef] [PubMed]
23. Bell, E.D.; Converse, M.; Mao, H.; Unnikrishnan, G.; Reifman, J.; Monson, K.L. Material Properties of Rat Middle Cerebral Arteries at High Strain Rates. *J. Biomech. Eng.* **2018**. [CrossRef]
24. Karunaratne, A.; Li, S.; Bull, A.M.J. Nano-scale Mechanisms Explain the Stiffening and Strengthening of Ligament Tissue with Increasing Strain Rate. *Sci. Rep.* **2018**, *8*. [CrossRef]
25. Anssari-Benam, A.; Bucchi, A.; Screen, H.R.C.; Evans, S.L. A Transverse Isotropic Viscoelastic Constitutive Model for Aortic Valve Tissue. *R. Soc. Open Sci.* **2017**. [CrossRef]
26. Anssari-Benam, A.; Tseng, Y.-T.; Holzapfel, G.A.; Bucchi, A. Rate-dependency of The Mechanical Behavior of Semilunar Heart Valves under Biaxial Deformation. *Acta Biomater.* **2019**, *88*, 120–130. [CrossRef] [PubMed]

bioengineering

Article

Adverse Hemodynamic Conditions Associated with Mechanical Heart Valve Leaflet Immobility

Fardin Khalili [1,2,*], Peshala P. T. Gamage [1], Richard H. Sandler [1,3] and Hansen A. Mansy [1]

1 Biomedical Acoustics Research Laboratory, University of Central Florida, 4000 Central Florida Blvd, Orlando, FL 32816, USA; peshala@knights.ucf.edu (P.P.T.G.); rhsandler@gmail.com (R.H.S.); hansen.mansy@ucf.edu (H.A.M.)

2 Department of Mechanical Engineering, Embry-Riddle Aeronautical University, 600 South Clyde Morris Blvd., Daytona Beach, FL 32114-3900, USA

3 College of Medicine, University of Central Florida, 6850 Lake Nona Blvd, Orlando, FL 32827, USA

* Correspondence: fardin.khalili@erau.edu; Tel.: +1-616-920-2661

Received: 24 July 2018; Accepted: 10 September 2018; Published: 16 September 2018

Abstract: Artificial heart valves may dysfunction, leading to thrombus and/or pannus formations. Computational fluid dynamics is a promising tool for improved understanding of heart valve hemodynamics that quantify detailed flow velocities and turbulent stresses to complement Doppler measurements. This combined information can assist in choosing optimal prosthesis for individual patients, aiding in the development of improved valve designs, and illuminating subtle changes to help guide more timely early intervention of valve dysfunction. In this computational study, flow characteristics around a bileaflet mechanical heart valve were investigated. The study focused on the hemodynamic effects of leaflet immobility, specifically, where one leaflet does not fully open. Results showed that leaflet immobility increased the principal turbulent stresses (up to 400%), and increased forces and moments on both leaflets (up to 600% and 4000%, respectively). These unfavorable conditions elevate the risk of blood cell damage and platelet activation, which are known to cascade to more severe leaflet dysfunction. Leaflet immobility appeared to cause maximal velocity within the lateral orifices. This points to the possible importance of measuring maximal velocity at the lateral orifices by Doppler ultrasound (in addition to the central orifice, which is current practice) to determine accurate pressure gradients as markers of valve dysfunction.

Keywords: computational fluid dynamics; bileaflet mechanical heart valve; adverse hemodynamics; transvalvular pressure gradients; turbulent shear stresses; blood damage; platelet activation

1. Introduction

Cardiovascular disease is the leading cause of death in the world [1]. There are more than 300,000 heart valves implanted annually worldwide [2,3], with approximately half of them being mechanical valves [4]. The bileaflet mechanical heart valves (BMHVs) is currently the most common valve given their durability and desirable hemodynamics [5]. However compared to bioprostheses, they are associated with more post-surgical complications such as thrombus and pannus formation, hemolysis, and platelet activation [6,7]. Improved understanding of mechanical valve hemodynamics may be vital for diagnostic, treatment and design improvements.

Several studies [6,8,9] investigated the etiology of insidious prosthetic valve dysfunction, showing that failure of mechanical heart valves is usually related to thrombus formation and tissue overgrowth. The time interval between the valve replacement and obstruction is very broad (from 6 weeks to 13 years) and some patients with significant prosthetic valve obstruction may be completely asymptomatic long before a diagnosis is made [6]. It is often difficult to distinguish between a normally functioning BMHV and a dysfunctional BMHV with mild severity, which unfortunately can still cause life-threatening sequela in the short-term [8]. Montorsi et al. [10] found that 35% of patients had normal Doppler study despite fluoroscopy showing significant restriction in one of the leaflets. They also concluded that the distinction between blocked and hypomobile leaflet is vital. Accordingly, a great deal of research has been performed on aortic and mitral heart valves in normal function and in various states of malfunction [6,9,11] ranging from slightly restricted opening to total occlusion of one leaflet including 25%, 50%, 75%, 100% dysfunctions [2,7,12,13]. Pibarot et al. [8] reported that the increase of Doppler gradients caused by dysfunction of the valve may underestimate the true hemodynamic changes [12]. Clinicians often opt for early surgical intervention since the surgical complication rate is relatively low while valve dysfunction can lead to rapid cardiovascular collapse even with minimal or absent symptoms [6]. But controversy remains whether patients with an obstructed valve should be managed by valve replacement [14], mechanical declotting [15] or nonsurgical thrombolysis [16].

Computational fluid dynamics (CFD), along with fluoroscopic or Doppler measurements, have the potential to provide clinically important insights by providing unprecedented hemodynamic detail for prosthetic heart valves [17]. For example, analysis of blood flow characteristics such as velocity, vortex formation, and turbulent stresses, especially around the valve hinge regions [18–21] can help identify conditions that may increase the risk of blood cell damage [22–24]. Critical turbulent shear stress thresholds of 400 $N{\cdot}m^{-2}$ [25] and 800 $N{\cdot}m^{-2}$ [26] for blood cell damage were reported. Studies also showed that high turbulent shear stress levels at the valve hinges and downstream of the valve can lead to thrombus formation and leaflets motion restriction [27,28]. This, in turn, may lead to a life-threatening dysfunction of one or both leaflets of BMHVs [12]. Fortunately, prompt recognition of valve dysfunction allows early treatment [8], and many potential complications can be prevented or minimized with careful medical management and periodic monitoring of valve function; e.g., blocked leaflets could be fully recovered when valve thrombosis is detected early [10]. CFD may also provide valuable information to speed up the design of implantable devices during the prototype development [29] and reduce the costs and risks associated with new heart valve designs [30]. Hence, analysis of flow dynamics and the resulting turbulence [31–34] and sounds [35–40] has been an active area of research.

The current computational study provides new quantitative information on blood flow characteristics, plus forces and moments acting on the leaflets of bileaflet mechanical heart valves at different levels of leaflet dysfunctionality during peak systolic flow. Model improvements compared to previous studies include: A more realistic aortic sinuses geometry (compared to References [41,42]), addition of the valve ring to the model (compared to References [43,44]), and creation of a 3-D model instead of a 2-D model (compared to References [2,13,45]). The study quantified important hemodynamic characteristics (such as principle stresses) that are not measurable using currently available standard diagnostic tools. This approach may provide a patient-specific tool for identification of adverse conditions that are associated with increased risk of hemolysis and thrombus formation [46,47], thereby potentially providing a more complete picture of the valve status useful in clinical management of patients with dysfunctional valves. The current CFD study focused on a geometric representative of leaflet dysfunction, which provided condition-specific hemodynamic changes. Patient-specific information can be obtained by carrying out similar CFD studies for actual geometries extracted from medical imaging modalities.

2. Materials and Methods

In this study, the computational domain was divided into four sequential regions in the flow direction: Upstream, BMHV, aortic sinuses and downstream. The heart valve geometry (Figure 1a) was chosen to be similar to previous studies [48,49]. A realistic geometry of the aortic sinuses was created since this is important for appropriate flow field analysis [50,51]. Another enhancement implemented in the current study (compared to some previous two-dimensional CFD studies) was to include the valve ring into the model. Figure 1b shows the asymmetric aortic sinuses geometry with inlet aortic root diameter of D_O = 23 mm, which was extracted from angiograms [52]. In this paper, the aortic root was modeled based on following parameters [52]: D_O = 22.3 mm is the diameter of aortic annulus, D_A = 27.7 mm is aortic diameter, D_B = 34.6 mm is the maximum projected sinus diameter, L_A = 22.3 mm is the length of the sinuses, and L_B = 7.6 mm is the distance between D_O and D_B (from the entrance of the aortic sinuses to the middle of the sinuses with the maximum projected sinus diameter), as described in Figure 1d. These parameters can be computed based on the aortic annulus diameter (D_O), which is the same as the size of the implanted mechanical heart valve. L_D = 100 mm is the length of the region downstream of the heart valve. Here, the BMHV is in the fully open position and divides the flow into three orifices: Two of them (top and bottom orifices) are roughly semicircular and the third (middle orifice) is approximately rectangular.

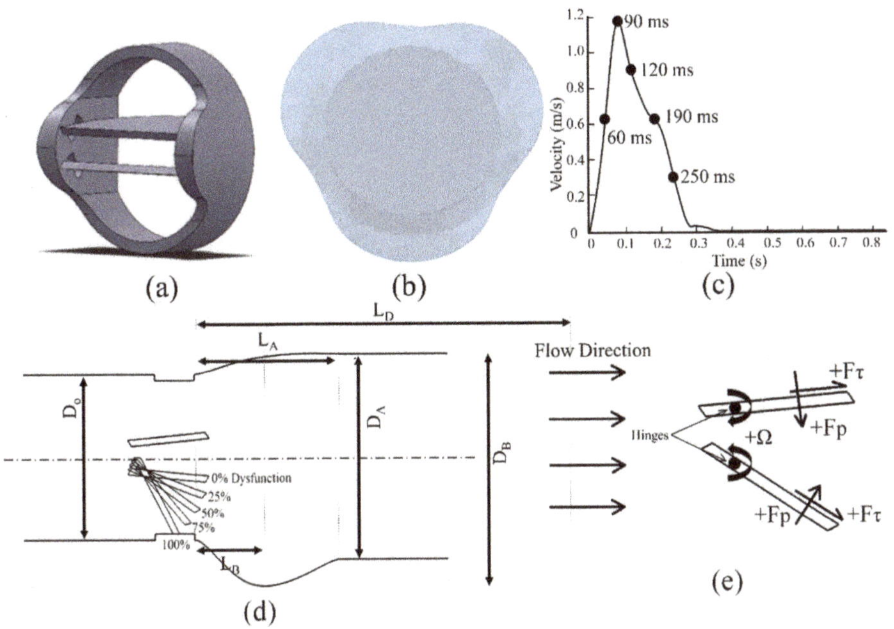

Figure 1. The geometry, inlet conditions and sign convention used in the current study: (**a**) Valve geometry; (**b**) Cross-section of the aortic root sinuses; (**c**) Inlet velocity profile; (**d**) Degrees of bottom leaflet dysfunction; and (**e**) Sign conventions for forces acting on the leaflets.

The CFD analysis was performed for a pulsatile flow through a three-dimensional BMHV during one cardiac cycle. The analyses were focused on the period from 60 to 250 ms, where the leaflets are expected to be fully open [30]. Some results concentrated on the peak systole (90 ms), as the highest flow fluctuations, pressure gradient, and turbulent stresses associated with high risk of blood damage and platelet activation could occur at this time. To reproduce a physiological flow waveform through the aortic heart valve, the following properties were obtained from recent and previous experimental and numerical studies [2,41,53]. The inlet velocity corresponded to cardiac output of 5 $L{\cdot}min^{-1}$ and heart rate of 70 bpm with a systolic phase duration of 0.3 s (Figure 1c). The peak inflow velocity was about 1.2 ms^{-1}. The density and dynamic viscosity of blood were set to ρ = 1080 $kg{\cdot}m^{-3}$ and μ = 0.0035 Pa·s, respectively. This corresponds to an inlet peak Reynolds number ($Re_{peak} = \frac{\rho U_{peak} d_{inlet}}{\mu}$) of 8516 and a Womersley number ($W_o = \frac{d}{2}\sqrt{\frac{\omega\rho}{\mu}}$) = 26.5; where, $\omega = \frac{2\pi}{T}$ = 17.21 $rad{\cdot}s^{-1}$, is the frequency of pulsatile flow and T = 0.866 s is the period.

In the current study, a normal functioning (i.e., 0% dysfunction) and a BMHV with different levels of dysfunction were simulated using a commercial CFD software package (STAR-CCM+, CD-Adapco,

Siemens PLM, Plano, TX, USA). Figure 1d shows the side cross section of the BMHV with a top functional leaflet and a bottom dysfunctional leaflet at 0, 25, 50, 75 and 100% levels of dysfunctionality (corresponding to a gradually decreasing effective orifice area (EOA)). In addition, Figure 1e shows the leaflet hinges as well as the direction of net pressure, shear forces (F_p and F_τ, respectively), and moments (Ω) acting on the leaflets. The positive direction of the F_p and moments acting on both leaflets are in the direction tending to open the leaflets.

The Wilcox's standard-Reynolds k-Omega turbulence model [43,54], which is known to perform well for internal flows, was used to simulate the flow during a complete cardiac cycle. The current and other studies [2,55] focus on the period from 60 to 250 ms, where the leaflets are expected to be fully open [30]. Hence, the dynamics of the leaflet opening and closure were not simulated as done in previous studies [2,30,43], which lowers computational cost. Therefore, the valve leaflets were assumed to remain fully open throughout the forward flow phase, which was considered reasonable because the opening and closing motions occur quickly compared to the total opening time. The unsteady simulation was performed with a time step of 0.5 ms and 25 iterations per time step. Numerical solution typically converged to residuals about $<10^{-4}$. Moreover, high quality polyhedral mesh was generated in the flow domain, especially in the heart valve and aortic sinuses regions (Figure 2). y^+ was maintained at less than 1 close to all walls including leaflet surfaces ($y^+ = 0.46$ at the peak flow).

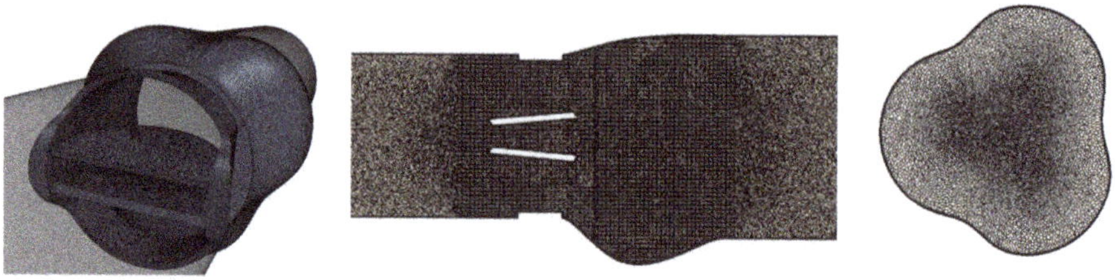

(a) **(b)** **(c)**

Figure 2. High quality mesh generated (**a**) close to the wall and leaflet surfaces; (**b**) in the flow domain; and (**c**) cross-sectional view of the mesh in the aortic root sinuses region downstream of the heart valve.

2.1. Numerical Uncertainty

Steady flow simulation was conducted to establish grid density prior to unsteady simulation. The uncertainty and error in the study was calculated following ASME recommendations [56]. Figure 3 shows velocity profile at the entrance of the aortic sinuses along with the corresponding error bars while Table 1 shows the discretization error of the maximum velocity value in the entire field. The fine-grid convergence index (GCI_{fine}) in Table 1 was 0.139% (excluding modeling errors [56]). In addition, the maximum discretization uncertainty was approximately 7% in the area close to the leaflets. These numerical uncertainties are comparable to previous studies [2].

Table 1. Calculation of discretization error.

ϕ = Maximum Velocity in the Entire Field (m/s)			
N_1; N_2; N_3	6,529,062; 2,598,513; 1,390,150		
r_{21} (Refinement factor of N_2/N_1)	1.35	$e^{21}{}_a$	0.11%
r_{32} (Refinement factor of N_3/N_2)	1.32	$e^{21}{}_{ext}$	0.11%
ϕ_1	2.523	$GCI^{21}{}_{fine}$	0.14%
ϕ_2	2.521	$\phi^{32}{}_{ext}$	2.515
ϕ_3	2.526	$e^{32}{}_a$	0.21%
P	2.289	$e^{32}{}_{ext}$	0.24%
$\phi^{21}{}_{ext}$	2.526	$GCI^{32}{}_{course}$	0.29%

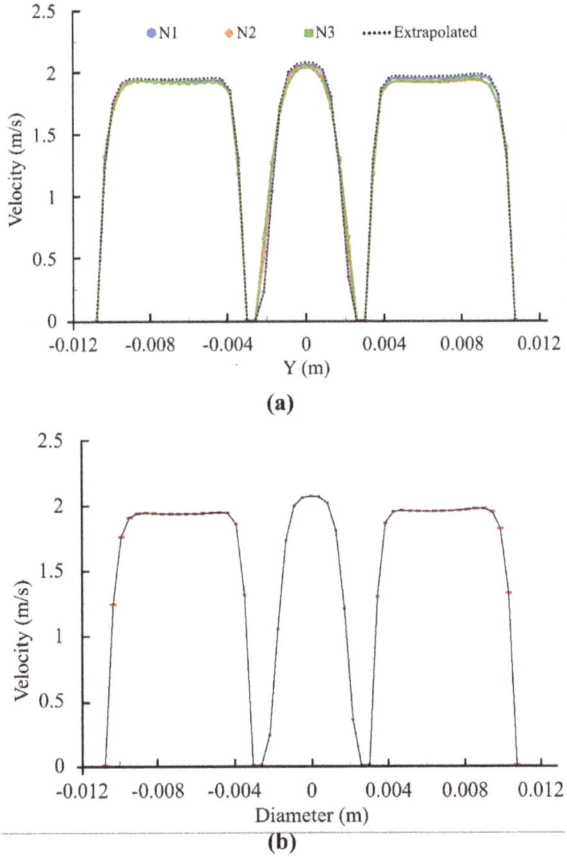

Figure 3. (**a**) Velocity profile at the entrance of the aortic sinuses for different grid solution; (**b**) Fine-grid solution with discretization error bars.

2.2. Validation

The normalized velocity profile along a line located 7 mm downstream of the healthy valve (at the peak systole) is shown in Figure 4a for a normal functioning valve. The velocity profiles obtained in previous studies that considered similar geometries and flow conditions [13,53] are also shown in the same figure. Here, normalized velocities are plotted to facilitate comparison with studies that reported normalized profiles [13]. The maximum velocities were compared for steady cardiac outputs of 5 and 7 L·min^{-1}. These velocities were 0.96 ms^{-1} and 1.35 ms^{-1} in the current study, respectively, which were comparable to maximum velocities of 1.0 ms^{-1} and 1.36 ms^{-1} reported in the previous study [13]. To quantify the difference between our computational results and the previous experimental results [53], the root-mean-square (RMS) of the velocity differences between the two studies were calculated. The RMS of the velocity difference was 6.58% of the maximum velocity, suggesting agreement between the results of the current study and measured values. The normalized velocity profile was also compared with two other experimental and computational studies at the trailing edge of the leaflet and 105 ms after the peak systole [7,43] (Figure 4b). The RMS of the velocity difference was <6% of the maximum velocity, suggesting agreement with these studies.

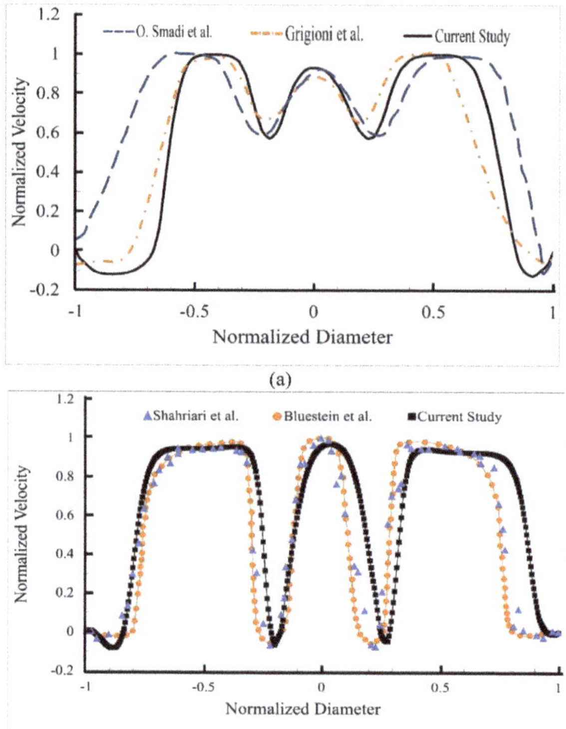

Figure 4. (**a**) Normalized velocity profiles at 7 mm downstream of the valve (at the peak systole) in the current study compared to previous experimental [53] and computational [13] studies. More agreement can be seen between the current and the experimental study; (**b**) Normalized velocity profiles at the trailing edge of the leaflets (105 ms after the peak systole) in the current study compared to previous experimental [43] and computational [7] studies.

3. Results and Discussions

Figure 5a shows a cross-sectional view of the velocity at t = 90 ms, where the color represents the magnitude and the short lines indicate direction. For 0% dysfunction (Figure 5(a1)), the flow was more uniform; especially compared to cases with dysfunctional leaflets (Figure 5(a2–a5)). Figure 5(a1) also shows a relatively small increase in velocity in the orifices and wake regions downstream of the leaflets as would be expected. As the bottom leaflet dysfunction took place, the velocity magnitude in the orifices increased. This is likely because of the narrowing of bottom orifice with dysfunction, which led to flow area reduction. Flow separation in the middle orifice was observed around the leading edge of the bottom leaflet for dysfunctionalities of 25–100% (Figure 5(a2–a5)). Separation also occurred close to the trailing edge of the top leaflet for 75% and 100% (Figure 5(a4,a5)). In addition, Figure 5a shows a trend of increasing separation bubble size with dysfunctionality. Although not clearly shown in the figure, vortex shedding was also observed. While Figure 5 shows information for t = 90 ms, flow structures were also examined for all times between 60 to 250 ms and were found similar to those shown in Figure 5.

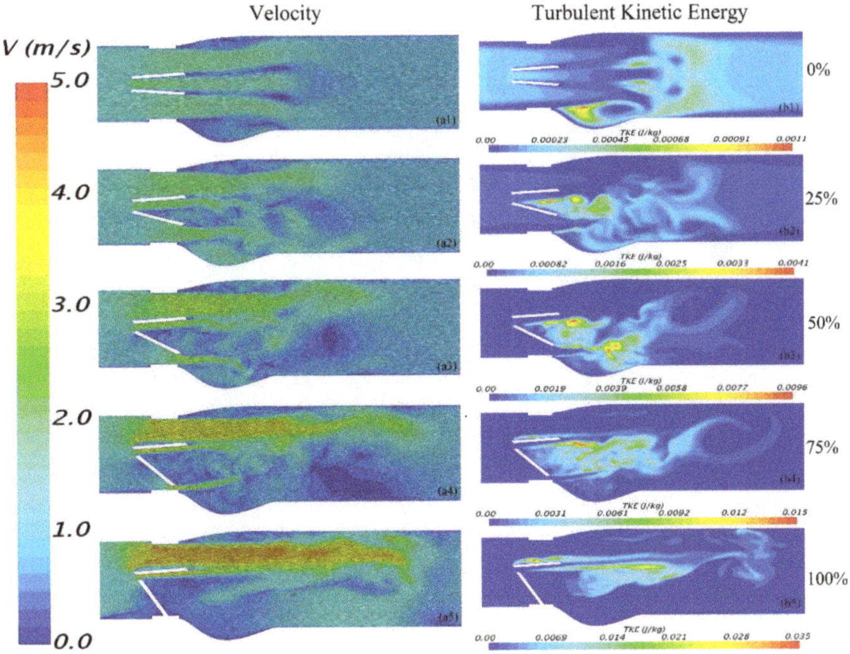

Figure 5. Velocity (a1–a5) and turbulent kinetic energy (b1–b5) at 90 ms for different degrees of lower leaflet dysfunction. There was a general trend of increased maximum velocity and TKE with increased dysfunction. (Note that the scale for TKE was increased with dysfunction).

Figure 5b shows the turbulent kinetic energy (TKE), which is an indicative of velocity fluctuations. TKE tended to increase with dysfunction and a region of higher TKEs (up to 150% compared to the healthy valve) around the top leaflet started to develop when dysfunction reached ≥75%.

Figure 6a shows the maximum velocities at the entrance of the aortic sinuses, which were comparable to a previous computational study in which the results for only three dysfunctional cases (0%, 50%, and 100%) were reported [2]. In the current study, the maximum velocity increased from 2.05 ms^{-1} to 4.49 ms^{-1} as dysfunction increased from 0% to 100%. The highest velocity elevation was likely associated with the jet that originates from the orifice between the healthy leaflet and the valve ring and not from the center orifice between the two leaflets. However, when velocity gradients are measured using Doppler, it is more common that that velocity at the center orifice is measured. The smaller peak velocities that may be detected at the center orifice can lead to false estimation of velocity and pressure gradients, which can translate into errors in in assessing the severity level of leaflet dysfunction [8].

It is also to be noted that the maximum transvalvular pressure gradient (TPG_{max}) can be computed from the maximal instantaneous velocity using the simplified Bernoulli equation ($TPG_{max} = 4v^2_{max}$) [12]. Figure 6b shows the maximum pressure gradient compared to the previous study [2] for different levels of dysfunction. Here, the TPG_{max} increased from 16.48 to 80.64 mmHg. The higher velocities and pressure gradients in the current study can be because of the smaller valve diameter and the addition of valve ring (which likely caused more flow obstruction).

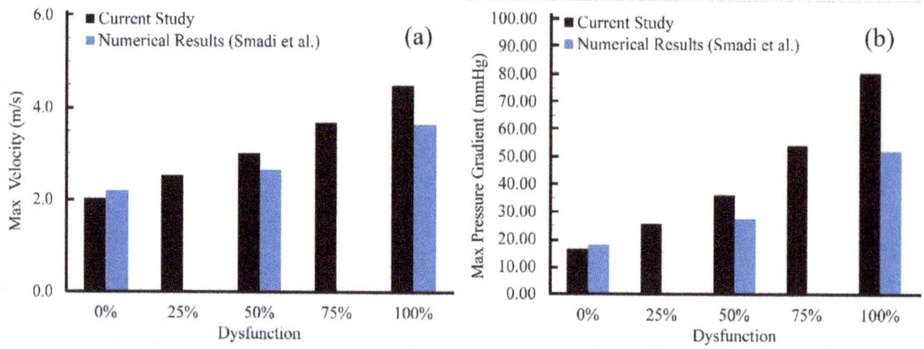

Figure 6. Comparison of the current study results with available data from a previous computational study [2]: (**a**) Maximum velocity at the entrance of the aortic sinuses, and (**b**) maximum pressure gradients across the valve computed from simplified Bernoulli equation. Both quantities continuously increased with dysfunction. While the trends were similar, differences may be due to the geometrical variations and the fact that the current study performed 3D compared to 2D simulation.

Figure 7 shows helicity isosurfaces at different times and dysfunction levels. Since helicity is proportional to the flow velocity and the vorticity, it indicates the potential for development of helical flow. The data in this figure showed that helicity increased with dysfunction and peaked around peak systolic velocity time. Figure 7 also suggested that intense vortical structures start to appear in the valve and sinus regions during the acceleration phase (e.g., 60 ms) before spreading downstream at later times. For leaflet dysfunction of ≥75%, lower helicity (compared to dysfunctionality of <75%) was observed in the dysfunctional leaflet side, which can be because the region downstream of that leaflet may contain lowered velocity and vorticity.

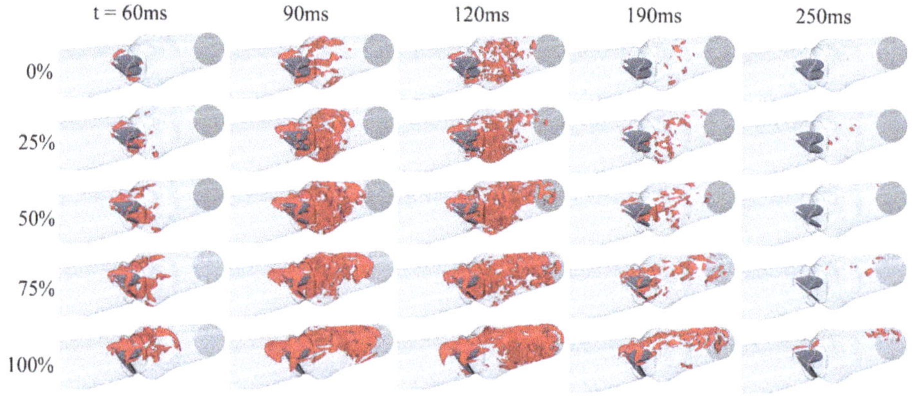

Figure 7. Helicity isosurfaces (isovalue = 414 m.s^{-2}) at different times and dysfunctions. A general increase in helicity was observed with dysfunction.

Several studies reported that the hemolysis (the breakage of a red blood cell membrane), can occur for turbulent shear stresses in the range from 400 to 5000 N·m^{-2} with exposure time as small as 10 ms [15,51]. In addition, these high turbulent shear stresses can lead to platelets activation, which increase the risk of platelet aggregation and blood clots formation [10,15]. Clots may detach and the resulting free-floating clot can block arteries leading to serious consequences such as embolism and stroke [26].

While stresses acting on the fluid occur in different directions, principal stresses are the highest. Three-dimensional principal stress analysis requires the computation of the full Reynolds stress tensor (T):

$$T = \begin{bmatrix} \sigma_{xx} & \tau_{xy} & \tau_{xz} \\ \tau_{yx} & \sigma_{yy} & \tau_{yz} \\ \tau_{zx} & \tau_{zy} & \sigma_{zz} \end{bmatrix} = \rho \begin{bmatrix} \overline{uu} & \overline{uv} & \overline{uw} \\ \overline{vu} & \overline{vv} & \overline{vw} \\ \overline{wu} & \overline{wv} & \overline{ww} \end{bmatrix} \quad (1)$$

where, u, v, and w are the velocity fluctuation components and, σ and τ represent normal and shear stresses, respectively. Popov [57] provides a detailed discussion of the calculation of three-dimensional maximum or principal stresses which involves the solution of the roots of the following third order equation:

$$\sigma^3 - I_1\sigma^2 + I_2\sigma - I_3 = 0 \quad (2)$$

where,

$$I_1 = \sigma_{xx} + \sigma_{yy} + \sigma_{zz} \quad (3)$$

$$I_2 = \sigma_{xx}\sigma_{yy} + \sigma_{yy}\sigma_{zz} + \sigma_{xx}\sigma_{zz} - \tau_{xy}{}^2 - \tau_{yz}{}^2 - \tau_{xz}{}^2 \quad (4)$$

$$I_3 = \sigma_{xx}\sigma_{yy}\sigma_{zz} + 2\tau_{xy}\tau_{yz}\tau_{xz} - \sigma_{xx}\tau_{yz}{}^2 - \sigma_{yy}\tau_{xz}{}^2 - \sigma_{zz}\tau_{xy}{}^2 \quad (5)$$

The three roots $\sigma_1 < \sigma_2 < \sigma_3$ of the above equation are the three principal normal stresses. The coefficients I_1, I_2 and I_3 are functions of the measured Reynolds stress tensor and are the three stress invariants of the Reynolds stress tensor. In addition, the maximum or principal shear stresses (τ_{ijP}) are linearly related to the normal stresses by the following equations:

$$\tau_{ijP} = \frac{\sigma_i - \sigma_j}{2}; \ \tau_{max} = \frac{\sigma_3 - \sigma_1}{2} \quad (6)$$

Figure 8 displays turbulent shear (τ_{max}) principal stresses for different levels of dysfunction at the peak systole. Since an increased risk of blood damage may occur for stresses exceeding 400 N·m^{-2}, only stresses in this range are shown. These results suggested that as the leaflet dysfunctionality increased, the principal turbulent shear stresses increased. More specifically for 0 %, 25%, 50%, 75%, and 100% dysfunction levels, the maximum principal shear stresses at peak systole were 420, 510, 760, 1155, and 1695 N·m^{-2}. In addition, the regions of elevated stresses grew with dysfunction and were concentrated around and downstream of the functional (top) leaflet where high jet velocity and stronger helical structures existed (Figures 5 and 7). These regions are of the particular interest since elevated turbulent stress levels are known to be associated with blood damage and thrombus formation. In addition, careful examination of Figure 8, indicates that the increase in the region with high principal stresses accelerates later (>50%) for the current model. While this ~50% threshold may vary with geometry, CFD will allow patient-specific analysis, which may further increase its utility. Future investigations of other realistic geometries may be performed to quantify this effect.

The highest principal turbulent stresses, however, occurred slightly after (100–120 ms) peak systole during the deceleration phase. Table 2 shows the highest principal turbulent stress values and their occurrence time. It can be seen that these values were somewhat higher (~4–14%) than those at peak systole. Comparing to previous experimental studies, lethal and sublethal damages of red cells can occur with turbulent shear stresses as low as 150 and 50 N·m^{-2}, respectively [46,58]. These levels can be significantly lower (1–10 N·m^{-2}) in the presence of foreign surfaces such as valve prostheses [59,60]. In addition, platelet activation can occur when turbulent shear stresses are in the range of 10–50 N·m^{-2} [46,47]. Studies also showed that high turbulent shear stresses at the valve hinges and downstream of the valve, for normal cases (valve leaflet with 0% dysfunction), can lead to thrombus formation and the leaflets' motion restriction [27,28]. This, in turn, may lead to a life-threatening dysfunction of one or both leaflets of BMHVs [12].

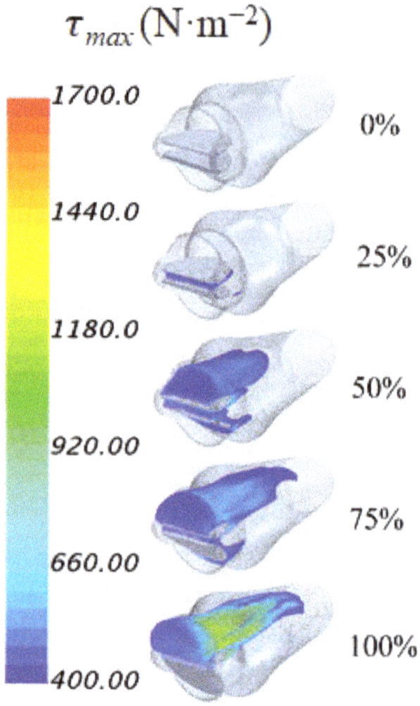

Figure 8. Principal shear stresses for different levels of dysfunction at the peak systole. Elevated levels of principal stresses were observed with dysfunction, which increase blood damage risks. Published cutoff stress value for damage is above 400 N·m^{-2} [25].

Table 2. Maximum Principal Shear Stresses.

Dysfunction	Max. Principal Shear Stress (N·m^{-2})	Time (s)
0%	440	0.102
25%	534	0.103
50%	832	0.112
75%	1276	0.112
100%	1972	0.119

Figure 9 shows the pressure distribution in the vicinity of the leaflets. The maximum pressure at the blocked leaflet increased with dysfunction. For dysfunctions higher than 50%, a region of high pressure developed at the bottom surface of the functional leaflet upstream the hinge, which would generate higher moments in the direction of leaflet opening.

It is important to document elevated forces and moments, as they would lead to higher reaction forces at the hinges (where thrombus tends to form), which may create more adverse conditions. Collection and analysis of this information can also aid in the development of improved valve designs. The net pressure and shear forces on the top and bottom leaflets for the full cardiac cycle are displayed in Figure 10. Results showed that increased dysfunctionality of one leaflet led to higher net forces on the functional and dysfunctional leaflets up to 200%, and 600%, respectively. Note that although the net pressure forces (F_p) on the top leaflet were negative (upward) for 75% and 100% dysfunctions, forces were acting upstream of the hinges (Figure 9d–e), which would result in positive moments (Figure 11a). Figure 10b shows the F_p on the bottom leaflet, which was positive for all cases. Net shear forces (F_τ) on the top and bottom leaflets (Figures 10c and 10d, respectively) were positive during the period under consideration for all levels of dysfunction except for the dysfunctional leaflet with 100% dysfunction. The change in the sign may be attributed to the large revered flow regions (Figures 5a and 9) that formed downstream of the leaflet, as resulted in positive moments on bottom leaflet (Figure 11b).

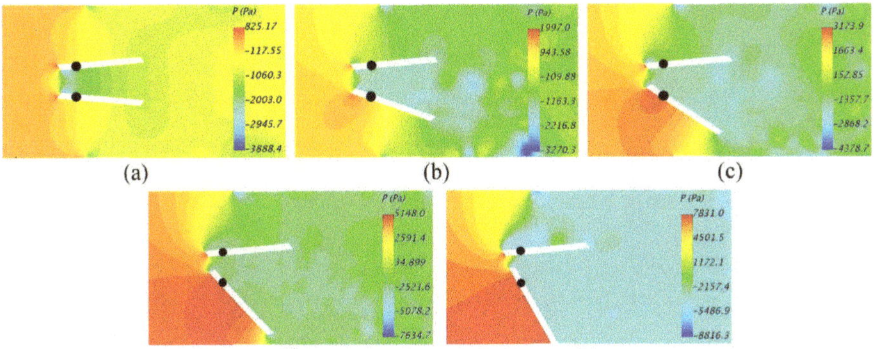

Figure 9. (**a**) 0%; (**b**) 25%; (**c**) 50%; (**d**) 75%; and (**e**) 100%. For dysfunction $\geq$ 75%, a region of high pressure developed at the bottom surface of the functional leaflet upstream of the hinge, which would generate moments that tend to keep that leaflet open.

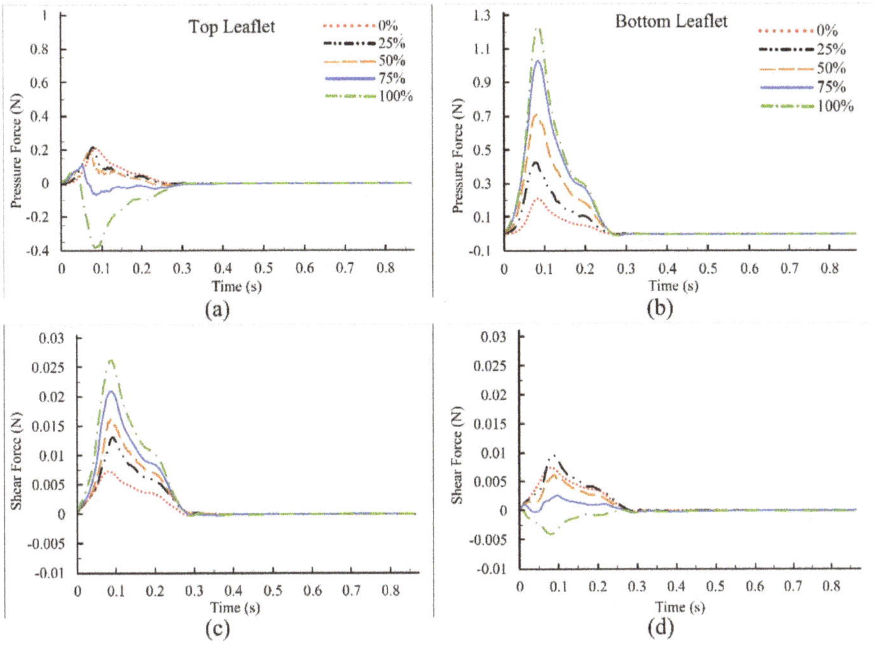

Figure 10. Net pressure and shear forces on leaflets: (**a**) F_p on top leaflet; (**b**) F_p on bottom leaflet; (**c**) F_τ on top leaflet; and (**d**) F_τ on bottom leaflet. The sign of some forces started to reverse at high levels of dysfunction. The legends are consistent for all four figures.

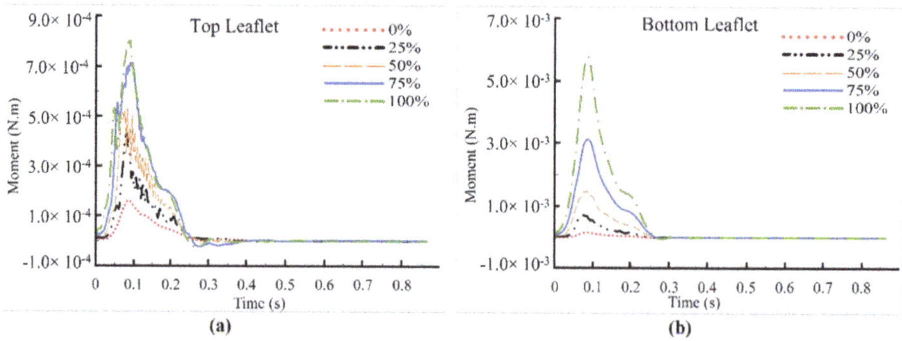

Figure 11. Net moments on: (**a**) Top leaflet, and (**b**) Bottom leaflet. The moments tended to be in the directions of leaflet opening. All moments increased with dysfunction. In most cases of dysfunction, the moments on the dysfunctional leaflet were higher (note the different scale for the dysfunctional leaflet).

Future CFD studies can explore new heart valve designs and structural materials and determine how blood-material interactions and hemodynamics can be affected by design changes [61] with the aim of reducing thrombo-embolic complications associated with these valves, which can lead to improved valve designs. For example, analysis of blood flow characteristics through a BMHV especially around the valve hinge regions can help identify conditions that may increase the risk of blood cell damage [22,23]. An investigation of the effect of the leaflet opening angles on the blood flow also suggested that the opening angle can highly affect the flow downstream of BMHV and that opening angles >80 degrees would be more effective in reducing flow resistance and vortical structures [62].

4. Conclusions

In this study, adverse hemodynamic conditions at peak systole due to incomplete leaflet opening of a bileaflet mechanical heart valve were investigated. A realistic 3-D geometry of the aortic sinuses and a complete model of a bileaflet mechanical heart valve including the valve ring were constructed. The results suggest that maximum blood velocities increased when the effective orifice area was reduced due to the increase of leaflet dysfunction, as expected. Leaflet immobility also appears to cause maximal velocity within the lateral orifices. This points to the possible importance of measuring maximal velocity at the lateral orifices by Doppler ultrasound (in addition to the central orifice which is current practice) to determine accurate pressure gradients as markers of valve dysfunction. Dysfunctionality also increased the transvalvular pressure gradient by up to 300%, which would increase the effort to produce the same cardiac output.

Results also suggested that the higher levels of dysfunction were accompanied with flow separation at the leaflet surfaces and growing eddies especially downstream of the valve in the aortic sinuses. Principal turbulent stresses for immobile leaflet increased up to 1695 $N{\cdot}m^{-2}$, which exceeds the threshold values for elevated risk of hemolysis and platelet activation and lead to potential development of thrombosis, especially around the normal leaflet. The region with high principal stresses (i.e., above threshold = 400 $N{\cdot}m^{-2}$) initially grew slowly (i.e., between 0 and 25% dysfunction), and then covered a significantly large region at higher dysfunctions (i.e., >50% of leaflet dysfunction), Figure 8, suggesting a possible need for closer monitoring of the patients with >50% of leaflet dysfunction. Dysfunctionality of one leaflet led to higher net forces on the leaflets (by up to 200%, and 600% for healthy and the dysfunctional leaflets, respectively). The resulting moments acting on the leaflets also increased with dysfunctionality (up to 550%, and 4000% for healthy and dysfunctional leaflets, respectively). These higher forces and moments can increase the reaction forces and stresses in the hinge region where vulnerability to thrombus and pannus formations tend to be high and can lead to more leaflet motion restriction.

Author Contributions: Methodology and Modeling, F.K., P.P.T.G.; Software, F.K., P.P.T.G.; Validation, F.K., P.P.T.G., Formal Analysis, F.K.; Investigation, F.K., H.A.M.; Writing-Original Draft Preparation, F.K.; Writing-Review & Editing, H.A.M., R.H.S.; Supervision, H.A.M., R.H.S.; Project Administration, F.K., H.A.M.; Funding Acquisition, H.A.M., R.H.S.

Funding: This Study was supported by NIH R01 EB012142, R43HL099053.

Conflicts of Interest: No conflict of interest to be declared.

References

1. WHO Fact sheet: Cardiovascular Diseases (CVDs). Available online: http://www.who.int/mediacentre/factsheets/fs317/en/ (accessed on 16 September 2018).
2. Smadi, O.; Hassan, I.; Pibarot, P.; Kadem, L. Numerical and experimental investigations of pulsatile blood flow pattern through a dysfunctional mechanical heart valve. *J. Biomech.* **2010**, *43*, 1565–1572. [CrossRef] [PubMed]
3. Khalili, F.; Gamage, P.P.T.; Mansy, H.A. Prediction of Turbulent Shear Stresses through Dysfunctional Bileaflet Mechanical Heart Valves using Computational Fluid Dynamics. In Proceedings of the 3rd Thermal and Fluids Engineering Conference (TFEC), Fort Lauderdale, FL, USA, 4–7 March 2018; pp. 1–9.
4. Emery, R.W.; Mettler, E.; Nicoloff, D.M. A new cardiac prosthesis: the St. Jude Medical cardiac valve: In vivo results. *Circulation* **1979**, *60*, 48–54. [CrossRef] [PubMed]
5. Gott, V.L.; Alejo, D.E.; Cameron, D.E. Mechanical Heart Valves: 50 Years of Evolution. *Ann. Thorac. Surg.* **2003**, *76*, 2230–2239. [CrossRef]
6. Deviri, E.; Sareli, P.; Wisenbaugh, T.; Cronje, S.L. Obstruction of mechanical heart valve prostheses: Clinical aspects and surgical management. *J. Am. Coll. Cardiol.* **1991**, *17*, 646–650. [CrossRef]
7. Shahriari, S.; Maleki, H.; Hassan, I.; Kadem, L. Evaluation of shear stress accumulation on blood components in normal and dysfunctional bileaflet mechanical heart valves using smoothed particle hydrodynamics. *J. Biomech.* **2012**, *45*, 2637–2644. [CrossRef] [PubMed]
8. Pibarot, P.; Dumesnil, J.G. Prosthetic heart valves: Selection of the optimal prosthesis and long-term management. *Circulation* **2009**, *119*, 1034–1048. [CrossRef] [PubMed]
9. Vesey, J.M.; Otto, C.M. Complications of prosthetic heart valves. *Curr. Cardiol. Rep.* **2004**, *6*, 106–111. [CrossRef] [PubMed]
10. Montorsi, P.; Cavoretto, D.; Alimento, M.; Muratori, M.; Pepi, M. Prosthetic mitral valve thrombosis: Can fluoroscopy predict the efficacy of thrombolytic treatment? *Circulation* **2003**, *108*, II79–II84. [CrossRef] [PubMed]
11. Fernandes, V.; Olmos, L.; Nagueh, S.F.; Quiñones, M.A.; Zoghbi, W.A. Peak early diastolic velocity rather than pressure half-time is the best index of mechanical prosthetic mitral valve function. *Am. J. Cardiol.* **2002**, *89*, 704–710. [CrossRef]
12. Baumgartner, H.; Schima, H.; Tulzer, G.; Kühn, P. Effect of stenosis geometry on the Doppler-catheter gradient relation in vitro: A manifestation of pressure recovery. *J. Am. Coll. Cardiol.* **1993**, *21*, 1018–1025. [CrossRef]
13. Smadi, O.; Fenech, M.; Hassan, I.; Kadem, L. Flow through a defective mechanical heart valve: A steady flow analysis. *Med. Eng. Phys.* **2009**, *31*, 295–305. [CrossRef] [PubMed]
14. Kinsley, R.H.; Antunes, M.J.; Colsen, P.R. Jude Medical valve replacement. An evaluation of valve performance. *J. Thorac. Cardiovasc. Surg.* **1986**, *92*, 349–360. [PubMed]
15. Venugopal, P.; Kaul, U.; Iyer, K.S.; Rao, I.M.; Balram, A.; Das, B.; Sampathkumar, A.; Mukherjee, S.; Rajani, M.; Wasir, H.S. Fate of thrombectomized Björk-Shiley valves. A long-term cinefluoroscopic, echocardiographic, and hemodynamic evaluation. *J. Thorac. Cardiovasc. Surg.* **1986**, *91*, 168–173. [PubMed]
16. Kurzrok, S.; Singh, A.K.; Most, A.S.; Williams, D.O. Thrombolytic therapy for prosthetic cardiac valve thrombosis. *J. Am. Coll. Cardiol.* **1987**, *9*, 592–598. [CrossRef]
17. Yoganathan, A.P.; He, Z.; Casey Jones, S. Fluid Mechanics of Heart Valves. *Annu. Rev. Biomed. Eng.* **2004**, *6*, 331–362. [CrossRef] [PubMed]

18. Klusak, E.; Bellofiore, A.; Loughnane, S.; Quinlan, N.J. High-Resolution Measurements of Velocity and Shear Stress in Leakage Jets From Bileaflet Mechanical Heart Valve Hinge Models. *J. Biomech. Eng.* **2015**, *137*, 111008. [CrossRef] [PubMed]
19. Ellis, J.T.; Travis, B.R.; Yoganathan, A.P. An In Vitro Study of the Hinge and Near-Field Forward Flow Dynamics of the St. Jude Medical®Regent™ Bileaflet Mechanical Heart Valve. *Ann. Biomed. Eng.* **2000**, *28*, 524–532. [CrossRef] [PubMed]
20. Khalili, F.; Mansy, H.A. Blood Flow through a Dysfunctional Mechanical Heart Valve. In Proceedings of the 38th Annual International Conference of the IEEE Engineering in Medicine and Biology Societ, Orlando, FL, USA, 17–20 August 2016.
21. Khalili, F.; Gamage, P.P.T.; Mansy, H.A. Hemodynamics of a Bileaflet Mechanical Heart Valve with Different Levels of Dysfunction. *J. Appl. Biotechnol. Bioeng.* **2017**, *2*. [CrossRef]
22. Min Yun, B.; Aidun, C.K.; Yoganathan, A.P. Blood Damage Through a Bileaflet Mechanical Heart Valve: A Quantitative Computational Study Using a Multiscale Suspension Flow Solver. *J. Biomech. Eng.* **2014**, *136*, 101009. [CrossRef] [PubMed]
23. Dumont, K.; Vierendeels, J.; Kaminsky, R.; Van Nooten, G.; Verdonck, P.; Bluestein, D. Comparison of the hemodynamic and thrombogenic performance of two bileaflet mechanical heart valves using a CFD/FSI model. *J. Biomech. Eng.* **2007**, *129*, 558–565. [PubMed]
24. Khalili, F. Fluid Dynamics Modeling and Sound Analysis of a Bileaflet Mechanical Heart Valve. Ph.D. Thesis, University of Central Florida, Orlando, FL, USA, May 2018.
25. Sallam, A.M.; Hwang, N.H. Human red blood cell hemolysis in a turbulent shear flow: Contribution of Reynolds shear stresses. *Biorheology* **1984**, *21*, 783–797. [CrossRef] [PubMed]
26. Lu, P.C.; Lai, H.C.; Liu, J.S. A reevaluation and discussion on the threshold limit for hemolysis in a turbulent shear flow. *J. Biomech.* **2001**, *34*, 1361–1364. [CrossRef]
27. Woo, Y.-R.; Yoganathan, A.P. In vitro pulsatile flow velocity and shear stress measurements in the vicinity of mechanical mitral heart valve prostheses. *J. Biomech.* **1986**, *19*, 39–51. [CrossRef]
28. Hasenkam, J.M. Studies of velocity fields and turbulence downstream of aortic valve prostheses in vitro and in vivo. *Dan. Med. Bull.* **1990**, *37*, 235. [PubMed]
29. Chandran, K.B. Role of Computational Simulations in Heart Valve Dynamics and Design of Valvular Prostheses. *Cardiovasc. Eng. Technol.* **2010**, *1*, 18–38. [CrossRef] [PubMed]
30. Kelly, S.G.D. Computational fluid dynamics insights in the design of mechanical heart valves. *Artif. Organs* **2002**, *26*, 608–613. [CrossRef] [PubMed]
31. Thibotuwawa, P.G.; Khalili, F.; Azad, M.; Mansy, H. Modeling Inspiratory Flow in a Porcine Lung Airway. *J. Biomech. Eng.* **2017**. [CrossRef]
32. Khalili, F.; Gamage, P.P.T.; Mansy, H.A. The Influence of the Aortic Root Geometry on Flow Characteristics of a Bileaflet Mechanical Heart Valve. *arXiv* **2018**, arXiv:1803.03362.
33. Khalili, F.; Gamage, P.P.T.; Mansy, H.A. Verification of Turbulence Models for Flow in a Constricted Pipe at Low Reynolds Number. In Proceedings of the 3rd Thermal and Fluids Engineering Conference (TFEC), Fort Lauderdale, FL, USA, 4–7 March 2018; pp. 1–10.
34. Tiari, S. An Experimental Study of Blood Flow in a Model of Coronary Artery with Single and Double Stenoses. In Proceedings of the 2011 18th Iranian Conference of Biomedical Engineering (ICBME), Tehran, Iran, 14–16 December 2011; pp. 33–36.
35. Taebi, A. Characterization, Classification, and Genesis of Seismocardiographic Signals, University of Central Florida. Ph.D. Thesis, University of Central Florida, Orlando, FL, USA, May 2018.
36. Taebi, A.; Mansy, H.A. Time-Frequency Distribution of Seismocardiographic Signals: A Comparative Study. *Bioengineering* **2017**, *4*, 32. [CrossRef] [PubMed]
37. Taebi, A.; Mansy, H.A. Analysis of Seismocardiographic Signals Using Polynomial Chirplet Transform and Smoothed Pseudo Wigner-Ville Distribution. In Proceedings of the 2017 IEEE Signal Processing in Medicine and Biology Symposium (SPMB), Philadelphia, PA, USA, 2 December 2017; IEEE: Philadelphia, PA, USA, 2017; pp. 1–6.

38. Taebi, A.; Solar, B.E.; Mansy, H.A. An Adaptive Feature Extraction Algorithm for Classification of Seismocardiographic Signals. In Proceedings of the IEEE SoutheastCon 2018, Tampa Bay Area, FL, USA, 19–22 April 2018; IEEE: Philadelphia, PA, USA, 2018; pp. 1–5.
39. Taebi, A.; Mansy, H.A. Grouping Similar Seismocardiographic Signals Using Respiratory Information. In Proceedings of the 2017 IEEE Signal Processing in Medicine and Biology Symposium (SPMB), Philadelphia, PA, USA, 2 December 2017; IEEE: Philadelphia, PA, USA, 2017; pp. 1–6.
40. Khalili, F.; Gamage, P.P.T.; Mansy, H.A. A coupled CFD-FEA study of sound generated in a stenosed artery and transmitted through tissue layers. In Proceedings of the IEEE SoutheastCon 2018, Tampa Bay Area, FL, USA, 19–22 April 2018; IEEE: St. Petersburg, Russia, 2018.
41. Fontaine, A.A.; Ellis, J.T.; Healy, T.M.; Hopmeyer, J.; Yoganathan, A.P. Identification of peak stresses in cardiac prostheses. A comparison of two-dimensional versus three-dimensional principal stress analyses. *ASAIO J.* **1996**, *42*, 154–163. [CrossRef] [PubMed]
42. Dasi, L.P.; Ge, L.; Simon, A.H.; Sotiropoulos, F.; Yoganathan, P.A. Vorticity dynamics of a bileaflet mechanical heart valve in an axisymmetric aorta. *Phys. Fluids* **2007**, *19*. [CrossRef]
43. Bluestein, D.; Rambod, E.; Gharib, M. Vortex shedding as a mechanism for free emboli formation in mechanical heart valves. *J. Biomech. Eng.* **2000**, *122*, 125–134. [CrossRef] [PubMed]
44. Ge, L.; Dasi, L.P.; Sotiropoulos, F.; Yoganathan, A.P. Characterization of hemodynamic forces induced by mechanical heart valves: Reynolds vs. viscous stresses. *Ann. Biomed. Eng.* **2008**, *36*, 276–297. [CrossRef] [PubMed]
45. Barannyk, O.; Oshkai, P. The Influence of the Aortic Root Geometry on Flow Characteristics of a Prosthetic Heart Valve. *J. Biomech. Eng.* **2015**, *137*, 51005. [CrossRef] [PubMed]
46. Bruss, K.H.; Reul, H.; Van Gilse, J.; Knott, E. Pressure drop and velocity fields at four mechanical heart valve prostheses. *Life Support Syst.* **1983**, *1*, 3–22. [PubMed]
47. Hung, T.C.; Hochmuth, R.M.; Joist, J.H.; Sutera, S.P. Shear-induced aggregation and lysis of platelets. *ASAIO J.* **1976**, *22*, 285–290.
48. Bluestein, D.; Einav, S.; Hwang, N.H.C. A squeeze flow phenomenon at the closing of a bileaflet mechanical heart valve prosthesis. *J. Biomech.* **1994**, *27*, 1369–1378. [CrossRef]
49. Fatemi, R.; Chandran, K.B. An in vitro comparative study of St. Jude Medical and Edwards-Duromedics bileaflet valves using laser anemometry. *J. Biomech. Eng.* **1989**, *111*, 298–302. [CrossRef] [PubMed]
50. Chandran, K.B.; Yearwood, T.L. Experimental study of physiological pulsatile flow in a curved tube. *J. Fluid Mech.* **1981**, *111*, 59–85. [CrossRef]
51. De Vita, F.; de Tullio, M.D.; Verzicco, R. Numerical simulation of the non-Newtonian blood flow through a mechanical aortic valve: Non-Newtonian blood flow in the aortic root. *Theor. Comput. Fluid Dyn.* **2016**, *30*, 129–138. [CrossRef]
52. Reul, H.; Vahlbruch, A.; Giersiepen, M.; Schmitz-Rode, T.; Hirtz, V.; Effert, S. The geometry of the aortic root in health, at valve disease and after valve replacement. *J. Biomech.* **1990**, *23*. [CrossRef]
53. Grigioni, M.; Daniele, C.; D'Avenio, G.; Barbaro, V. The influence of the leaflets' curvature on the flow field in two bileaflet prosthetic heart valves. *J. Biomech.* **2001**, *34*, 613–621. [CrossRef]
54. Wilcox, D.C. *Turbulence Modeling for CFD (Second Edition)*; Springer: New York, NY, USA, 1998; ISBN 9781928729082.
55. Nobili, M.; Morbiducci, U.; Ponzini, R.; Del Gaudio, C.; Balducci, A.; Grigioni, M.; Maria Montevecchi, F.; Redaelli, A. Numerical simulation of the dynamics of a bileaflet prosthetic heart valve using a fluid-structure interaction approach. *J. Biomech.* **2008**, *41*, 2539–2550. [CrossRef] [PubMed]
56. Celik, I.B.; Ghia, U.; Roache, P.J. Procedure for estimation and reporting of uncertainty due to discretization in {CFD} applications. *J. Fluids Eng.-Trans. ASME* **2008**, *130*.
57. Popov, E.P. *Engineering Mechanics of Solids*; Prentice Hall: Englewood Cliffs, NJ, USA, 1990.
58. Sutera, S.P.; Mehrjardi, M.H. Deformation and fragmentation of human red blood cells in turbulent shear flow. *Biophys. J.* **1975**, *15*, 1–10. [CrossRef]
59. Blackshear, P.L. Hemolysis at prosthetic surfaces. *Chem. Biosurf.* **1972**, *2*, 523–561.
60. Mohandas, N.; Hochmuth, R.M.; Spaeth, E.E. Adhesion of red cells to foreign surfaces in the presence of flow. *J. Biomed. Mater. Res.* **1974**, *8*, 119–136. [CrossRef] [PubMed]

61. Bark, D.L.; Vahabi, H.; Bui, H.; Movafaghi, S.; Moore, B.; Kota, A.K.; Popat, K.; Dasi, L.P. Hemodynamic Performance and Thrombogenic Properties of a Superhydrophobic Bileaflet Mechanical Heart Valve. *Ann. Biomed. Eng.* **2017**, *45*, 452–463. [CrossRef] [PubMed]
62. King, M.J.; David, T.; Fisher, J. Three-dimensional study of the effect of two leaflet opening angles on the time-dependent flow through a bileaflet mechanical heart valve. *Med. Eng. Phys.* **1997**, *19*, 235–241. [CrossRef]

bioengineering

MDPI

Article

Biomechanical Restoration Potential of Pentagalloyl Glucose after Arterial Extracellular Matrix Degeneration

Sourav S. Patnaik [1], Senol Piskin [1,2], Narasimha Rao Pillalamarri [1], Gabriela Romero [3], G. Patricia Escobar [4], Eugene Sprague [4] and Ender A. Finol [1,*]

1 Department of Mechanical Engineering, The University of Texas at San Antonio, One UTSA Circle, San Antonio, TX 78249, USA
2 Department of Mechanical Engineering, Koc University, Rumelifeneri Kampusu, Istanbul 34450, Turkey
3 Chemical Engineering Program, Department of Biomedical Engineering, The University of Texas at San Antonio, San Antonio, TX 78249, USA
4 Department of Medicine, University of Texas Health San Antonio, San Antonio, TX 78229, USA
* Correspondence: ender.finol@utsa.edu; Tel.: +1-(210)-458-8058

Received: 1 May 2019; Accepted: 30 June 2019; Published: 3 July 2019

Abstract: The objective of this study was to quantify pentagalloyl glucose (PGG) mediated biomechanical restoration of degenerated extracellular matrix (ECM). Planar biaxial tensile testing was performed for native (N), enzyme-treated (collagenase and elastase) (E), and PGG (P) treated porcine abdominal aorta specimens (n = 6 per group). An Ogden material model was fitted to the stress–strain data and finite element computational analyses of simulated native aorta and aneurysmal abdominal aorta were performed. The maximum tensile stress of the N group was higher than that in both E and P groups for both circumferential (43.78 ± 14.18 kPa vs. 10.03 ± 2.68 kPa vs. 13.85 ± 3.02 kPa; $p = 0.0226$) and longitudinal directions (33.89 ± 8.98 kPa vs. 9.04 ± 2.68 kPa vs. 14.69 ± 5.88 kPa; $p = 0.0441$). Tensile moduli in the circumferential direction was found to be in descending order as N > P > E (195.6 ± 58.72 kPa > 81.8 ± 22.76 kPa > 46.51 ± 15.04 kPa; $p = 0.0314$), whereas no significant differences were found in the longitudinal direction ($p = 0.1607$). PGG binds to the hydrophobic core of arterial tissues and the crosslinking of ECM fibers is one of the possible explanations for the recovery of biomechanical properties observed in this study. PGG is a beneficial polyphenol that can be potentially translated to clinical practice for preventing rupture of the aneurysmal arterial wall.

Keywords: pentagalloyl glucose; aneurysm; enzyme; biomechanics; aorta

1. Introduction

The etiology of abdominal aortic aneurysm (AAA) development is believed to be multi-factorial, in that (i) the pathology is initiated at the molecular level (protease- and enzyme-related); (ii) it builds up to the tissue level through extracellular matrix (ECM) and structural changes; and (iii) it manifests as geometrical-, biomechanical-, and blood flow-related alterations in the abdominal aorta, resulting in rupture if left untreated [1–3]. Of the numerous etiological theories of AAA pathology, the degraded ECM theory is the widely accepted one, as human AAA specimens usually exhibit a reduction in elastin content and elastin crosslinking, and an increase in collagen crosslinking [4]. Increased elastase activity leads to disorganized and tortuous elastin fibers [5], which represents a compromised organization of load bearing proteins, resulting in reduced aortic elasticity [4,6], and further weakening of the aortic wall. With the deficiency in elastin, collagen dominates the ECM [7]. Disease progression is characterized by an increase in matrixmetalloproteinase (MMP) activity, which subsequently yields elevated wall stress and concomitantly higher wall stress to strength ratios [4,6–8]. In addition, most aneurysms exhibit an intraluminal thrombus (ILT), which is also a source of proteolytic activity [9], increased wall weakening [10], and a preferential site for rupture [11]. This multifaceted presentation of the disease makes the discovery of potential pharmacological targets a complex one (i.e., it has to consider the biological factors, the biomechanical environment, and the presence of ILT as a potential transport barrier).

Anti-inflammatory or matrix metalloproteinase inhibiting chemicals are the primary choice for stabilizing the aortic extracellular matrix (ECM). We envision the use of pentagalloyl glucose (PGG), a multifunctional polyphenol [12,13], as a potential pharmacological agent for AAA suppression [14]. PGG has been shown to bind to elastin and collagen, and stabilizes the ECM [4,5]. PGG has multiple phenolic hydroxyl groups that have high affinity towards the hydrophobic regions of the tissues [15] and can bind to proline-rich proteins such as elastin and collagen by surface adsorption mechanisms [16]. Isenberg and colleagues applied PGG periadventitially to the abdominal aorta of adult male Sprague–Dawley rats, previously exposed to $CaCl_2$-mediated aortic elastin injury, and found that early inhibition of aneurysm and stabilization of elastin lamellae is possible [17]. Since then, multiple studies have investigated nanoparticle-based PGG delivery to the site of AAA [18,19]. Sinha et al. [20] reported that addition of PGG to rat aneurysmal smooth muscle cells increased lysyl oxidase production, enhanced elastin crosslinking, and assisted in lowering MMP-2 levels. Using a rat $CaCl_2$ model of AAA, Thirugnanasambandam and colleagues showed that PGG was able to mitigate the inflammatory response, lower the MMP-2,9 levels, and prevent biomechanical stress build up on the aortic wall [21]. PGG has been applied as a preventive measure in porcine AAA models by Kloster and co-authors [22]; they found that PGG was able to lower the abnormal dilation of the abdominal aorta to a certain extent.

AAA porcine models based on elastase–collagenase combination are uncommon [23–27], but their combined effect produces maximal damage to the ECM and pronounced inflammatory infiltration in vivo. Conversely, in vitro studies utilizing enzymatic digestion of porcine aortic tissues have been reported widely [28–33]. The objective of the present work is to quantify biomechanical changes in ex-vivo porcine abdominal aortas after treatment with PGG. Biaxial mechanical testing was performed on native, elastase, and collagenase treated (to mimic the presence of AAA), and PGG-treated enzyme degraded porcine abdominal aorta specimens. We hypothesize that enzyme treated specimens will exhibit loss of biomechanical strength compared to native or PGG-treated tissue specimens. We report on the ability of PGG to restore the biomechanical properties of the porcine abdominal aorta after enzymatic damage (elastase + collagenase).

2. Materials and Methods

2.1. Biomechanical Testing

Three porcine abdominal aorta tracts (Yorkshire mixed breed, 125–250 lb, 6–9 months) were obtained from a local abattoir and all excess connective tissue removed. Approximately 7 mm-long cylindrical rings were dissected from the tracts and further utilized for biomechanical testing as shown in Figure 1.

To evaluate the restorative potential of PGG, specimens were tested consecutively in their native state (N), followed by a simulated aneurysmal condition (E), and then treated with PGG (P). The simulated aneurysmal condition was achieved by treating the specimen for 1 h in an enzyme solution of 1.5 mg/mL purified elastase and 0.5 mg/mL purified collagenase at 37 °C [31] (Worthington Biochemical Corporation, Lakewood, NJ, USA). The treatment with PGG consisted of a 12-hour incubation in 0.6 mg/mL PGG at 4 °C after enzymatic treatment [22] (Sigma-Aldrich Inc., St. Louis, MO, USA).

Specimens of an approximate size of 7 × 7 mm were prepared for biaxial testing as shown in Figure 1 and a suture knot (6-0 Silk—Ethicon Inc., Somerville, NJ, USA) was tied to the upper right-hand corner of the specimen to maintain the orientation throughout the study. Prior to testing, the wall thickness of each specimen was measured using digital caliper (Mitutoyo America Corporation,

Aurora, IL, USA). Four fiducial markers were placed on the specimens using cyanoacrylate glue (Loctite Professional Super Glue Liquid, Henkel, Germany) for tracking local deformation, and the specimens were secured with metallic hooks as per the specified orientation (Figure 1).

A CellScale© Biotester (CellScale Biomaterials Testing, Ontario, Canada) was utilized for biaxial mechanical testing of the specimens. Briefly, specimens were preloaded up to 2 g, preconditioned ten times in the physiological range (0–5% strain), and ultimately stretched equi-biaxially up to ~50% tensile strain over 45 seconds followed by unloading to its reference state [34–36]. Using a constant strain rate, the following strain-based protocol was performed ($\lambda_C : \lambda_L$)—1:1, 0.5:1, and 1:0.5—where λ_C *and* λ_L represent stretch ratios in the circumferential and longitudinal directions, respectively. LabJoy 8.1 software (Waterloo Instruments Inc., Ontario, Canada) was utilized to collect the data at 30 Hz and images were captured at the rate of 1 Hz using the Biotester's overhead CCD camera. All biomechanical tests were performed in a 37 °C saline bath.

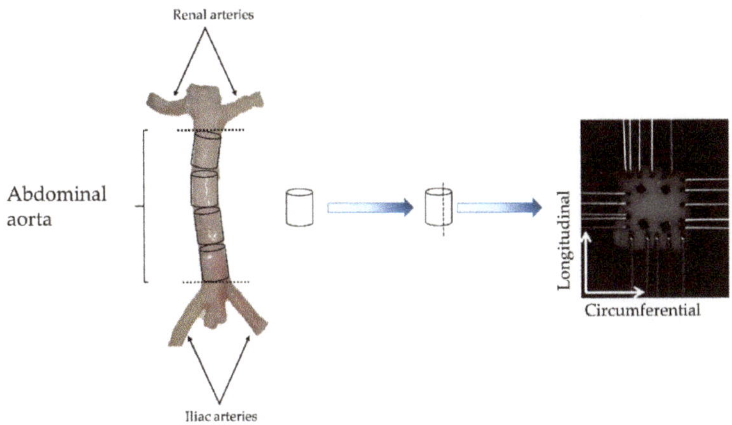

Figure 1. Exemplary schematic for specimen procurement. Each specimen consisted of a cylindrical ring approximately 7 mm long, which were subject to planar biaxial tensile testing. N = 6 specimens per group were tested using a CellScale BioTester® while submerging the specimen in saline solution at 37 °C.

2.2. Data Analysis

Force and displacement data were exported (LabJoy 8.1) and further processed to generate stress–strain curves for each specimen in the three experimental groups (N, E, and P) using MATLAB (R2018, The MathWorks Inc., Natick, MA, USA). Shear components of biaxial deformation were assumed negligible [37,38].

Stretch ratios in the circumferential and longitudinal directions are given by Equation (1),

$$\lambda_C = \frac{l_C}{l_{C0}}, \lambda_L = \frac{l_L}{l_{L0}}, \tag{1}$$

where l_{C0} and l_{L0} are the undeformed specimen lengths in mm, and l_C and l_L are the deformed specimen lengths in mm, for the circumferential and longitudinal directions, respectively.

The Green strain tensor components were calculated from the respective stretch ratios following Equation (2),

$$\epsilon_C = \frac{1}{2}\left(\lambda_C^2 - 1\right), \epsilon_L = \frac{1}{2}\left(\lambda_L^2 - 1\right) \tag{2}$$

where ϵ_C and ϵ_L are the Green strains in circumferential and longitudinal directions, respectively.

The Cauchy stress (σ) was calculated by dividing the force by the cross-sectional area of each specimen (width multiplied by thickness), as indicated by Equation (3),

$$\sigma_C = \frac{F_C * \lambda_C}{A_C}, \ \sigma_L = \frac{F_L * \lambda_C}{A_L}, \quad (3)$$

where F_C and F_L are forces in Newton, and A_C and A_L are the specimen cross-sectional areas in mm^2, for the circumferential and longitudinal directions, respectively.

The tensile moduli (TM), defined as the slope of the upper linear portion of the stress–strain curves that best represents the linear elastic region of the material (see Figure 2A) for the circumferential and longitudinal directions, were calculated using a pointwise linear regression in the upper 10% of the strain range. The strain energy, determined by area under the stress–strain curve (AUC), and maximum stress (σ_{max}) were also calculated for both tissue orientations (see Figure 2A).

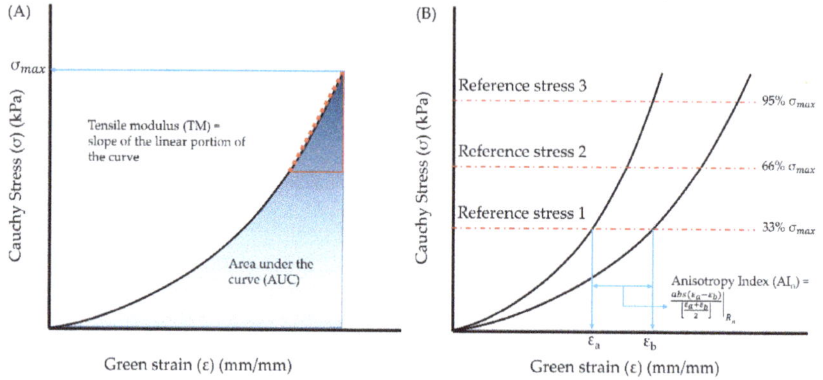

Figure 2. Biomechanical parameters evaluated from the stress–strain curves generated from biaxial mechanical testing of porcine abdominal aorta specimens. (**A**) Schematic for calculation of area under the curve (AUC) and maximum tensile stress (σ_{max}). (**B**) Procedure for calculation of anisotropy index (AI) at each reference stress level, as reported in [39,40]. The reference stresses for each specimen were calculated by estimating 33rd, 66th, and 95th percentiles of σ_{max} in each stress–strain curve.

Enzymatic damage introduces some degree of changes to the ECM and in many cases, a change in anisotropic behavior was observed in porcine arterial tissues [30]. To measure these changes in the mechanical anisotropy, we calculated the anisotropy index (AI, illustrated in Figure 2B) according to Equation (4) [39,40],

$$Anisotropy\ Index\ (AI_n) = \left. \frac{abs(\varepsilon_a - \varepsilon_b)}{\left[\frac{\varepsilon_a + \varepsilon_b}{2}\right]} \right|_{R_n} \quad (4)$$

AI was calculated for each group at reference stresses (R_n) corresponding to the 33rd, 66th, and 95th percentiles of the maximum stress (σ_{max}) in each orientation (i.e., AI_1, AI_2, and AI_3, respectively). A perfectly isotropic material will have an AI of zero, whereas tissues and most biological materials exhibit non-zero AI values.

To quantify recovery, we normalized the biomechanical parameters obtained from the E and P groups to their N matching counterparts according to Equation (5),

$$\text{Normalized } \varnothing_{C \text{ or } L} = \frac{\varnothing|_E}{\varnothing_N}; \frac{\varnothing|_P}{\varnothing_N} \quad (5)$$

where $\varnothing$ = biomechanical parameter such as σ_{max}, TM, or AUC. C and L stands for circumferential or longitudinal. N, E, P are the three experimental groups. The aforementioned biomechanical data collected from the specimens were further utilized for constitute modeling and as input for finite element modeling.

2.3. Constitutive Modeling

To characterize the material behavior of the specimens, a first-order incompressible hyperelastic Ogden material model [41–43], Equation (6), was fitted to the experimental stress–strain data,

$$\Psi = \frac{m_1}{c_1}\left(\lambda_1^{c_1} + \lambda_2^{c_1} + \lambda_3^{c_1} - 3\right) \quad (6)$$

where c_1 and m_1 are constants, and λ_i are the principal stretches. For planar biaxial tension, there are no stretch data in the 3rd direction. Therefore, λ_3 was obtained by applying an incompressibility condition (i.e., the *determinant of* $\boldsymbol{F} = 1$, where $\boldsymbol{F}$ is the deformation gradient tensor), as expressed by Equation (7).

$$\lambda_3 = \frac{1}{\lambda_1 \cdot \lambda_2} \quad (7)$$

We have assumed that there are no shear components during the planar biaxial tension. The ANSYS (Ansys, Inc., Canonsburg, PA, USA) biaxial curve fitting tool (a non-linear least squares algorithm) was utilized to generate the best subset of material constants that can minimize the differences of the sum of the squares between the experimental data and the constitutive model. The tool uses the Levenberg–Marquardt algorithm to solve the non-linear least squares problem. This method requires a set of initial values for each parameter of the material model (i.e., c_1 and m_1), as many other optimization algorithms. The error calculation is performed using a normalized error instead of an absolute error. We calculated residual errors for all native and aneurysm samples, but do not report them in the manuscript since the models have an excellent goodness of fit, as described in Section 3.2.

2.4. Finite Element Modeling

The material constants from the Ogden model fitting from each group (N, E, or P) were further utilized for computational modeling following previously established protocols [21,44]. Briefly, idealized models of a native abdominal aorta (NAA) and an aneurysmal abdominal aorta (AAA) were created using ANSYS® SpaceClaim (SpaceClaim Corporation, Concord, MA, USA). Figure 3 shows the geometries of the models and their dimensions, based on the work reported by Azar et al. [44]. The inner and outer surfaces of the geometries were meshed with 2D triangle elements using Gmsh open source software [45]. The surface meshes were converted into volume meshes using TetGen [46] by generating linear tetrahedral elements. FEBio® Preview was utilized to setup the volumetric meshes for both models [47]. The Ogden material properties obtained with the native tissue properties (N) were assigned to the NAA model, whereas the Ogden material properties obtained with the enzymatic (E) and PGG tissue properties (P) were assigned to the AAA model. Both ends of the models were fixed for all degrees of freedoms. An average of systolic and diastolic pressures (100 mmHg) was applied homogenously at the intraluminal surface of the models [21]. A quasi-static structural analysis was performed with the open source finite element analysis (FEA) solver FEBio [47]. The first principal stress [48] generated by the FEA simulations was postprocessed with FEBio PostView [47] to quantify the differences in in silico wall stress distributions due to changes in material properties (N group vs. E and P groups).

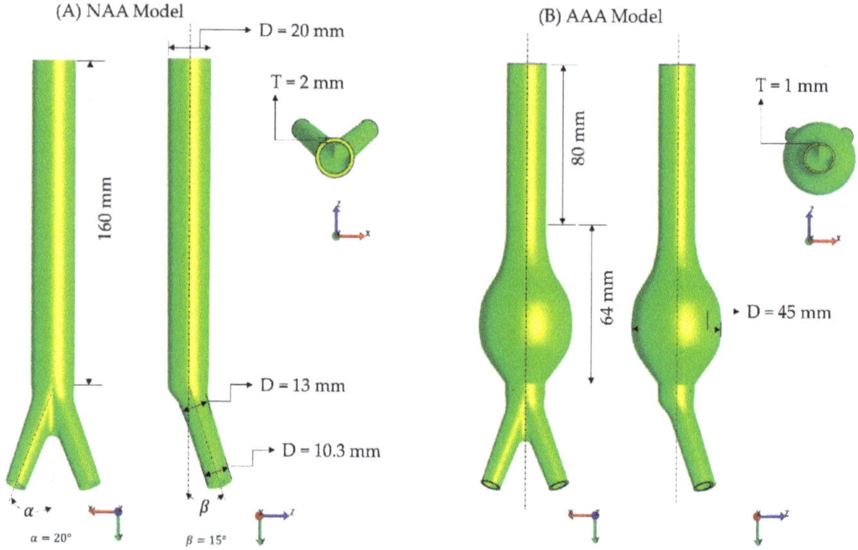

Figure 3. Geometric models and their dimensions for the (**A**) native abdominal aorta (NAA) and (**B**) aneurysmal abdominal aorta (AAA).

2.5. Statistical Analysis

Data were reported as mean ± standard error of mean (SEM). The same porcine arterial specimens (N) underwent enzymatic treatment (E) and followed by PGG treatment (P), a repeated-measures ANOVA was performed to elucidate the differences across the biomechanical data (σ_{max}, TM, AI, and AUC). Sphericity was assumed for the data and pairwise comparisons were performed using Tukey's test with results considered significant when $p < 0.05$. All analyses were performed using SPSS (IBM Corp., Armonk, NY, USA).

3. Results

3.1. Biomechanical Testing

Biomechanical parameters such as σ_{max} and AUC were found to be significantly different across the groups for both tissue orientations ($p < 0.05$), whereas TM was significantly different for only the circumferential direction. The anisotropy indices (AI_1, AI_2, and AI_3), derived from the three reference stresses, were found to be non-zero, but conserved across the three groups ($p = 0.2702$, $p = 0.0813$, and $p = 0.1425$, respectively). All biomechanical parameters calculated from the three biaxial testing protocols ($\lambda_C : \lambda_L$ – 1:1, 0.5:1, and 1:0.5) are listed in Table 1.

The maximum tensile stress of the N group was higher than in the E and P groups for both circumferential (43.78 ± 14.18 kPa vs. 10.03 ± 2.68 kPa vs. 13.85 ± 3.02 kPa; $p = 0.0226$) and longitudinal directions (33.89 ± 8.98 kPa vs. 9.04 ± 2.68 kPa vs. 14.69 ± 5.88 kPa; $p = 0.0441$), as shown in Figure 4A,B. Likewise, the tensile moduli was found to be in descending order as N > P > E for the circumferential direction (195.6 ± 58.72 kPa > 81.8 ± 22.76 kPa > 46.51 ± 15.04 kPa; $p = 0.0314$), as illustrated in Figure 4C, whereas no significant differences were found in the longitudinal direction ($p = 0.1607$). Strain energy, represented by AUC, was nearly four times greater for the N group than the E or P groups in the circumferential direction (6.48 ± 2.22 kPa vs. 1.55 ± 0.34 kPa or 1.56 ± 0.26 kPa; $p = 0.0224$), as shown in Figure 4E. For the longitudinal direction, AUC was nearly three times greater for the N group than the E or P groups (4.77 ± 1.04 kPa vs. 1.45 ± 0.42 kPa or 1.35 ± 0.32 kPa; $p = 0.0034$), as illustrated in Figure 4F.

Table 1. Biomechanical parameters calculated from biaxial tensile testing of porcine abdominal aorta specimens for the three experimental groups (native (N), enzyme-treated (collagenase and elastase) (E), and pentagalloyl glucose (PGG) (P)). Values are reported as mean ± SEM.

Testing Protocol	Biomechanical Parameters (kPa)	N	E	P	*p*-Value
$\lambda_C : \lambda_L = 1{:}1$	$\sigma_{C,max}$	43.78 ± 14.18	10.03 ± 2.66	13.85 ± 3.02	0.0226 [a]
	$\sigma_{L,max}$	33.89 ± 8.98	9.04 ± 2.97	14.69 ± 5.88	0.0441 [a]
	TM_C	195.6 ± 58.72	46.51 ± 15.04	81.8 ± 22.76	0.0314 [a]
	TM_L	168.0 ± 51.53	39.75 ± 15.56	101.6 ± 50.87	n.s.
	AUC_C	6.48 ± 2.22	1.55 ± 0.34	1.56 ± 0.26	0.0224 [a,b]
	AUC_L	4.77 ± 1.04	1.45 ± 0.42	1.35 ± 0.32	0.0034 [a,b]
$\lambda_C : \lambda_L = 0.5{:}1$	$\sigma_{C,max}$	7.6 ± 1.35	2.65 ± 0.54	1.51 ± 0.14	0.0004 [a,b]
	$\sigma_{L,max}$	18.66 ± 3.78	5.71 ± 1.92	4.05 ± 1.04	0.0013 [a,b]
	TM_C	67.46 ± 12.94	22.19 ± 5.63	16.86 ± 2.73	0.0011 [a,b]
	TM_L	101.9 ± 30.18	26.11 ± 10.26	31.61 ± 12.91	0.0273 [a]
	AUC_C	0.63 ± 0.12	0.24 ± 0.04	0.1 ± 0.01	0.0006 [a,b]
	AUC_L	2.84 ± 0.41	0.93 ± 0.3	0.48 ± 0.07	<0.0001 [a,b]
$\lambda_C : \lambda_L = 1{:}0.5$	$\sigma_{C,max}$	19.84 ± 5.14	5.84 ± 1.44	3.1 ± 0.73	0.0034 [a,b]
	$\sigma_{L,max}$	6.99 ± 2.71	2.28 ± 1.299	0.8 ± 0.08	<0.0001 [a,b]
	TM_C	125.3 ± 38.93	30.08 ± 9.42	22.37 ± 6.28	0.0102 [a,b]
	TM_L	60.95 ± 12.84	18.15 ± 4.11	7.6 ± 1.38	0.0007 [a,b]
	AUC_C	2.64 ± 0.57	0.91 ± 0.18	0.37 ± 0.09	0.011 [a,b]
	AUC_L	0.59 ± 0.08	0.19 ± 0.05	0.06 ± 0.007	<0.0001 [a,b]

[a] denotes significant pairwise differences across N and E groups and [b] denotes significant pairwise differences across N and P groups (repeated measures ANOVA—sphericity assumed). n.s.: not significant.

Normalized biomechanical parameters were calculated for the E and P groups to demonstrate the biomechanical recovery of the degenerated ECM owing to PGG treatment. Figure 5 shows these parameters for both tissue orientations. For the circumferential orientation, normalized σ_{max} and TM were found to be 43.8% and 58.6% greater for P group than the E group, respectively (Figure 5A,C). However, the normalized AUC exhibited a minimal increase of approximately 18.4% from the E group to the P group (Figure 5E). Following a similar trend, normalized σ_{max} and normalized TM were higher by 54.0% and 72.4%, respectively, for group P vs. group E in the longitudinal direction (Figure 5B,D). A small increase, of approximately 13.6%, was observed in the normalized AUC of the P group compared to the E group in the longitudinal direction (Figure 5F).

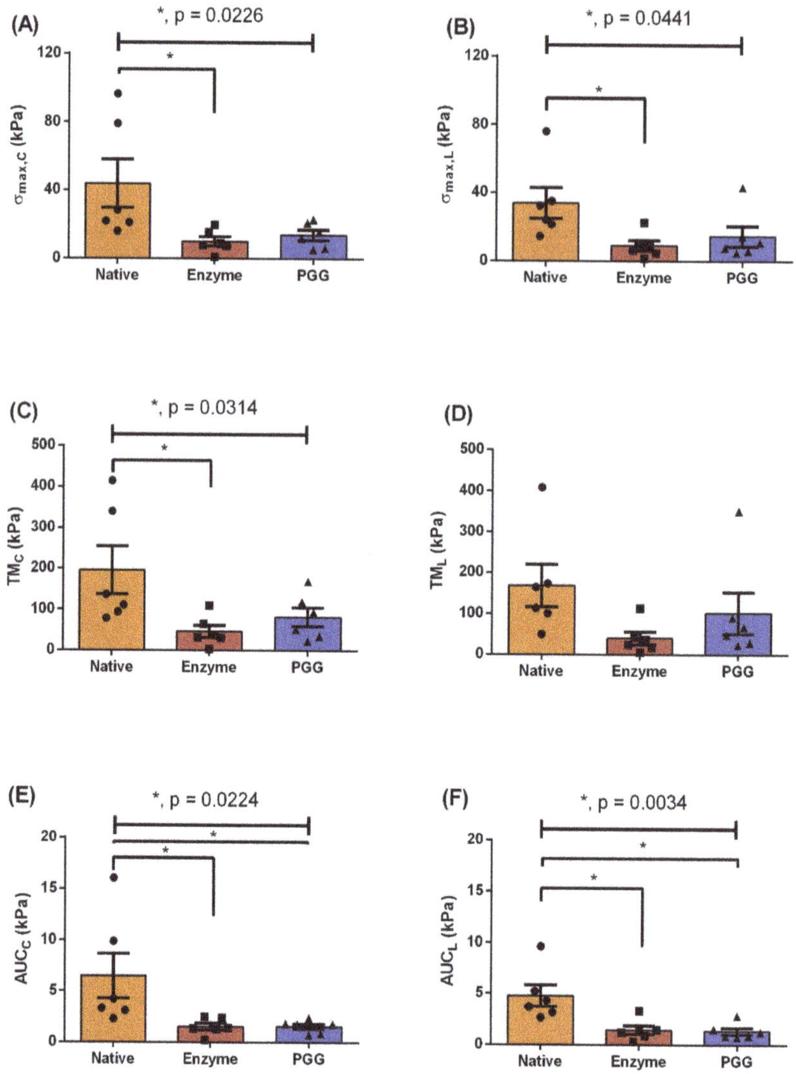

Figure 4. Biomechanical parameters for the native (N), elastase-digested (E), and PGG-treated (P) porcine abdominal aorta specimens, which are represented as mean ± SEM. Maximum stress (kPa) is displayed in both circumferential ($\sigma_{max,C}$) (**A**) and longitudinal orientations ($\sigma_{max,L}$) (**B**). (**C**,**D**) tensile moduli (TM) (kPa) and (**E**,**F**) area under the curve (AUC) (kPa) for both circumferential and longitudinal directions, respectively. *denotes significance across the groups ($p < 0.05$).

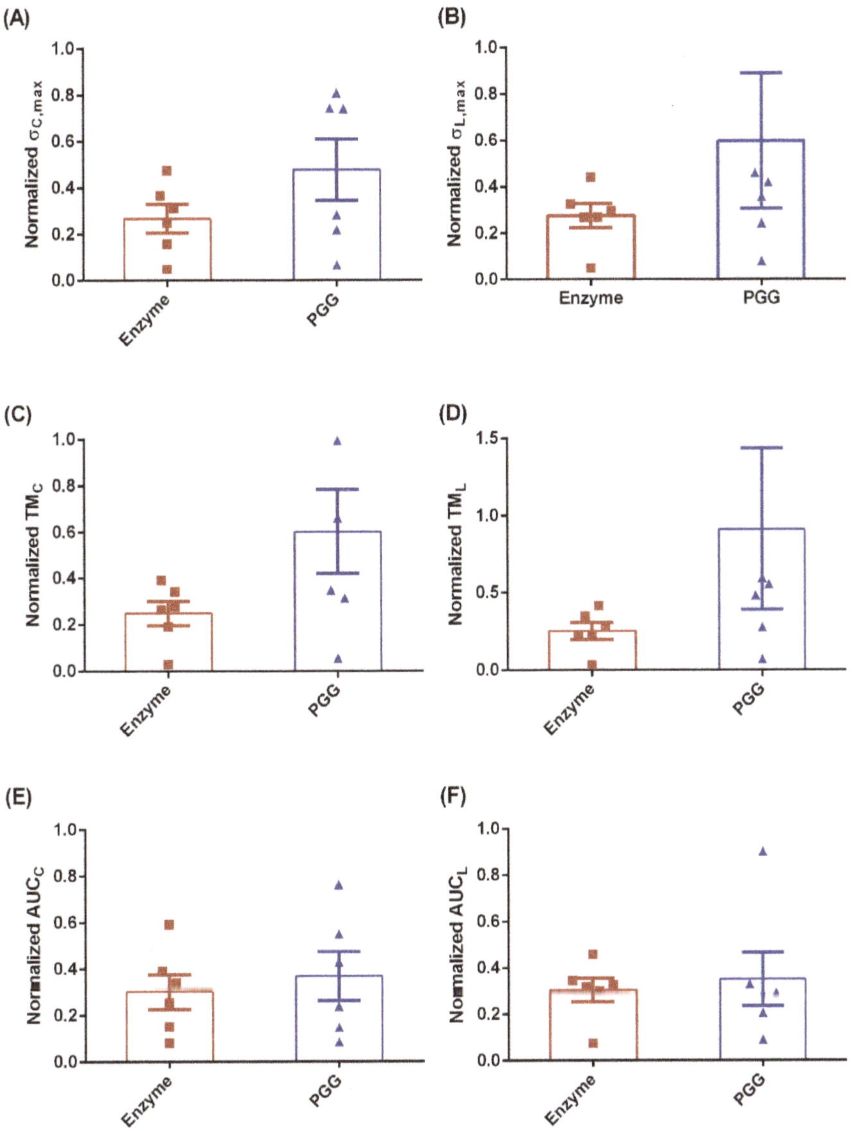

Figure 5. An assessment of the recovery of biomechanical properties by normalizing the enzymatic- and PGG-based biomechanical parameters to their native equivalents. We observe an increasing trend in σ_{max} (**A**,**B**) and TM (**C**,**D**) for both circumferential and longitudinal orientations, respectively. AUC for both orientations exhibited similar but limited increase in strain energy (**E**,**F**), thereby indicating a conservation of elastic energy.

3.2. Constitutive Modeling and Finite Element Analyses

Following the Ogden constitutive relation, the phenomenological behavior was represented for each specimen of the N, E, and P groups (with exemplary stress–strain curves shown in Figure 6 and their respective material constants reported in Table 2). Good correlations ($R^2 > 0.99$) between experimental and theoretical data were observed for the three experimental groups (Figure 6).

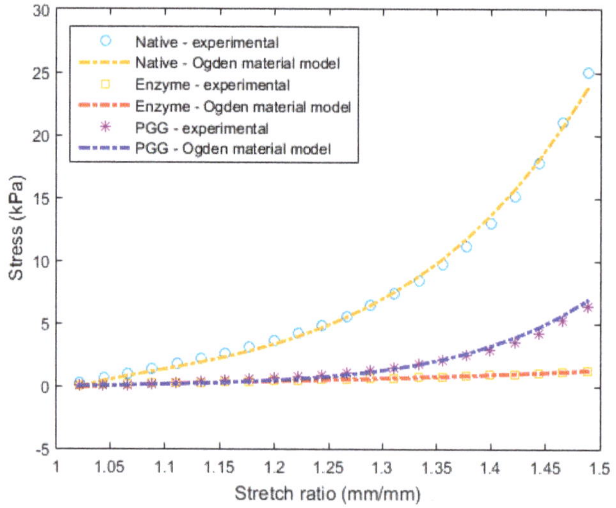

Figure 6. Ogden model fitting with the corresponding experimental stress–strain curves for an exemplary specimen of the native (N), enzyme (E), and PGG treatment (P) groups.

Table 2. Ogden material model constants (mean ± SEM) for the three tissue types.

Group	m_1 (kPa)	c_1 (–)
N	0.96 ± 0.025	10.06 ± 0.38
E	0.37 ± 0.06	8.53 ± 0.67
P	0.09 ± 0.03	13.57 ± 0.87

The wall stress (calculated at the mid-section or sac of the geometries) obtained for each FEA model is summarized in Table 3. Figure 7 illustrates the spatial distribution of wall stress for the idealized FEA abdominal aorta models (native and AAA) based on the Ogden constitutive relations derived from stress–strain curves of the three experimental groups. Colorimetric surface plots of the wall stress in the normal aorta shows a uniform stress distribution until the aortic bifurcation (Figure 7A). Similar to the experimental data, the FEA models reveal that the maximum wall stress was in the order of N > P > E (35 ± 4.0 kPa vs. 16 ± 0.5 vs. 13 ± 1.0 kPa; $p = 0.0002$). The E and P models exhibited maximum wall stresses that were, respectively, 62.6% and 53.7% lower than the N models (Table 3). The average and minimum wall stresses at the sac region of the PGG-treated model were 1.3 and 1.7 times greater than the enzyme-treated model, respectively. However, these stresses of the PGG-treated model were almost 2.6 and 4 times lower than the native model stresses, respectively.

Table 3. Wall stress (mean ± SEM) computed in the mid-section of the FEA geometries corresponding to the three Ogden material models.

Group	Maximum Wall Stress (kPa)	Average Wall Stress (kPa)	Minimum Wall Stress (kPa)
N	35 ± 4.0	26 ± 4.0	20 ± 4.0
E	13 ± 1.0	8.0 ± 0.4	3.0 ± 0.2
P	16 ± 0.5	10 ± 0.5	5.0 ± 0.3
p-value	0.0002 [ab]	0.0003 [ab]	0.0002 [ab]

[a] denotes significant pairwise differences across N and E groups and [b] denotes significant pairwise differences across N and P groups (repeated measures ANOVA—sphericity assumed).

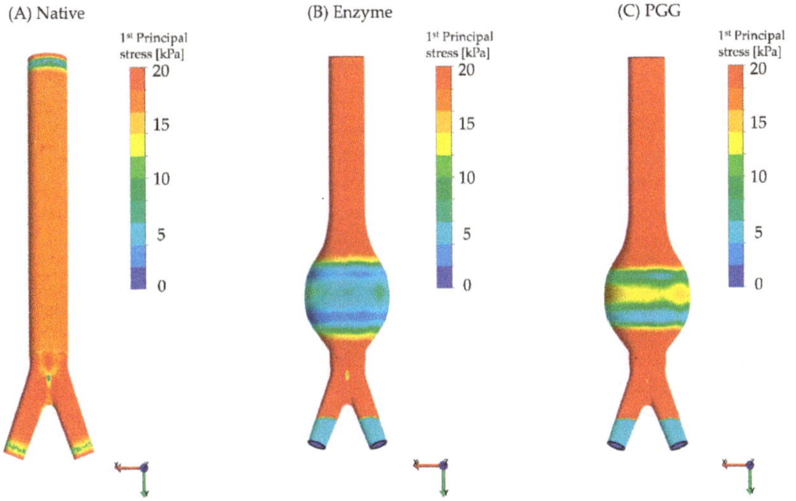

Figure 7. Wall stress spatial distribution for three exemplary FEA models (anterior view): (**A**) Native, (**B**) enzyme and (**C**) PGG treated. The Ogden hyperelastic constitutive material was utilized for the FEA simulations and a 100 mmHg intraluminal pressure was applied to the models. The upper limit of the stress legend was lowered to emphasize the differences across the three groups at the midsection.

4. Discussion

This investigation is a "proof-of-concept" contribution that highlights the beneficial crosslinking properties of PGG—specifically with respect to the degenerated arterial ECM, which is a common finding in AAA. Experimentally, we infer that PGG leads to crosslinking between the ECM proteins that improve the biomechanical strength of enzymatically degraded tissues in vitro (Table 1). To simulate the potential application of this finding, idealized finite element models were created to estimate changes in the stress build up on the aneurysmal wall (due to enzymatic damage) with and without PGG treatment (fibrillar crosslinking). The application of PGG after enzymatic degradation yielded some degree of biomechanical recovery—both experimentally and computationally (Figures 4 and 5 and Tables 2 and 3). The computational models were utilized to demonstrate changes in stress distribution owing to geometry (e.g., an aneurysmal expansion) and the effect of the constitutive material model under simulated intraluminal pressure. The primary contributions of this work are the quantification of the biomechanical restoration potential of PGG and the inference of this finding on the binding of PGG to the arterial ECM.

4.1. Biomechanical Restoration Potential of PGG

The three types of tissue specimens in this study were used to represent a healthy aorta, an aneurysm pathology, and a PGG-treated aneurysmal condition, respectively. By using a mixture of enzymes (collagenase and elastase), we successfully compromised the porcine abdominal aorta ECM integrity (see Supplementary Material), which was evident by its reduced biomechanical strength (Figure 4A–F and Table 1). An hour-long digestion of arterial tissue, similar to Gundiah et al. [30], was sufficient to reduce the structural integrity of elastin and collagen. In this process, the aorta may have become more permeable to PGG influx (see Supplementary Material). The stresses in the PGG group were higher than in the enzyme-digested group, but lower than in the native group (Table 1). For clarification, the increased stresses observed in the PGG group did not yield a "stiffening" of the aorta.

AAA rupture typically occurs when the ECM fiber distribution is altered [4,5], which results in increased stresses that exceed the strength of the diseased arterial wall. In addition, the localized concentration of stresses has been postulated as one of the primary causes of AAA rupture [1].

Noteworthy is that there was an increase in stiffness of the PGG-treated specimens (compared to the enzyme-treated specimens), but not comparable to the stiffness of the native aorta specimens (Figures 4, 5 and 7). The FEA results showed an average wall stress along the dilated portion of the PGG-treated aneurysmal model that was nearly 1.7 times greater than the enzymatic model, and the average stresses for both models were significantly less (nearly 2.6 times and 3.2 times, respectively) than the native model (Figure 7 and Table 3). From the experimental data, maximum stresses in the circumferential and longitudinal directions were found to decrease for the E group compared to the N group (by 77.1% and 73.3%, respectively) owing to the enzymatic cleavage of the native elastin and collagen crosslinks (Figure 4A,B). Following PGG treatment, possibly due to PGG crosslinking activity, the maximum stress for group P increased by 1.4 times in the circumferential direction and 1.6 times in the longitudinal direction compared to the E group. A similar outcome was obtained for the tensile modulus; however, only the increase in circumferential TM was significant for the PGG-treated group (almost 1.8 times greater than the enzyme digested group; Figure 4C,D). Isenberg et al. [49] reported that porcine ascending aorta specimens, treated with 0.15% PGG solution for four days, exhibited a reduced distensibility (or elastic modulus) compared to their untreated native counterparts undergoing uniaxial tension. They also suggest that this reduction in distensibility is a result of the PGG-elastin binding mechanism. However, their work does not discuss the interaction of collagen with PGG, which is one of the proline-rich compounds that selectively binds with this polyphenol [14,50]. Although our results exhibit a similar trend in elastic moduli (Figure 5C,D), they are not directly comparable to the study by Isenberg and co-authors. For example, our specimens underwent (i) biaxial tension and (ii) enzymatic degradation prior to PGG treatment (12 h at 4 °C) in contrast to Isenberg et al.'s direct PGG treatment for a longer period (four days; no temperature reported).

Native aortic tissue exhibits anisotropic mechanical behavior due to the circumferentially oriented elastin and collagen network. We observed a non-significant anisotropy in the three experimental groups. Upon elastase treatment, arterial tissues lose some of its anisotropic characteristics [30] and concordantly, the collagen fiber arrangement is also disrupted [29]. Within 6 h of enzyme treatment, there is pronounced softening of the mechanical characteristics [47] (similar to Figure 4A–F and Table 1) and by 96 h the arterial specimens behave as a collagen-scaffold-type material [51]. The cleavage of ECM fibers by elastase and collagenase exposes some of the hydrophobic cores in the elastin fiber network. These exposed hydrophobic sites or residues are favorable for PGG attachment [17,49], as this polyphenolic compound is known to "lock" or affix the orientation of fibrous structural proteins resulting in reduced residual stresses in arteries [49]. We infer that the restoration of structural integrity and improvement of the degenerated ECM's biomechanical characteristics (Figure 4A–F and Table 1), originates from the intricate hydrophobic bonds forged by PGG [13,52,53] that specifically crosslink arterial elastin and collagen [14,17].

Under typical in vivo intraluminal loads, ECM fibers in blood vessels are engaged and assumed to gradually straighten out from their "crimped" state. With increasing age (or pathology), arterial fiber arrangement is altered, leading to a change in the overall strain energy, and progressively toward a stiffened arterial matrix [54]. In our study, enzymatic and PGG treatments produced transitional change in the native microstructural fiber architecture of the porcine abdominal aorta. The strain energy (AUC) of the PGG-treated specimens exhibited some degree of recovery in both tissue directions, although it was insignificant compared to σ_{max} or TM (Figure 5A–E). The experimental data was well represented by the hyperelastic Ogden model (Equation (6)) [41,42]; the suitability of the Ogden hyperelastic model for blood vessel mechanics is not uncommon [42,43,55]. Due to its dependence on large strain behavior, the Ogden phenomenological model fit the biaxial stress–strain curves across the three groups alike (Figure 6). To the best of our knowledge, this is the first study that compares the computational biomechanical analysis of native, enzyme-degraded, and PGG-treated abdominal aortic tissues (Figure 7A–C).

The use of computational models with idealized geometries allowed us to simulate potential changes in wall stress distribution due to ECM modifications—either due to a pathological or a regenerative condition. FEA simulations revealed the substantial effect of the abdominal aorta geometry and the constitutive material properties on the wall stress distributions (Figure 7A–C). In addition, we noted a 62.6% and 53.6% reduction in maximum wall stress at the AAA sac of the enzyme and PGG models, respectively, compared to the native ones. The idealized native and aneurysmal models utilized in our study were similar to those of Azar et al. [44]; however, the material parameters for our FEA study were derived directly from the biaxial experiments for all native, enzyme, and PGG-treated specimens. Similar to [44], the wall stress distribution of the native aorta model was uniform until the aortic bifurcation (Figure 7A). However, the stresses reported in [44] are higher than those of our study (average wall stress—110 kPa vs. 26 kPa; maximum wall stress—760 kPa vs. 35 kPa). The intraluminal pressure applied in our FEA simulations was approximately 20 mmHg less than that of [44], which could account for the differences in the wall stresses, in addition to the different constitutive material model. Strain and stress distributions in idealized AAA geometries using the Ogden material model are presented in [43]. The stress distributions in the sac region of these geometries are similar to our models as the maximum stress is found at the bulge (as in Figure 7). They have reported a maximum von Mises stress of 135 kPa, which is less than that reported in [44], but greater than our maximum Cauchy stress, likely due to the different materials used and the wall thickness of the models. Niestrawska et al. [56] also used an idealized geometry with uniform wall thickness. They report on the maximum circumferential and longitudinal Cauchy stresses at the maximum AAA diameter for three different assumptions of fiber dispersions: non-rotationally symmetric dispersion, transversely isotropic dispersion, and isotropic. Although we have used a different constitutive equation that does not include fiber orientation, the location of the maximum stress at the AAA sac matches that of [56].

While the uniform wall thickness of our AAA models (groups E and P) was half of the native model, the maximum wall stress of the N group was 2.7 and 2.2 times greater than the E and P groups, respectively, due to the difference in their corresponding material properties (Table 3). Similarly, the average and minimum wall stresses of the E group (3.3 times and 6.4 times, respectively) and P group (2.5 times and 4.3 times, respectively) were lower than the N group stresses. A change in the biomechanical properties, due to PGG crosslinking, can be evidently inferred from the exhibited stress recovery in the PGG FEA models. The in silico models (similar to the experimental data) showed an increase in wall stress for the PGG-treated AAA compared to the enzymatically digested AAA. However, they also exhibited lower stresses than the native abdominal aorta (Table 3).

4.2. Binding of PGG to Degenerated Arterial ECM

PGG attaches to hydrophobic portions of the proteins by surface adsorption [16], selectively binding to the elastic lamellae [17]. The disrupted aneurysmal ECM could favor the permeation of this polyphenol. Other than ECM stability, the successful inhibition of aneurysmal growth by PGG in AAA rat $CaCl_2$ models in vivo is likely due to (i) its ability to be a radical scavenger [12–14], (ii) lower inflammatory responses [12,17,21,57–59], (iii) acting as a calcium antagonist (blocks inositol 1,4,5-trisphosphate receptors) [60], and (iv) reducing MMP activity [18–21,58]. In general, PGG has been shown to be less toxic than tannic acid- and glutderaldehyde-based treatments [49]. While the present work in some measure replicates the periadventitial [17–19] or intraluminal routes [22] of PGG administration for aneurysm suppression, our goal was to quantify the biomechanical characterization of the PGG-ECM protein binding. In rodent AAA studies, with the exception of Isenburg et al. [17] and Thirugnanasambandam et al. [21], the administration of PGG to the abdominal aorta was systemic and the aneurysm inducing-$CaCl_2$ injury was created periadventitially [18,19]. Conversely, Kloster et al. [22] applied PGG intraluminally to their porcine animal model, after a combined application of elastase and balloon based-mechanical expansion of the abdominal aorta. However, it is uncertain if PGG works better with $CaCl_2$- or elastase-based AAA models. Binding of polyphenols with proline-rich proteins is favorable for pH in the range 3.8–6.0 [50], which potentially makes it difficult for PGG to be transported and distributed through circulatory routes (blood pH is 7.4). This may be a possible explanation for why most AAA animal studies are based on a localized (i.e., not systemic) PGG administration.

The concentration of PGG utilized in our in vitro investigation (0.6 mg/mL), was relatively higher than in the previously reported rodent studies that also utilize this unique polyphenol to prevent pathological aortic dilation [17–19,21]. However, it was lower than the in vivo PGG-saline formulations used by Kloster et al. (0.6 vs. 2.5 or 5 mg/mL) [22]. This high concentration reported in [21] is a possible explanation for their claim of "stiffened arterial system" as a potential side effect of PGG treatment. Moreover, there is a known difference in polyphenol activity due to variation in in vivo vs. in vitro experimental settings. Mechanistically, PGG binds to the enzyme cleaved sites of the arterial ECM and the crosslink formation between the hydrophobic cores of the elastin helps in the stabilization of the elastic lamellae. This has shown to suppress any further aortic dilation in rat $CaCl_2$ aneurysm models [17–19,21]. Our study supports the fact that the change in aortic biomechanical properties is a direct result of this hydrophobic binding of PGG with proline-rich proteins such as elastin and collagen, even though these arterial ECM proteins were partially degraded or cleaved by a collagenase and elastase enzyme treatment. PGG has shown to lower the biomechanical stresses in rodent aneurysm models [21]; however, it is unclear if the increased wall stress, as experienced in all aneurysm models compared to native abdominal aortas, can lead to reversal of the PGG-protein binding behavior. Furthermore, it is unclear if PGG binding efficiency is affected by the degree of arterial ECM damage.

4.3. Limitations

Our in vitro study has several limitations and does not completely replicate the periadventitial [17–19] or intraluminal routes [22] of PGG administration, so a direct comparison with known elastase AAA models is not possible. Further, the simulated aneurysmal matrix (group E) is not an exact replica of the complex ECM degradation observed in human AAA [7,34,61]. The shear components of deformation were considered negligible during the biaxial tensile experiments, similar to a previous study of the human abdominal aorta by Vande Geest et al. [37] Nevertheless, shear calculations from planar biaxial testing of soft tissues are a complex and controversial topic [38,62–64]. Biological tissue specimens begin degrading within a few hours of incubation in saline at room temperature (21–24 °C). It was challenging to maintain the porcine abdominal aortic specimens in a PGG solution at 37 °C for days or weeks, due to the amount of bacterial growth and its associated tissue deterioration that would take place in the solution. Inclusion of other chemicals, such as anti-bacterial agents in the PGG solution, may interfere with the overall chemical reactions and could possibly delay the reaction potential of the polyphenolic compound. In addition, we would like to clarify that enzyme treated arterial specimens underwent more rapid degradation compared to the native specimens. Even after thorough saline washes, the enzymatic degradation of the extracellular matrix continues in vitro (as it is possible for the enzymes to permeate the arterial tissue and continue the degradation process) and this degradation is maximum at body temperature (37 °C). As reported by Gundiah et al. [30], one hour of enzyme digestion is sufficient to yield significant changes in the biomechanics in vitro. At 37 °C, the remnant enzymes in the arterial tissue matrix could have further damaged the structural proteins and led to an altered biomechanical state. Therefore, specimen incubation with PGG at 37 °C, being ideal testing conditions, could have produced more arterial matrix damage than expected in vivo. Further, we found no structural differences between specimens that underwent 12 h vs. 48 h of incubation with PGG solution. Consequently, to minimize changes in the tissue microstructure due to incubation at 37 °C, we opted for 12-hour PGG treatments at 4 °C.

The permeation of PGG in the native arterial ECM is a largely unexplored matter. For example, the permeability of the aneurysmal ECM is likely affected by the poroelastic properties of intraluminal thrombus, and/or potential biophysical interaction of PGG with the thrombus, thereby leading to several unknown queries [9–11,61]. Our computational models are also subject to several important limitations. We did not use a multilayer geometry or a multilayer constitutive material model (e.g., an arterial wall composed of an adventitia, media, and endothelium). Although the native and aneurysm materials show orthotropic behavior, we have implemented a first-order isotropic material model. Using a transversely isotropic (or orthotropic) or a multilayer Holzapfel–Gasser–Ogden model may improve the accuracy of the stress estimations of the present work. The FEA models lacked subject-specificity, although the use of idealized geometries in lieu of non-invasive imaging is properly justified. The intraluminal pressure loading for the FEA simulations was assumed static and spatially homogenous rather than pulsatile. The interaction between blood flow and the vessel wall was also ignored. The wall in the FEA models was impermeable and had a uniform thickness. Many of the limitations in the computational models are mitigated by the used of idealized geometries, acknowledging that the goal of the FEA simulations was to analyze the effect of constitutive material properties while maintaining the abdominal aorta geometry as a control.

5. Conclusions

Due to the absence of adequate non-surgical treatment options for AAA, one possible alternative is to translate the novel PGG-based treatment from the established rodent models to prospective large animal models, and ultimately to clinical trials. In a clinical setting, a PGG-based treatment would be aimed at preventing the progressive increase in aneurysm size and the eventual rupture of the abdominal aorta. From the present work, we can infer that PGG treatment of enzyme-digested porcine aortas leads to stabilization of the arterial ECM and restores some of the tissues' mechanical characteristics. Future investigations will focus on the tissue microstructural changes that may occur due to PGG treatment and the potential translation of this work toward an in vivo application.

Supplementary Materials: The following are available online at http://www.mdpi.com/2306-5354/6/3/58/s1, Figure S1: Histological characterization of porcine arterial specimens from native (A), enzyme treated (B), and PGG treated (C) aortic specimens. Specimens were stained with Verhoeff–Van Gieson stain (VVG) for identification of elastin fibers. Images were taken at 10× magnification using a Leica® microscope (Buffalo Grove, IL, USA); the scale bar size is 297.5 µm for the three frames.

Author Contributions: Conceptualization: S.S.P. and E.A.F.; Data curation: S.S.P., S.P., N.R.P., and G.P.E.; Formal analysis: S.S.P. and S.P.; Funding acquisition: E.A.F.; Investigation: S.S.P., S.P., N.R.P., and G.P.E.; Methodology: S.S.P., S.P., G.R., and E.A.F.; Project Administration: E.A.F.; Resources: E.A.F.; Software: S.S.P. and S.P.; Supervision: G.R. and E.A.F.; Validation: S.S.P., G.R., E.S., and E.A.F.; Visualization: S.S.P. and SP; Writing—original draft: S.S.P., SP, N.R.P., and E.A.F.; Writing—reviewing and editing: S.S.P., S.P., N.R.P., G.R., G.P.E., E.S., and E.A.F.

Funding: This work has been supported by a U.S. National Institutes of Health Award (R01HL121293) and an American Heart Association Collaborative Sciences Award (16CSA28480006). The use of the ANSYS biaxial curve fitting tool is gratefully acknowledged through an educational licensing agreement with Ansys Inc.

Conflicts of Interest: The authors declare no conflict of interest. The funders had no role in the design of the study; in the collection, analyses, or interpretation of data; in the writing of the manuscript, or in the decision to publish the results.

References

1. Samarth, S.R.; Santanu, C.; Judy, S.; Christopher, B.W.; Satish, C.M.; Ender, A.F.; Jose, F.R. Biological, geometric and biomechanical factors influencing abdominal aortic aneurysm rupture risk: A comprehensive review. *Recent Pat. Med. Imaging* **2013**, *3*, 44–59.
2. Raut, S.S.; Chandra, S.; Shum, J.; Finol, E.A. The role of geometric and biomechanical factors in abdominal aortic aneurysm rupture risk assessment. *Ann. Biomed. Eng.* **2013**, *41*, 1459–1477. [CrossRef] [PubMed]
3. Golledge, J.; Norman, P.E.; Murphy, M.P.; Dalman, R.L. Challenges and opportunities in limiting abdominal aortic aneurysm growth. *J. Vasc. Surg.* **2017**, *65*, 225–233. [CrossRef] [PubMed]
4. Carmo, M.; Colombo, L.; Bruno, A.; Corsi, F.R.M.; Roncoroni, L.; Cuttin, M.S.; Radice, F.; Mussini, E.; Settembrini, P.G. Alteration of elastin, collagen and their cross-links in abdominal aortic aneurysms. *Eur. J. Vasc. Endovasc. Surg.* **2002**, *23*, 543–549. [CrossRef] [PubMed]
5. White, J.V.; Haas, K.; Phillips, S.; Comerota, A.J. Adventitial elastolysis is a primary event in aneurysm formation. *J. Vasc. Surg.* **1993**, *17*, 371–380. [CrossRef]
6. Cohen, J.R.; Mandell, C.; Chang, J.B.; Wise, L. Elastin metabolism of the infrarenal aorta. *J. Vasc. Surg.* **1988**, *7*, 210–214. [CrossRef]

7. Tanios, F.; Gee, M.W.; Pelisek, J.; Kehl, S.; Biehler, J.; Grabher-Meier, V.; Wall, W.A.; Eckstein, H.H.; Reeps, C. Interaction of biomechanics with extracellular matrix components in abdominal aortic aneurysm wall. *Eur. J. Vasc. Endovasc. Surg.* **2015**, *50*, 167–174. [CrossRef]
8. Thompson, R.W.; Baxter, B.T. MMP inhibition in abdominal aortic aneurysms. Rationale for a prospective randomized clinical trial. *Ann. N. Y. Acad. Sci.* **1999**, *878*, 159–178. [CrossRef]
9. Swedenborg, J.; Eriksson, P. The intraluminal thrombus as a source of proteolytic activity. *Ann. N. Y. Acad. Sci.* **2006**, *1085*, 133–138. [CrossRef]
10. Vorp, D.A.; Lee, P.C.; Wang, D.H.; Makaroun, M.S.; Nemoto, E.M.; Ogawa, S.; Webster, M.W. Association of intraluminal thrombus in abdominal aortic aneurysm with local hypoxia and wall weakening. *J. Vasc. Surg.* **2001**, *34*, 291–299. [CrossRef]
11. Roy, J.; Labruto, F.; Beckman, M.O.; Danielson, J.; Johansson, G.; Swedenborg, J. Bleeding into the intraluminal thrombus in abdominal aortic aneurysms is associated with rupture. *J. Vasc. Surg.* **2008**, *48*, 1108–1113. [CrossRef] [PubMed]
12. Zhang, J.; Li, L.; Kim, S.H.; Hagerman, A.E.; Lu, J. Anti-cancer, anti-diabetic and other pharmacologic and biological activities of penta-galloyl-glucose. *Pharm. Res.* **2009**, *26*, 2066–2080. [CrossRef] [PubMed]
13. Cao, Y.; Himmeldirk, K.B.; Qian, Y.; Ren, Y.; Malki, A.; Chen, X. Biological and biomedical functions of Penta-O-galloyl-D-glucose and its derivatives. *J. Nat. Med.* **2014**, *68*, 465–472. [CrossRef] [PubMed]
14. Patnaik, S.S.; Simionescu, D.T.; Goergen, C.J.; Hoyt, K.; Sirsi, S.; Finol, E.A. Pentagalloyl Glucose and Its Functional Role in Vascular Health: Biomechanics and Drug-Delivery Characteristics. *Ann. Biomed. Eng.* **2019**, *47*, 39–59. [CrossRef] [PubMed]
15. Luck, G.; Liao, H.; Murray, N.J.; Grimmer, H.R.; Warminski, E.E.; Williamson, M.P.; Lilley, T.H.; Haslam, E. Polyphenols, astringency and proline-rich proteins. *Phytochemistry* **1994**, *37*, 357–371. [CrossRef]
16. Dobreva, M.A.; Frazier, R.A.; Mueller-Harvey, I.; Clifton, L.A.; Gea, A.; Green, R.J. Binding of pentagalloyl glucose to two globular proteins occurs via multiple surface sites. *Biomacromolecules* **2011**, *12*, 710–715. [CrossRef] [PubMed]
17. Isenburg, J.C.; Simionescu, D.T.; Starcher, B.C.; Vyavahare, N.R. Elastin stabilization for treatment of abdominal aortic aneurysms. *Circulation* **2007**, *115*, 1729–1737. [CrossRef] [PubMed]
18. Nosoudi, N.; Chowdhury, A.; Siclari, S.; Karamched, S.; Parasaram, V.; Parrish, J.; Gerard, P.; Vyavahare, N. Reversal of Vascular Calcification and Aneurysms in a Rat Model Using Dual Targeted Therapy with EDTA- and PGG-Loaded Nanoparticles. *Theranostics* **2016**, *6*, 1975–1987. [CrossRef]
19. Nosoudi, N.; Chowdhury, A.; Siclari, S.; Parasaram, V.; Karamched, S.; Vyavahare, N. Systemic Delivery of Nanoparticles Loaded with Pentagalloyl Glucose Protects Elastic Lamina and Prevents Abdominal Aortic Aneurysm in Rats. *J. Cardiovasc. Transl. Res.* **2016**, *9*, 445–455. [CrossRef] [PubMed]
20. Sinha, A.; Nosoudi, N.; Vyavahare, N. Elasto-regenerative properties of polyphenols. *Biochem. Biophys. Res. Commun.* **2014**, *444*, 205–211. [CrossRef] [PubMed]
21. Thirugnanasambandam, M.; Simionescu, D.T.; Escobar, P.G.; Sprague, E.; Goins, B.; Clarke, G.D.; Han, H.-C.; Amezcua, K.L.; Adeyinka, O.R.; Goergen, C.J.; et al. The Effect of Pentagalloyl Glucose on the Wall Mechanics and Inflammatory Activity of Rat Abdominal Aortic Aneurysms. *J. Biomech. Eng.* **2018**, *140*, 084502. [CrossRef] [PubMed]
22. Kloster, B.O.; Lund, L.; Lindholt, J.S. Inhibition of early AAA formation by aortic intraluminal pentagalloyl glucose (PGG) infusion in a novel porcine AAA model. *Ann. Med. Surg.* **2016**, *7*, 65–70. [CrossRef] [PubMed]
23. Hynecek, R.L.; DeRubertis, B.G.; Trocciola, S.M.; Zhang, H.; Prince, M.R.; Ennis, T.L.; Kent, K.C.; Faries, P.L. The creation of an infrarenal aneurysm within the native abdominal aorta of swine. *Surgery* **2007**, *142*, 143–149. [CrossRef] [PubMed]
24. Sadek, M.; Hynecek, R.L.; Goldenberg, S.; Kent, K.C.; Marin, M.L.; Faries, P.L. Gene expression analysis of a porcine native abdominal aortic aneurysm model. *Surgery* **2008**, *144*, 252–258. [CrossRef] [PubMed]
25. Czerski, A.; Bujok, J.; Gnus, J.; Hauzer, W.; Ratajczak, K.; Nowak, M.; Janeczek, M.; Zawadzki, W.; Witkiewicz, W.; Rusiecka, A. Experimental methods of abdominal aortic aneurysm creation in swine as a large animal model. *J. Physiol. Pharmacol.* **2013**, *64*, 185–192. [PubMed]
26. Hauzer, W.; Czerski, A.; Zawadzki, W.; Gnus, J.; Ratajczak, K.; Nowak, M.; Janeczek, M.; Witkiewicz, W.; Niespielak, P. The effects of aneurysm repair using an aortic prosthesis on the electrical parameters of the muscular layer of the abdominal aorta. *J. Physiol. Pharmacol.* **2014**, *65*, 853–858. [PubMed]

27. Lysgaard Poulsen, J.; Stubbe, J.; Lindholt, J.S. Animal Models Used to Explore Abdominal Aortic Aneurysms: A Systematic Review. *Eur. J. Vasc. Endovasc. Surg.* **2016**, *52*, 487–499. [CrossRef]
28. Kratzberg, J.A.; Walker, P.J.; Rikkers, E.; Raghavan, M.L. The effect of proteolytic treatment on plastic deformation of porcine aortic tissue. *J. Mech. Behav. Biomed. Mater.* **2009**, *2*, 65–72. [CrossRef]
29. Chow, M.J.; Choi, M.; Yun, S.H.; Zhang, Y. The effect of static stretch on elastin degradation in arteries. *PLoS ONE* **2013**, *8*, e81951. [CrossRef]
30. Gundiah, N.; Babu, A.R.; Pruitt, L.A. Effects of elastase and collagenase on the nonlinearity and anisotropy of porcine aorta. *Physiol. Meas.* **2013**, *34*, 1657–1673. [CrossRef]
31. Riches, K.; Angelini, T.G.; Mudhar, G.S.; Kaye, J.; Clark, E.; Bailey, M.A.; Sohrabi, S.; Korossis, S.; Walker, P.G.; Scott, D.J.; et al. Exploring smooth muscle phenotype and function in a bioreactor model of abdominal aortic aneurysm. *J. Transl. Med.* **2013**, *11*, 208. [CrossRef] [PubMed]
32. Zeinali-Davarani, S.; Chow, M.J.; Turcotte, R.; Zhang, Y. Characterization of biaxial mechanical behavior of porcine aorta under gradual elastin degradation. *Ann. Biomed. Eng.* **2013**, *41*, 1528–1538. [CrossRef] [PubMed]
33. Schriefl, A.J.; Schmidt, T.; Balzani, D.; Sommer, G.; Holzapfel, G.A. Selective enzymatic removal of elastin and collagen from human abdominal aortas: Uniaxial mechanical response and constitutive modeling. *Acta Biomater.* **2015**, *17*, 125–136. [CrossRef] [PubMed]
34. Vande Geest, J.P.; Sacks, M.S.; Vorp, D.A. The effects of aneurysm on the biaxial mechanical behavior of human abdominal aorta. *J. Biomech.* **2006**, *39*, 1324–1334. [CrossRef] [PubMed]
35. Matthews, P.B.; Azadani, A.N.; Jhun, C.S.; Ge, L.; Guy, T.S.; Guccione, J.M.; Tseng, E.E. Comparison of porcine pulmonary and aortic root material properties. *Ann. Thorac. Surg.* **2010**, *89*, 1981–1988. [CrossRef] [PubMed]
36. O'Leary, S.A.; Kavanagh, E.G.; Grace, P.A.; McGloughlin, T.M.; Doyle, B.J. The biaxial mechanical behaviour of abdominal aortic aneurysm intraluminal thrombus: Classification of morphology and the determination of layer and region specific properties. *J. Biomech.* **2014**, *47*, 1430–1437. [CrossRef] [PubMed]
37. Vande Geest, J.P.; Sacks, M.S.; Vorp, D.A. Age dependency of the biaxial biomechanical behavior of human abdominal aorta. *J. Biomech. Eng.* **2004**, *126*, 815–822. [CrossRef]
38. Macrae, R.A.; Miller, K.; Doyle, B.J. Methods in Mechanical Testing of Arterial Tissue: A Review. *Strain* **2016**, *52*, 380–399. [CrossRef]
39. Lee, J.M.; Ku, M.; Haberer, S.A. The bovine pericardial xenograft: III. Effect of uniaxial and sequential biaxial stress during fixation on the tensile viscoelastic properties of bovine pericardium. *J. Biomed. Mater. Res.* **1989**, *23*, 491–506. [CrossRef]
40. Langdon, S.E.; Chernecky, R.; Pereira, C.A.; Abdulla, D.; Lee, J.M. Biaxial mechanical/structural effects of equibiaxial strain during crosslinking of bovine pericardial xenograft materials. *Biomaterials* **1999**, *20*, 137–153. [CrossRef]
41. Ogden, R.W. Large Deformation Isotropic Elasticity: On the Correlation of Theory and Experiment for Compressible Rubberlike Solids. *Proc. R. Soc. A Math. Phys. Eng. Sci.* **1972**, *320*, 567–583. [CrossRef]
42. Ogden, R.W. *Non-Linear Elastic Deformations*; Dover Publications: Mineola, NY, USA, 1997.
43. Callanan, A.; Morris, L.G.; McGloughlin, T.M. Finite element and photoelastic modelling of an abdominal aortic aneurysm: A comparative study. *Comput. Methods Biomech. Biomed. Eng.* **2012**, *15*, 1111–1119. [CrossRef] [PubMed]
44. Azar, D.; Ohadi, D.; Rachev, A.; Eberth, J.F.; Uline, M.J.; Shazly, T. Mechanical and geometrical determinants of wall stress in abdominal aortic aneurysms: A computational study. *PLoS ONE* **2018**, *13*, e0192032. [CrossRef] [PubMed]
45. Geuzaine, C.; Remacle, J.F. Gmsh: A 3-D finite element mesh generator with built-in pre- and post-processing facilities. *Int. J. Numer. Methods Eng.* **2009**, *79*, 1309–1331. [CrossRef]
46. Si, H. TetGen, a Delaunay-Based Quality Tetrahedral Mesh Generator. *ACM Trans. Math. Softw.* **2015**, *41*, 1–36. [CrossRef]
47. Maas, S.A.; Ellis, B.J.; Ateshian, G.A.; Weiss, J.A. FEBio: Finite elements for biomechanics. *J. Biomech. Eng.* **2012**, *134*, 011005. [CrossRef] [PubMed]
48. Chauhan, S.S.; Gutierrez, C.A.; Thirugnanasambandam, M.; De Oliveira, V.; Muluk, S.C.; Eskandari, M.K.; Finol, E.A. The Association Between Geometry and Wall Stress in Emergently Repaired Abdominal Aortic Aneurysms. *Ann. Biomed. Eng.* **2017**, *45*, 1908–1916. [CrossRef] [PubMed]

49. Isenburg, J.C.; Karamchandani, N.V.; Simionescu, D.T.; Vyavahare, N.R. Structural requirements for stabilization of vascular elastin by polyphenolic tannins. *Biomaterials* **2006**, *27*, 3645–3651. [CrossRef] [PubMed]
50. Charlton, A.J.; Baxter, N.J.; Khan, M.L.; Moir, A.J.; Haslam, E.; Davies, A.P.; Williamson, M.P. Polyphenol/peptide binding and precipitation. *J. Agric. Food Chem.* **2002**, *50*, 1593–1601. [CrossRef] [PubMed]
51. Chow, M.J.; Mondonedo, J.R.; Johnson, V.M.; Zhang, Y. Progressive structural and biomechanical changes in elastin degraded aorta. *Biomech. Model. Mechanobiol.* **2013**, *12*, 361–372. [CrossRef]
52. Baxter, N.J.; Lilley, T.H.; Haslam, E.; Williamson, M.P. Multiple interactions between polyphenols and a salivary proline-rich protein repeat result in complexation and precipitation. *Biochemistry* **1997**, *36*, 5566–5577. [CrossRef] [PubMed]
53. Bennick, A. Interaction of plant polyphenols with salivary proteins. *Crit. Rev. Oral Biol. Med.* **2002**, *13*, 184–196. [CrossRef] [PubMed]
54. Zulliger, M.A.; Stergiopulos, N. Structural strain energy function applied to the ageing of the human aorta. *J. Biomech.* **2007**, *40*, 3061–3069. [CrossRef] [PubMed]
55. Owen, B.; Bojdo, N.; Jivkov, A.; Keavney, B.; Revell, A. Structural modelling of the cardiovascular system. *Biomech. Model. Mechanobiol.* **2018**, *17*, 1217–1242. [CrossRef] [PubMed]
56. Niestrawska, J.A.; Ch Haspinger, D.; Holzapfel, G.A. The influence of fiber dispersion on the mechanical response of aortic tissues in health and disease: A computational study. *Comput. Methods Biomech. Biomed. Eng.* **2018**, *21*, 99–112. [CrossRef] [PubMed]
57. Kang, D.G.; Moon, M.K.; Choi, D.H.; Lee, J.K.; Kwon, T.O.; Lee, H.S. Vasodilatory and anti-inflammatory effects of the 1,2,3,4,6-penta-O-galloyl-beta-D-glucose (PGG) via a nitric oxide-cGMP pathway. *Eur. J. Pharmacol.* **2005**, *524*, 111–119. [CrossRef] [PubMed]
58. Mendonca, P.; Taka, E.; Bauer, D.; Cobourne-Duval, M.; Soliman, K.F. The attenuating effects of 1,2,3,4,6 penta-O-galloyl-beta-d-glucose on inflammatory cytokines release from activated BV-2 microglial cells. *J. Neuroimmunol.* **2017**, *305*, 9–15. [CrossRef] [PubMed]
59. Mendonca, P.; Taka, E.; Bauer, D.; Reams, R.R.; Soliman, K.F.A. The attenuating effects of 1,2,3,4,6 penta-O-galloyl-β-d-glucose on pro-inflammatory responses of LPS/IFNγ-activated BV-2 microglial cells through NFƙB and MAPK signaling pathways. *J. Neuroimmunol.* **2018**, *324*, 43–53. [CrossRef]
60. Lu, Y.; Deng, Y.; Liu, W.; Jiang, M.; Bai, G. Searching for calcium antagonists for hypertension disease therapy from Moutan Cortex, using bioactivity integrated UHPLC-QTOF-MS. *Phytochem. Anal.* **2019**, *30*, 456–463. [CrossRef]
61. Adolph, R.; Vorp, D.A.; Steed, D.L.; Webster, M.W.; Kameneva, M.V.; Watkins, S.C. Cellular content and permeability of intraluminal thrombus in abdominal aortic aneurysm. *J. Vasc. Surg.* **1997**, *25*, 916–926. [CrossRef]
62. Holzapfel, G.A.; Ogden, R.W. On planar biaxial tests for anisotropic nonlinearly elastic solids. A continuum mechanical framework. *Math. Mech. Solids* **2008**, *14*, 474–489. [CrossRef]
63. Sommer, G.; Haspinger, D.; Andra, M.; Sacherer, M.; Viertler, C.; Regitnig, P.; Holzapfel, G.A. Quantification of Shear Deformations and Corresponding Stresses in the Biaxially Tested Human Myocardium. *Ann. Biomed. Eng.* **2015**, *43*, 2334–2348. [CrossRef] [PubMed]
64. Zhang, W.; Feng, Y.; Lee, C.H.; Billiar, K.L.; Sacks, M.S. A generalized method for the analysis of planar biaxial mechanical data using tethered testing configurations. *J. Biomech. Eng.* **2015**, *137*, 064501. [CrossRef] [PubMed]

bioengineering

MDPI

Review

Current Understanding of the Biomechanics of Ventricular Tissues in Heart Failure

Wenqiang Liu [1] and Zhijie Wang [1,2,*]

1 School of Biomedical Engineering, Colorado State University, Fort Collins, CO 80523, USA; Wenqiang.Liu@colostate.edu
2 Department of Mechanical Engineering, Colorado State University, Fort Collins, CO 80523, USA
* Correspondence: zhijie.wang@colostate.edu

Received: 4 November 2019; Accepted: 18 December 2019; Published: 20 December 2019

Abstract: Heart failure is the leading cause of death worldwide, and the most common cause of heart failure is ventricular dysfunction. It is well known that the ventricles are anisotropic and viscoelastic tissues and their mechanical properties change in diseased states. The tissue mechanical behavior is an important determinant of the function of ventricles. The aim of this paper is to review the current understanding of the biomechanics of ventricular tissues as well as the clinical significance. We present the common methods of the mechanical measurement of ventricles, the known ventricular mechanical properties including the viscoelasticity of the tissue, the existing computational models, and the clinical relevance of the ventricular mechanical properties. Lastly, we suggest some future research directions to elucidate the roles of the ventricular biomechanics in the ventricular dysfunction to inspire new therapies for heart failure patients.

Keywords: myocardium; stiffness; viscoelastic property; anisotropy; fibrosis

1. Introduction

Despite the advances in modern management, heart failure (HF) leads to high mortality and morbidity in the United States. More than 5 million Americans have HF, and around 550,000 new cases occur every year [1,2]. It is shown that the lifetime risk for developing HF at the age of 40 years old is around 20%, and the risk of HF increases with aging. As the number of elderly (≥65 years old) is expected to grow to 70.3 million in 2030, the prevalence of HF will continue to increase [1,3–5]. Economically, HF is the leading cause of hospitalization [2], with more than $33 billion in expenses annually in the United States. [1]; and in developed countries, the burden of HF is likely to keep increasing [4,6].

Ventricle dysfunction is the most common cause of heart failure, including left-sided HF with preserved ejection fraction (HFpEF) and reduced ejection fraction (HFrEF), as well as right-sided HF secondary to pulmonary hypertension and congenital heart disease (CHD) [7–14]. The malfunction of the myocardium in these diseases can occur in the left ventricle (LV), right ventricle (RV), or both ventricles (biventricular HF). It is known that the LV and RV have distinct embryological, geometrical, and structural properties [15–17], and the mechanism of RV failure is likely to be different than that of LV failure [18]. However, compared with LV failure, RV failure has been less understood, and it remains unclear if the two ventricles present similar mechanical behaviors or adaptations in the pathogenesis of ventricular dysfunction.

It is generally accepted that the mechanical property of the myocardium is an important determinant of the ventricular function [19,20]. Indeed, changes in the ventricular mechanical properties during the HF progression have been reported in numerous studies for both LVs and RVs. The alteration of the extracellular environment can result in the dysfunctions of cardiac cells, and thus the overall organ function is impaired, which forms a vicious cycle in the maladaptive remodeling of the ventricle.

Therefore, it is critical to unravel the roles of the tissue biomechanics in the ventricular dysfunction to inspire new therapies for HF patients. In this review, we will summarize the methodologies of the mechanical measurement of ventricle free walls, as well as the current understanding of ventricular mechanical properties including the tissue viscoelasticity, the existing computational models, and the clinical relevance of the biomechanical properties of ventricles. In particular, we have discussions on the right ventricle and the dynamic mechanical properties of the tissue—viscoelasticity, both of which have received less attention in the current research on cardiac biomechanics. Finally, future directions are suggested to advance the understanding of the biomechanical mechanisms of the heart failure in systemic and pulmonary circulations.

2. Characterization of the Mechanical Behavior of Ventricles

2.1. *Ex Vivo Measurements*

The ventricular free wall is known as an anisotropic and viscoelastic material, which means it has different mechanical behaviors in different directions and presents both elastic and viscous features in dynamic deformations (Figure 1). Depending on the mechanical behavior to measure, the mechanical tests can be uniaxial or biaxial (for anisotropic behavior), static or dynamic (for elastic or viscoelastic behavior), and in different testing conditions (e.g., bath medium, temperature, preconditioning protocol, removal of residue stress). To obtain the viscoelastic properties, either stress relaxation/creep tests or cyclic tensile mechanical tests can be used (Figure 2, detailed discussions in 2.1.2 and 2.1.3). Then, the viscous behavior is quantified to capture the time-, strain rate-, or frequency-dependent character [18]. In this review, we will focus on the macroscopic mechanical measurements, and thus the experimental methods using atomic force microscopy (AFM) or length-tension tests on the isolated cardiac muscle (e.g., papillary muscle) or cardiomyocytes are not included.

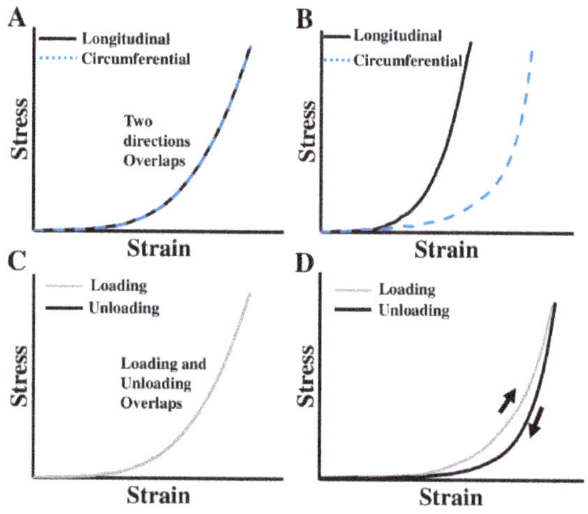

Figure 1. (**A**)–(**B**) Stress–strain curves obtained from different directions in isotropic and anisotropic materials, respectively; (**C**)–(**D**) stress–strain curves obtained from loading and unloading periods of cyclic deformation in nonlinear elastic and viscoelastic materials, respectively.

2.1.1. Preconditioning and Residual Stress Measurement

As needed in other biological tissues' mechanical tests, preconditioning is often performed prior to the data acquisition to ensure a constant and accurate mechanical behavior of the tissue [21]. This procedure has been described in the mechanical tests of cardiac tissues [17,22–29]. The number of preconditioning cycles in the biaxial/uniaxial tests varied among 5–10 cycles for the animal (canine, bovine, and murine) myocardium [17,22,24–29], whereas Sommer et al. and Fatemifar et al. showed that, after 3–5 cycles, the human heart tissue reached stable biaxial behavior [23,30]. Owing to the viscoelastic nature of the tissue, a sufficient resting period should be given between the tests. It is suggested that ten times of the previous mechanical testing period is appropriate for the tissue to be free from the 'memory' of former deformations [31,32].

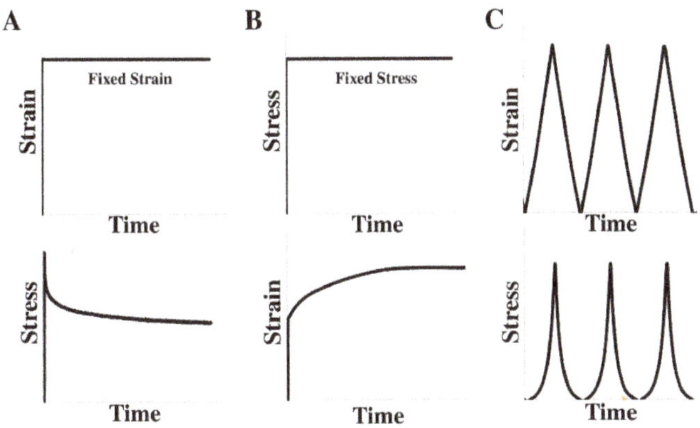

Figure 2. Different mechanical tests for the viscoelastic properties measurement. (**A**) Stress relaxation test, (**B**) creep test, and (**C**) displacement-controlled cyclic tensile mechanical test. The upper panels illustrate the mechanical inputs and the lower panels illustrate the mechanical responses of the material in these tests.

Residual stress is the stress that remains in the tissue after all external loads are removed [33]. The presence of residual stress in myocardium has been observed in both large animal (porcine) and small animal (rat) ventricles [34–37]. The exact cause of residual stress in biological tissues is not fully clear, but the different growth rates at different layers or directions of the tissue are likely the reason [36]. Residual stress is generally considered 'beneficial' to the tissue. From the study of opening angle in an arterial ring, it is found that the presence of residual stress leads to a homogenous distribution of the circumferential wall stress through the vessel thickness [38]. For myocardium, Shi et al. measured the residual stress by a curling angle characterization and found that the residual stress protected the ventricle wall by reducing myocardial stress during LV diastolic expansion [34]. The measurement of residual stress in myocardium is seldom seen in ex vivo mechanical tests and future experimental studies may consider to include such measurement.

2.1.2. Uniaxial and Biaxial Tensile Mechanical Tests

Uniaxial and biaxial mechanical tests are the most common methods to investigate the ventricular mechanical property after tissue harvest (Figure 3). While the uniaxial mechanical test offers a quicker and easier examination of the material mechanical property, the biaxial mechanical test better mimics the in vivo loading conditions and provides more comprehensive measurements of the anisotropic mechanical behavior [17,19,20,22,39–41]. Both methods have been used in prior studies of LV and RV mechanical properties [17,20,22–30,39–46] (please see Table 1 for a summary of these studies).

Furthermore, when the entire cycle of stress–strain data is used (i.e., including loading and unloading curves), the ventricular viscoelastic behavior can be derived from the hysteresis stress–strain loop (Figure 4). However, the biaxial measurement of viscoelasticity is less common than the elasticity measurement and only sporadic studies have examined canine [39], porcine [46], and human ventricles [23]. Recently, the viscoelasticity of neonatal porcine LVs and RVs was obtained using the cyclic uniaxial mechanical tests. The myocardial hysteresis was quantified by the ratio of the area enclosed in the hysteresis loop over the area beneath the loading curve, but the elasticity of these ventricles was not quantified [46]. To our knowledge, the first human myocardium viscoelastic behavior quantified by biaxial testing was reported by Sommer et.al. [23]. Increased stress and hysteresis area were evident with increased stretch rate (from 3 mm/min to 30 mm/min), but no viscoelastic property (e.g., elasticity or viscosity) was quantified from these biaxial tests.

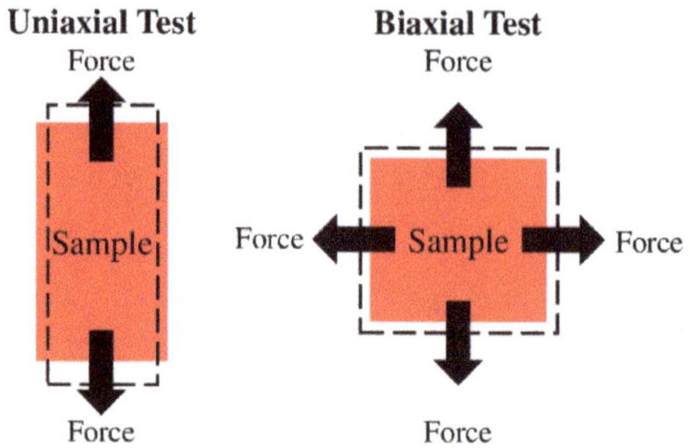

Figure 3. Diagrams of the uniaxial (**left**) and biaxial (**right**) tissue mechanical tests. Dashed rectangles illustrate the deformed configurations of the sample after the mechanical stretch.

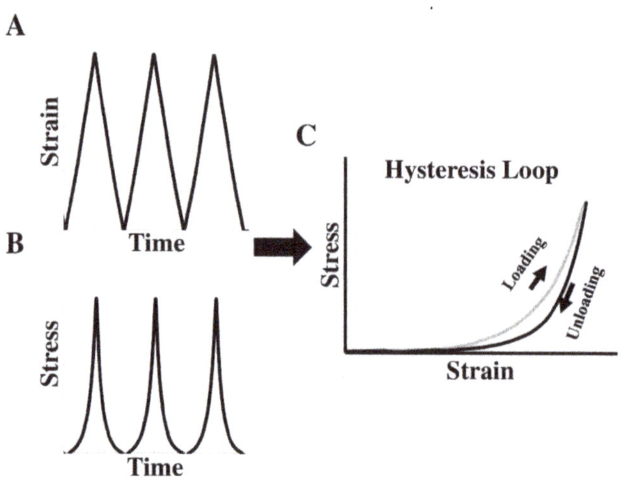

Figure 4. Hysteresis stress–strain loop obtained from the cyclic tensile mechanical tests. Triangle or sinusoidal mechanical loadings are typically applied during the cyclic stretches. (**A**)–(**B**) Representative strain and stress curves as a function of time in the tensile mechanical test; (**C**) representative hysteresis loop derived from the synchronized stresses and strains in (**A**) and (**B**).

2.1.3. Stress Relaxation and Creep Tests

Stress relaxation and creep tests are traditional methods to measure the viscoelasticity of soft tissues such as tendon, cartilage, and heart valves [47–53]. The stress relaxation test is the recording of a time-dependent stress reduction under a fixed strain/stretch, whereas the creep test is the recording of a time-dependent strain increase under a constant stress/load (Figure 2) [18]. These methods have been applied to the myocardium or papillary muscle [23,54–58], although slightly different testing protocols and conditions were adopted (please see Table 2).

Table 1. Summary of the prior biaxial/uniaxial tensile mechanical tests performed in ventricular tissues. The experimental details on testing methods and conditions are listed as well. Viscoelastic mechanical studies are marked with * in the Methods. CPS: Cardioplegia Solution; BDM: 2,3-butanedione monoxime; PBS: Phosphate-buffered saline.

Sample	Method	Axial Definition	Preconditioning Cycles	Strain Range/Rate	Bath Medium	Temperature	Immerse Condition
Canine RV [17]	Biaxial	Main fiber direction	10	30% /	Water with recycle required oxygenated carioplegic solution	Room temperature	Immersed
Rat RV [20]	Biaxial	Outflow tract	/	/ /	Modified Kreb's solution with 2,3-butanedione 2-monoxime and oxygen	/	Immersed
Bovine LV/RV [22]	Biaxial and uniaxial	Main fiber direction	5	20% 0.10.75 cm/s	Saline with O_2 and CO_2 (pH = 7.4)	Physiological range	Immersed
Human LV/RV [23]	Biaxial* and Triaxial	Main fiber direction	4	20% Quasi-static	CPS with 20 mM BDM	37 °C	Immersed
Canine LV [24]	Biaxial	Main fiber direction	≥7	/ 50s/cycle	Modified Kreb's Ringers solution with a ~ 10 mM potassium, O_2, and CO_2 (pH = 7.4)	30 °C	Float
Canine LV [25]	Biaxial	Main fiber direction	5–7	20% 0.05 or 0.1Hz	Bath containing the oxygenated solution	Room temperature	Immersed
Canine LV [26]	Biaxial	Main fiber direction	7–10	5%–27% 0.1 Hz	Oxygenated cardioplegic solution	Room temperature	Immersed
Rabbit LV [27]	Biaxial	Main fiber direction	Several	/	BDM–Krebs solution	/	Immersed
Ovine LV [28]	Biaxial	/	10	20%–25% 0.5 Hz	Isotonic cardioplegic solution (pH:7.4)	20 °C	Immersed
Murine RV [29]	Biaxial	Outflow tract	10	5–25 kPa /	Modified Kreb's solution with BDM	Room temperature	Immersed
Human LV/RV [30]	Biaxial and uniaxial	Main fiber direction	5	40% ~6 mm/min	Phosphate-buffered saline (PBS)	37 °C	Immersed
Canine LV [39]	Biaxial and uniaxial*	Main fiber direction	9	/ 0.0025–0.25 mm/s	Tyrode solution with O_2 and CO_2 (pH:7.4)	29.5–30.5 °C	Float
Rat RV [40]	Biaxial	Outflow tract	/	/ /	Modified Kreb's solution with BDM and oxygen	Room temperature	Immersed
Canine LV/RV [43]	Biaxial	Apex to base	/	/ /	Oxygenated solution	Room temperature	Immersed
Ovine LV/RV [44]	Biaxial	Main fiber direction	10	40% 8 s per cycle	Saline bath	37 °C	Immersed
Rat LV [45]	Biaxial and uniaxial	/	10	/ 0.5mm/s	PBS	37 °C	Submerged
Porcine LV/RV [46]	Biaxial and uniaxial*	Main fiber direction	/	/ 0.5mm/s	PBS	37 °C	Submerged

Table 2. Summary of the prior studies with stress relaxation or creep tests on ventricular tissues. LV, left ventricle; RV, right ventricle. CPS: Cardioplegia Solution; BDM: 2,3-butanedione monoxime; PBS: Phosphate-buffered saline; KHB: Krebs-Henseleit buffer.

Sample	Method	Ramp Speed	Stretch Level	Duration	Bath Condition
Human LV/RV [23]	Stress relaxation	100 mm/min	10%	5 min	CPS with 20 mM BDM at 37 °C
Rabbit LV papillary muscle [54]	Stress relaxation and creep	/	/	5 min	Ringer–Lacke solution with O_2, CO_2, pH = 7.38
Cats, Rabbits papillary muscle; Frog and Turtle LV [55]	Stress relaxation and creep	/	20%, 30%	/	Tyrode solution with O_2, CO_2, pH = 7.3, at 24 °C (for papillary muscles); Modified PBS solution at pH = 7.3 (for LVs)
Chicken embryonic heart [56]	Stress relaxation	Fast linear	10%, 20%, 40%	10 min	Oxygenated KHB–CPS at 35 °C
Chicken LV/RV [57]	Stress relaxation	1000% axial strain/s	5%, 10%, 20%, 30%	5 min	Oxygenated KHB–CPS at 35 °C
Cat LV papillary muscle [58]	Stress relaxation	/	/	/	Oxygenated Kreb's–Ringer's solution at 20 °C

2.2. *In Vivo Measurements*

The ex vivo measurement discussed above can provide better controls of the experimental conditions (e.g., the strain rage, cardiac muscle tone) and eliminate the interference of physiological factors (e.g., blood pressure, heart rate, hormone levels) in the mechanical properties of the ventricle tissues. However, the ex vivo tests require tissue removal and are often limited by the contractile state of cardiomyocytes (passive only) and the configuration of the tissue (non-physiological stretches in the biaxial directions). Therefore, the in vivo measurements could provide useful information of the tissue mechanical behavior that is absent in the ex vivo conditions.

2.2.1. The Elasticity Measurement

At the whole-organ level, pressure–volume (PV) loop measurement (Figure 5) is the gold standard to assess the ventricle performance invasively by inserting a PV catheter into the ventricle lumen [19]. The end-diastolic pressure–volume relation (EDPVR) derived from the steady-state PV loops is often used to represent the ventricular passive stiffness. Similarly, diastolic stiffness can also be estimated by the ratio of end-diastolic pressure (EDP) to end-diastolic volume (EDV) [59,60]. In addition, chamber compliance, which is the ratio of ventricular volume change over pressure change during a cardiac cycle ($\Delta V/\Delta P$), has also been used to describe the ventricle stiffness [61–63]. Another type of ventricular elastance, end-systolic pressure-volume relation (ESPVR or Ees), can be derived from a serious PV loops during a temporal vena cava occlusion or estimated by other formulas with a single beat technique. Ees is considered a measure of load-independent contractility of the ventricle [64]. However, this parameter has also been viewed as an index of *systolic* stiffness of the ventricle [59,60].

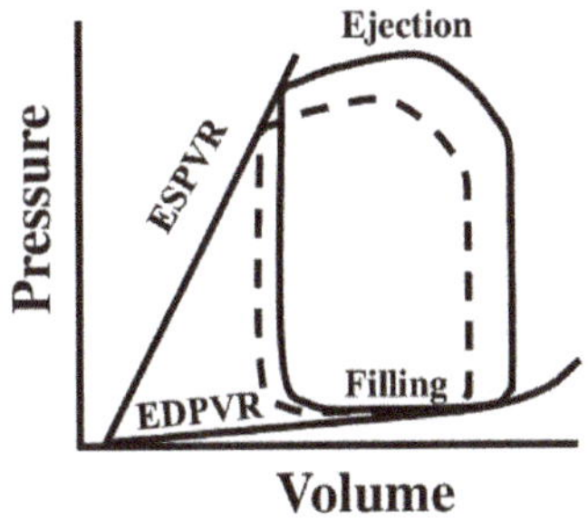

Figure 5. Diagram of the pressure–volume (PV) loop obtained from cardiac catheterization. The loop in solid line denotes a steady-state PV loop, whereas the loop in dotted line denotes a transient loop obtained by brief vena cava occlusion to reduce the ventricle filling. ESPVR: end-systolic pressure-volume relationship; EDPVR: end-diastolic pressure-volume relationship.

Non-invasively, cardiovascular magnetic resonance (CMR) and speckle-tracking echocardiography (STE) are alternative methods to measure the myocardial performance [65–67]. Depend on the imaging technique, 3D geometry is reconstructed and the strain and strain rate are then calculated as the indicators of ventricle stiffness. The in vivo 3D strain analyses can be achieved by applying a so-called hyperelastic warping method to various types of medical images such as cine CMR or echocardiography, from which global or regional myocardial strain can be calculated [68–70]. The hyperelastic warping method is a deformable image registration technique, which uses a deformable finite element mesh to register the target image to the reference image. The reference image is typically selected as the image at the end-diastole [71,72]. Then, the 3D deformation of the ventricular geometry can be derived over a cardiac cycle, and the strains in different directions (longitudinal, circumferential, and radial) are calculated [70–73]. This technique is powerful because it enables the measurement of the myocardial strain temporally and spatially, and both ventricles can be examined at the same time to further investigate the ventricular interactions in HF patients. These strain measurements could potentially offer new diagnostic or prognostic indices for LV or RV dysfunction [70,74]. However, it should be noted that the strain is essentially a measure of relevant deformation of the ventricular chamber, and such deformation is affected by both the passive stiffness and active contraction of the ventricular wall. Therefore, it is not a direct measurement of ventricular stiffness.

The direct non-invasive measurement of ventricular stiffness (e.g., elastic modulus) can be obtained by magnetic resonance elastography (MRE) [75,76]. MRE is a phase contrast magnetic resonance imaging (MRI) technique. The underlying principle of this imaging method is based on the fact that the different stiffness of a material generates different shear wave length. With an induction of shear waves in the tissue region of interest, the waves are encoded in the phase of MR image and the wave images can be converted to the stiffness maps with temporal and spatial information included. MRE has been investigated in animals and a couple of clinical studies to study the effect of myocardial infarction, aging, hypertension, or hypertrophic cardiomyopathy on cardiac stiffness. A good review of MRE in cardiovascular tissues is given by Khan et. al [75]. However, although the methodology has been validated in animals with the gold standard PV loop, the elastic moduli reported in human subjects (<12 kPa) are much lower than the values reported in animals or in ex vivo measurements (in hundreds of kPa) [75]. Thus, more work is warranted in this area.

Finally, with the combination of medical imaging and computational modeling such as finite element methods, it is also possible to estimate the ventricular material properties using 'inverse modeling' [77–82]. These computational methods are briefly reviewed in the works of [83–85].

2.2.2. The Viscoelasticity Measurement

The viscoelasticity of ventricles has been occasionally reported with the measurement of cyclic stress–strain relations. Some early studies measured the viscoelasticity of the LV from healthy canine and human hearts by individual measurements of pressure and volume in vivo [86–88]. Briefly, cardiac catheterization was performed and a micromanometer was introduced into the LV to measure the pressure. In the meanwhile, echocardiogram was performed and the endocardial diameter and the posterior wall thickness were recorded. These data were synchronized and further used to calculate the meridional wall stress and midwall strain during the diastolic phase. Viscoelastic properties in the 'passive' state of the LV were then derived from the nonlinear stress–strain curve using an empirical model of viscoelasticity. Interestingly, these studies were all published in the late 1970s and early 1980s, and there is no further investigation of the in vivo measurement of ventricular viscoelasticity.

2.3. *Basic Behavior of Ventricles—Tissue with Anisotropy and Viscoelasticity*

2.3.1. Anisotropic Behavior of Ventricles

The characterization of the anisotropic behavior of ventricles is highly dependent on the definition of the biaxial coordinate system. To date, there are two main types of coordinate systems: the main fiber and cross-fiber coordinate system [17,30,44], and the outflow tract and cross-outflow tract coordinate system [20,29,40]. Using the former coordinate system, it is consistently observed that the tissue behaves stiffer in the fiber direction compared with the cross-fiber direction [22]. However, the degrees of anisotropy in the ventricles are not consistent among observations. Sacks et al. reported that the canine RV had greater anisotropy than the LV [17]. Similarly, Ahmad et al. found that the neonatal porcine RV had significantly greater anisotropy than the LV in different anatomic regions [89]. However, Javani et al. reported that the ovine LV was more anisotropic than the RV [44]. Ghaemi et al. reported that both LV and RV were anisotropic, but there was no comparison between these chambers [22]. Therefore, there is no consensus about the difference in anisotropic behavior between a healthy LV and RV. The discrepancies may depend on the age and species of samples, methods of tissue selection and preparation and testing protocols. Besides, it has been noted that the determination of the main fiber direction is challenging and could induce variations in the anisotropic behavior as well [17].

Using the second coordinate system, Valdez-Jasso et al. found that the rat RV had greater stiffness in the outflow tract direction compared with the cross-outflow tract direction [29], and Hill et al. found that the degree of rat RV anisotropy increased in the pressure overload state [40].

2.3.2. Viscoelastic Behavior of Ventricles

The viscoelastic property of a material is manifested by the non-overlapping of loading and unloading stress–strain curves over an entire cycle [18]. Such behavior has been observed for both LV and RV tissues [18,39,90], which implies that the ventricular elasticity (or stiffness) is dependent on the strain rate, and there is energy loss during the cyclic deformation owing to the viscous property of the ventricle. Particularly, Ahmad et al. found that the neonatal porcine LV had greater viscoelasticity than the RV, and both ventricles exhibited greater viscoelasticity at the mean-fiber direction compared with the cross-fiber direction [46]. Sommer et al. measured the viscoelastic property of various diseased human LVs, and their findings also showed a larger hysteresis in the mean-fiber direction than the cross-fiber direction [23]. Our own recent study in ovine RVs showed that the chronic pressure overload increased hysteresis (viscosity) in both directions (unpublished data (Figure 6)).

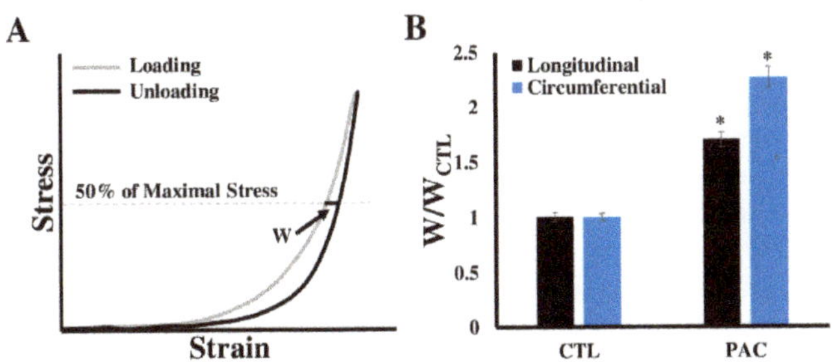

Figure 6. Change in the right ventricle (RV) viscosity after three-month pressure elevation in adult sheep. Pressure elevation was induced by pulmonary artery constriction (PAC). (**A**) Viscosity is defined as the loop width (w) at the 50% of the maximal stress of the loop; (**B**) Loop width normalized by the average loop width of the control RVs in the individual direction. CTL: control; PAC: pulmonary artery constriction. * $p < 0.05$ vs. CTL in the same direction.

2.4. *Computational Modeling of Ventricular Biomechanics*

Both empirical models and constitutive models have been applied to characterize the nonlinear, biaxial mechanical behavior of ventricles. Because of the nonlinear, 'J'-shaped stress–strain curve, the use of an exponential component is common in empirical models. However, these models provide little information on the relations of physical quantities or physiological conditions of the tissue, and thus constitutive models are developed to better describe the myocardium tissue mechanics [42]. With certain assumptions (hyperelasticity, incompressibility, homogeneity, and so on), a strain energy function is defined to relate the mechanical loadings (stress) to the geometry changes (strain). The determination of the strain energy function is the key in constitutive models. On the basis of the model parameters included in the strain energy function, different materials' properties can then be derived. A thorough review of the modeling for tissues biaxial mechanical properties can be found in the works of [41,42].

Classic empirical models to describe the tissue viscoelasticity are composed of springs and dashpots that represent the elastic and viscous behaviors, respectively. The two basic models of these are also known as the Maxwell model (consisting of a spring and a dashpot in series) and Kelvin–Voight model (consisting of a spring and a dashpot arranged in parallel). Different combinations of the spring and dashpot elements have been used to describe the ventricle and papillary muscle viscoelasticity. For example, a spring connected to two Maxwell elements in parallel was used to form a 1D viscoelastic model for the papillary muscle of the LV [58]. An elastic term and a viscous term in parallel were used to describe the viscoelasticity of the LV in different conditions [39,87,88,91,92]. In the constitutive models of ventricular viscoelasticity, a finite element analysis with orthotropic viscoelastic model has been used to describe the passive myocardium viscoelastic behavior [93]. Another option to represent the viscoelastic behavior is by the hereditary (or convolution) integral with a strain-dependent Prony series, which has been found to successfully capture the strain- and time-dependent behavior in *non-cardiovascular* tissues [51,94–96]. A nice review of constitutive models of cardiac tissue viscoelasticity can be found in the literature [93,97,98].

3. Biomechanical Changes of Ventricles in Heart Failure Development

Heart failure is associated with extensive remodeling of the tissue involving changes in extracellular matrix (ECM) (e.g., fibrosis or accumulation of collagen), recruitment of inflammatory cells (e.g., macrophage infiltration), upregulated oxidative stress (e.g., increased ROS), and altered metabolic activity (e.g., increased glycolysis) [99–102]. These changes not only lead to the malfunction of various cells in the myocardium, but also result in the impairment in the mechanical and hemodynamic functions of the organ. Because of our focus on the biomechanical behavior of the ventricle in this review, we will restrict our discussions to the extracellular matrix (ECM) proteins (particularly collagen) as they are the main determinant of mechanical properties including viscoelasticity [18,92,103].

The myocardium ECM consists of proteins such as collagen, elastin, fibronectin, proteoglycan, and laminin. Among these molecules, collagen is the most abundant ECM protein in the adult heart, with at least five different types of collagen (I, III, IV, V, and VI) that have been identified [104]. Types IV and V collagen are mostly found in the basement membrane of the cardiomyocytes, and types I and III collagen are the main constituents in the ECM: type I collagen represents 75%~80% of total collagen content and type III collagen represents approximately 15%~20% of the total collagen [100]. The collagen metabolism, that is, the balance of collagen synthesis and degradation, is regulated by the mechanical loadings (i.e., pressure-overload, volume-overload) and leads to rapid changes in cardiac ECM and mechanical properties [105–107].

Ventricular fibrosis (i.e., collagen accumulation) is frequently observed in cardiac remodeling in both LV failure and RV failure [104,108,109], and the cessation of the accumulation or cross-linking of collagen has been shown to reverse the maladaptive remodeling and improve ventricular function [107,110,111]. However, the story about collagen accumulation is not as simple as firstly viewed if more aspects are considered. For example, in the late stage of HF with LV dilation and wall thinning, conflicting results are given in collagen metabolism: some report that (type I) collagen is degraded and the extent of collagen cross-linking is reduced [106,112,113], whereas other report elevated collagen content or cross-linking [111,114]. In response to pressure overload, the findings on LV collagen deposition are not consistent either: increased collagen [115], decreased collagen [116,117], and no change in collagen [118] in the ventricles were all reported. During the progression of RV dysfunction in pulmonary hypertension, the total collagen was increased with respect to time, but the percentage of collagen cross-linking was decreased [61]. This suggests that the role of collagen content and cross-linking in RV dysfunction may be different. Overall, the variations in collagen deposition depending on the etiology or the specific phase of the heart disease development suggest that collagen metabolism is a key factor contributing to the heterogeneity of the heart failure. Therefore, further examination of the collagen metabolism in LV/RV failure progression is required.

While these previous studies investigated the role of fibrosis in the HF progression, the link of collagen deposition to the mechanical changes is another open area of research. Some biomechanical studies have quantified both biaxial mechanical properties and collagen/myo-fiber orientation in the ventricle (mouse RV, infarcted LV) [23,29,45]. However, how the collagen orientation or total amount is correlated with the ventricular anisotropy or elasticity remains unknown. We recently exposed the ovine RVs to pressure overload using a pulmonary artery constriction model. The chronic remodeling of the RV led to increased collagen deposition. More interestingly, we observed a larger increase in type III collagen than in type I collagen (unpublished data) (Figure 7). Further investigations on the structure–function relations of the ventricles in different physiological conditions will provide more insights into the role of fibrosis in heart failure development.

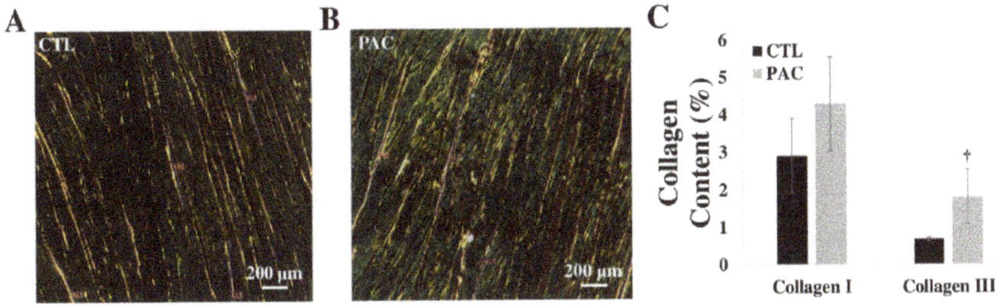

Figure 7. Collagen accumulation in hypertensive ovine RVs. (**A**)–(**B**) Representative histology images of the Picro Sirius Red staining of RVs in control and hypertensive groups, respectively; (**C**) Increase in type III collagen in the hypertensive RVs. CTL: control; PAC: pulmonary artery constriction. † $p = 0.05$ vs. CTL.

4. Clinical Relevance of Ventricular Mechanical Alterations

4.1. Significance of Ventricular Stiffening in Heart Failure

In chronic heart diseases, the myocardial structure and morphology changes lead to the stiffening of the ventricles [7–9,119–121]. These mechanical changes are considered as the changes in the passive mechanical behavior of the tissue, which is often related to the diastolic dysfunction [87,122]. The stiffening of the ventricle impedes the filling of blood during diastole, and thus leads to an increased filling pressure (EDP) at the same chamber volume. This is a key mechanism for the progression of LV dysfunction, particularly in heart failure with preserved ejection fraction (HFpEF) [122,123]. Recently, it was further demonstrated that the increase in passive stiffness proceeds the LV diastolic dysfunction [124]. Consequently, reducing LV stiffness has become one therapeutic target for HFpEF patients [125]. Ventricular stiffening also occurs in other conditions such as hypertension, aging, and hypertrophic cardiomyopathy [59,75], and the former two conditions are well-known risk factors of heart failure.

In addition, the increased passive stiffness could result in an increase in stiffness during the systolic contraction, which is why the age-related increases in Ees (ESPVR, the elastance at systole) and EDPVR are correlated, regardless of the changes in arterial load [60]. While Ees is considered as a measure of ventricular contractility, it is possible that the systolic function of the ventricle is affected by the passive stiffness. Indeed, reduced LV strains in the longitudinal and circumferential directions have been reported in HFpEF patients compared with normal and hypertensive heart disease patients, which indicates the stiffening of the LV. Furthermore, these strains were correlated to the LV systolic function (ejection fraction), but not the diastolic function (E′ or E/E′), suggesting a link of the LV strain (indicator of stiffness) with the systolic performance [126]. However, whether and how the systolic function is altered by the increased passive stiffness in different etiologies of LV failure remains largely unexplored.

Finally, the stiffening of LV could impact on the pulmonary circulation as well. Pulmonary edema and elevation in pulmonary venous pressures are observed as a result of the backward transmission of elevated left-sided pressures into the pulmonary circulation. This leads to the development of post-capillary pulmonary hypertension (PH), which is commonly found in HFpEF patients [127]. Therefore, both ventricles become dysfunctional, and this is probably why HFpEF is a more challenging type of heart failure to manage.

RV stiffening is consistently observed in a variety of PH etiologies as well as left-sided heart failure. Using non-invasive echocardiography, reduced RV longitudinal strains have been reported in pre-capillary PH (pulmonary arterial hypertension) patients and PH patients with other etiologies [128–130]. Increased RV stiffness was frequently reported in the preclinical studies of PH via the ex vivo tissue mechanical tests [20,29,40]. However, the impact of RV stiffening in the ventricular performance is rarely investigated. Recently, a correlation of RV longitudinal elastic modulus and the end-diastolic volume (EDV) was found in rodent RVs during PH development [20]. This is the first study to correlate the RV mechanics to the hemodynamic function of the organ. In another study of patient-specific biventricular constitutive modeling, a ratio of RVEDV/LVEDV was found to increase with increased RV free wall stiffness in PH patients, and this new index was strongly and inversely correlated with the RV peak contractility [131]. A following study from the same group suggested that this index can be used to estimate RV contractility [74]. Therefore, in both the left and right sides of the heart, the passive mechanical behavior is linked to the diastolic function as well as the contractility of the ventricle. This suggests that the improvement in the tissue mechanics may be a therapeutic target for heart failure patients.

4.2. Significance of Altered Ventricular Viscoelasticity in Heart Failure

The viscoelastic properties of the ventricle can impact the in vivo function. To date, the discussion of the relevance of ventricular viscoelasticity is mainly restricted to the diastolic function. Firstly, because the viscoelastic property is strain-rate dependent and because the early and later diastole have different filling rates, the diastolic function of the ventricle is time-dependent [86,87,132,133]. Furthermore, evidence has shown that the viscoelasticity of the ventricle changes from normal to diseased states. Increased viscosity of the LV has been reported in different types of patients (severe aortic regurgitation, congestive cardiomyopathy with preserved and reduced ejection fraction) with dilated, hypertrophy LVs [86]. Our preliminary data in pressure-overloaded ovine RVs also showed an increased viscosity in both biaxial directions compared with the healthy RVs (Figure 6). While these data indicate a change of tissue viscoelasticity in HF progression, the exact role of the viscous property in the ventricular function is not well understood.

5. Future Directions

It is well accepted that the passive mechanical properties of the ventricle are important for the diastolic function, and thus heart diseases with a change in myocardial mechanical properties are often associated with diastolic dysfunction [134–138]. However, if and how much of the systolic function is affected by the passive mechanical properties remain unclear. Second, the energy consumption of the tissue could also be affected by the mechanical properties of the tissue (e.g., viscosity) as the cyclic deformation involves energy storage, release, and dissipation. It is thus necessary to explore the energy consumption at the tissue level and how the use of energy at the organ/tissue level is related to the metabolism of individual cardiomyocytes. Overall, the comprehensive understanding of the relationship between the mechanical behavior and the ventricle performance awaits further investigations. Third, the research on ventricular viscoelasticity has been limited in the current literature. Future studies should characterize the viscoelastic properties of the ventricles at different physiological and pathological conditions and elucidate the role of acellular and cellular components in tissue viscoelastic properties. Finally, the RV, known as the 'forgotten chamber', has been less investigated compared with the left compartment. The understanding of the mechanical properties of the RV and their changes in RV failure progression will deepen the insights of the pathogenesis of RV failure or biventricular failure.

Author Contributions: Writing—original draft, W.L. and Z.W.; Writing—review and editing, W.L. and Z.W.; Funding acquisition—Z.W.; Project administration, Z.W. All authors have read and agreed to the published version of the manuscript.

Funding: Z.W. was supported by the institutional start-up funds from Department of Mechanical Engineering at Colorado State University.

Conflicts of Interest: The authors declare no conflict of interest.

References

1. Wayne, R.; Katherine, F.; Karen, F.; Alan, G.; Kurt, G.; Nancy, H.; Susan, M.H.; Michael, H.; Virginia, H.; Brett, K.; et al. Heart Disease and Stroke Statistics—2008 Update. *Circulation* **2008**, *117*, e25–e146.
2. Douglas, D.S.; Emelia, J.B.; Gregg, C.F.; Harlan, M.K.; Daniel, L.; George, A.M.; Jagat, N.; Eileen, S.S.; James, B.Y.; Yuling, H. Prevention of Heart Failure. *Circulation* **2008**, *117*, 2544–2565.
3. Lloyd-Jones, D.M.; Larson, M.G.; Leip, E.P.; Beiser, A.; D'Agostino, R.B.; Kannel, W.B.; Murabito, J.M.; Vasan, R.S.; Benjamin, E.J.; Levy, D. Lifetime Risk for Developing Congestive Heart Failure. *Circulation* **2002**, *106*, 3068–3072. [CrossRef] [PubMed]
4. Ho, K.K.L.; Pinsky, J.L.; Kannel, W.B.; Levy, D. The epidemiology of heart failure: The Framingham Study. *J. Am. Coll. Cardiol.* **1993**, *22*, A6–A13. [CrossRef]
5. Schocken, D.D.; Arrieta, M.I.; Leaverton, P.E.; Ross, E.A. Prevalence and mortality rate of congestive heart failure in the United States. *J. Am. Coll. Cardiol.* **1992**, *20*, 301–306. [CrossRef]
6. Levy, D.; Kenchaiah, S.; Larson, M.G.; Benjamin, E.J.; Kupka, M.J.; Ho, K.K.L.; Murabito, J.M.; Vasan, R.S. Long-Term Trends in the Incidence of and Survival with Heart Failure. *N. Engl. J. Med.* **2002**, *347*, 1397–1402. [CrossRef] [PubMed]
7. Yancy Clyde, W.; Mariell, J.; Biykem, B.; Javed, B.; Casey Donald, E.; Drazner Mark, H.; Fonarow Gregg, C.; Geraci Stephen, A.; Tamara, H.; Januzzi James, L.; et al. ACCF/AHA Guideline for the Management of Heart Failure. *Circulation* **2013**, *128*, e240–e327.
8. Dhingra, A.; Garg, A.; Kaur, S.; Chopra, S.; Batra, J.S.; Pandey, A.; Chaanine, A.H.; Agarwal, S.K. Epidemiology of Heart Failure with Preserved Ejection Fraction. *Curr. Heart Fail. Rep.* **2014**, *11*, 354–365. [CrossRef]
9. Owan, T.E.; Hodge, D.O.; Herges, R.M.; Jacobsen, S.J.; Roger, V.L.; Redfield, M.M. Trends in Prevalence and Outcome of Heart Failure with Preserved Ejection Fraction. *N. Engl. J. Med.* **2006**, *355*, 251–259. [CrossRef]
10. Konstam, M.A.; Kiernan, M.S.; Bernstein, D.; Bozkurt, B.; Jacob, M.; Kapur, N.K.; Kociol, R.D.; Lewis, E.F.; Mehra, M.R.; Pagani, F.D.; et al. On behalf of the American Heart Association Coun-cil on Clinical Cardiology; Council on Cardiovas-cular Disease in the Young; and Council on Cardiovascular Surgery and Anesthesia. *Circulation* **2018**, *137*, 578–622.
11. Lahm, T.; Douglas, I.S.; Archer, S.L.; Bogaard, H.J.; Chesler, N.C.; Haddad, F.; Hemnes, A.R.; Kawut, S.M.; Kline, J.A.; Kolb, T.M.; et al. Assessment of right ventricular function in the research setting: Knowledge gaps and pathways forward an official American thoracic society research statement. *Am. J. Respir. Crit. Care Med.* **2018**, *198*, e15–e43. [CrossRef] [PubMed]
12. Köhler, D.; Arnold, R.; Loukanov, T.; Gorenflo, M. Right Ventricular Failure and Pathobiology in Patients with Congenital Heart Disease—Implications for Long-Term Follow-Up. *Front. Pediatr.* **2013**, *1*, 37. [CrossRef] [PubMed]
13. Haddad, F.; Hunt, S.A.; Rosenthal, D.N.; Murphy, D.J. Right ventricular function in cardiovascular disease, part I: Anatomy, physiology, aging, and functional assessment of the right ventricle. *Circulation* **2008**, *117*, 1436–1448. [CrossRef] [PubMed]
14. Voelkel, N.F.; Quaife, R.A.; Leinwand, L.A.; Barst, R.J.; Mcgoon, M.D.; Meldrum, D.R.; Dupuis, J.; Long, C.S.; Rubin, L.J.; Smart, F.W.; et al. Right Ventricular Function and Failure Report of a National Heart, Lung, and Blood Institute Working Group on Cellular and Molecular Mechanisms of Right Heart Failure The Normal Right Ventricle The Right Ventricle in Pulmonary Hypertension Special Report. *Circulation* **2006**, *114*, 1883–1891. [CrossRef]
15. Golob, M.; Moss, R.L.; Chesler, N.C. Cardiac tissue structure, properties, and performance: A materials science perspective. *Ann. Biomed. Eng.* **2014**, *42*, 2003–2013. [CrossRef] [PubMed]
16. Bellofiore, A.; Chesler, N.C. Methods for measuring right ventricular function and hemodynamic coupling with the pulmonary vasculature. *Ann. Biomed. Eng.* **2013**, *41*, 1384–1398. [CrossRef]
17. Sacks, M.S.; Chuong, C.J. Biaxial Mechanical Properties of Passive Right Ventricular Free Wall Myocardium. *J. Biomech. Eng.* **1993**, *115*, 202–205. [CrossRef]
18. Wang, Z.; Golob, M.J.; Chesler, N.C. Viscoelastic Properties of Cardiovascular Tissues. In *Viscoelastic and Viscoplastic Materials*; Golob, M.J., Ed.; IntechOpen: Rijeka, Croatia, 2016; p. 7. ISBN 978-953-51-2603-4.
19. Nguyen-Truong, M.; Wang, Z. Biomechanical properties and mechanobiology of cardiac ECM. In *Advances in Experimental Medicine and Biology*; Springer: Cham, Switzerland, 2018; Volume 1098, pp. 1–19.

20. Jang, S.; Vanderpool, R.R.; Avazmohammadi, R.; Lapshin, E.; Bachman, T.N.; Sacks, M.; Simon, M.A. Biomechanical and Hemodynamic Measures of Right Ventricular Diastolic Function: Translating Tissue Biomechanics to Clinical Relevance. *J. Am. Hear. Assoc. Cardiovasc. Cerebrovasc. Dis.* **2017**, *6*, e006084. [CrossRef]
21. Fung, Y.C.; Fronek, K.; Patitucci, P. Pseudoelasticity of arteries and the choice of its mathematical expression. *Am. J. Physiol. Circ. Physiol.* **1979**, *237*, H620–H631. [CrossRef]
22. Ghaemi, H.; Behdinan, K.; Spence, A.D. In vitro technique in estimation of passive mechanical properties of bovine heart: Part I. Experimental techniques and data. *Med. Eng. Phys.* **2009**, *31*, 76–82. [CrossRef]
23. Sommer, G.; Schriefl, A.J.; Andrä, M.; Sacherer, M.; Viertler, C.; Wolinski, H.; Holzapfel, G.A. Biomechanical properties and microstructure of human ventricular myocardium. *Acta Biomater.* **2015**, *24*, 172–192. [CrossRef] [PubMed]
24. Yin, F.C.P.; Strumpf, R.K.; Chew, P.H.; Zeger, S.L. Quantification of the mechanical properties of noncontracting canine myocardium under simultaneous biaxial loading. *J. Biomech.* **1987**, *20*, 577–589. [CrossRef]
25. Humphrey, J.D.; Strumpf, R.K.; Yin, F.C.P. Determination of a Constitutive Relation for Passive Myocardium: I. A New Functional Form. *J. Biomech. Eng.* **1990**, *112*, 333–339. [CrossRef] [PubMed]
26. Novak, V.P.; Yin, F.C.P.; Humphrey, J.D. Regional mechanical properties of passive myocardium. *J. Biomech.* **1994**, *27*, 403–412. [CrossRef]
27. Lin, D.H.S.; Yin, F.C.P. A Multiaxial Constitutive Law for Mammalian Left Ventricular Myocardium in Steady-State Barium Contracture or Tetanus. *J. Biomech. Eng.* **1998**, *120*, 504–517. [CrossRef] [PubMed]
28. Gupta, K.B.; Ratcliffe, M.B.; Fallert, M.A.; Edmunds, L.H.; Bogen, D.K. Changes in passive mechanical stiffness of myocardial tissue with aneurysm formation. *Circulation* **1994**, *89*, 2315–2326. [CrossRef] [PubMed]
29. Valdez-Jasso, D.; Simon, M.A.; Champion, H.C.; Sacks, M.S. A murine experimental model for the mechanical behaviour of viable right-ventricular myocardium. *J. Physiol.* **2012**, *590*, 4571–4584. [CrossRef]
30. Fatemifar, F.; Feldman, M.D.; Oglesby, M.; Han, H.-C. Comparison of Biomechanical Properties and Microstructure of Trabeculae Carneae, Papillary Muscles, and Myocardium in the Human Heart. *J. Biomech. Eng.* **2018**, *141*, 021007. [CrossRef]
31. Ooi, C.Y.; Wang, Z.; Tabima, D.M.; Eickhoff, J.C.; Chesler, N.C. The role of collagen in extralobar pulmonary artery stiffening in response to hypoxia-induced pulmonary hypertension. *Am. J. Physiol. Circ. Physiol.* **2010**, *299*, H1823–H1831. [CrossRef]
32. Lakes, R.S. *Viscoelastic Solids*; CRC Press Revivals; CRC Press: Boca Raton, FL, USA, 1998; ISBN 9781351355650.
33. Fung, Y.C. What are the residual stresses doing in our blood vessels? *Ann. Biomed. Eng.* **1991**, *19*, 237–249. [CrossRef]
34. Shi, X.; Liu, Y.; Copeland, K.M.; McMahan, S.R.; Zhang, S.; Butler, J.R.; Hong, Y.; Cho, M.; Bajona, P.; Gao, H. Epicardial prestrained confinement and residual stresses: A newly observed heart ventricle confinement interface. *J. R. Soc. Interface* **2019**, *16*, 20190028. [CrossRef] [PubMed]
35. Omens, J.H.; Fung, Y.C. Residual strain in rat left ventricle. *Circ. Res.* **1990**, *66*, 37–45. [CrossRef] [PubMed]
36. Genet, M.; Rausch, M.K.; Lee, L.C.; Choy, S.; Zhao, X.; Kassab, G.S.; Kozerke, S.; Guccione, J.M.; Kuhl, E. Heterogeneous growth-induced prestrain in the heart. *J. Biomech.* **2015**, *48*, 2080–2089. [CrossRef] [PubMed]
37. Jöbsis, P.D.; Ashikaga, H.; Wen, H.; Rothstein, E.C.; Horvath, K.A.; McVeigh, E.R.; Balaban, R.S. The visceral pericardium: Macromolecular structure and contribution to passive mechanical properties of the left ventricle. *Am. J. Physiol. Heart Circ. Physiol.* **2007**, *293*, H3379–H3387. [CrossRef]
38. Hoskins, P.R.; Lawford, P.V.; Doyle, B.J. Cardiovascular Biomechanics. *Cardiovasc. Biomech.* **2017**, 1–462.
39. Demer, L.L.; Yin, F.C. Passive biaxial mechanical properties of isolated canine myocardium. *J. Physiol.* **1983**, *339*, 615–630. [CrossRef]
40. Hill, M.R.; Simon, M.A.; Valdez-Jasso, D.; Zhang, W.; Champion, H.C.; Sacks, M.S. Structural and Mechanical Adaptations of Right Ventricular Free Wall Myocardium to Pulmonary-Hypertension Induced Pressure Overload. *Ann. Biomed. Eng.* **2014**, *42*, 2451–2465. [CrossRef]
41. Sacks, M. Biaxial Mechanical Evaluation of Planar Biological Materials. *J. Elast. Phys. Sci. Solids* **2000**, *61*, 199.
42. Holzapfel, G.A.; Ogden, R.W. Constitutive modelling of passive myocardium: A structurally based framework for material characterization. *Philos. Trans. R. Soc. A Math. Phys. Eng. Sci.* **2009**, *367*, 3445–3475. [CrossRef]
43. Humphrey, J.D.; Strumpf, R.K.; Yin, F.C. Biaxial mechanical behavior of excised ventricular epicardium. *Am. J. Physiol. Circ. Physiol.* **1990**, *259*, H101–H108. [CrossRef]

44. Javani, S.; Gordon, M.; Azadani, A.N. Biomechanical Properties and Microstructure of Heart Chambers: A Paired Comparison Study in an Ovine Model. *Ann. Biomed. Eng.* **2016**, *44*, 3266–3283. [CrossRef] [PubMed]
45. Sirry, M.S.; Butler, J.R.; Patnaik, S.S.; Brazile, B.; Bertucci, R.; Claude, A.; McLaughlin, R.; Davies, N.H.; Liao, J.; Franz, T. Characterisation of the mechanical properties of infarcted myocardium in the rat under biaxial tension and uniaxial compression. *J. Mech. Behav. Biomed. Mater.* **2016**, *63*, 252–264. [CrossRef] [PubMed]
46. Ahmad, F.; Prabhu, R.J.; Liao, J.; Soe, S.; Jones, M.D.; Miller, J.; Berthelson, P.; Enge, D.; Copeland, K.M.; Shaabeth, S.; et al. Biomechanical properties and microstructure of neonatal porcine ventricles. *J. Mech. Behav. Biomed. Mater.* **2018**, *88*, 18–28. [CrossRef] [PubMed]
47. Ramo, N.L.; Troyer, K.L.; Puttlitz, C.M. Viscoelasticity of spinal cord and meningeal tissues. *Acta Biomater.* **2018**, *75*, 253–262. [CrossRef]
48. Ramo, N.L.; Puttlitz, C.M.; Troyer, K.L. The development and validation of a numerical integration method for non-linear viscoelastic modeling. *PLoS ONE* **2018**, *13*, e0190137. [CrossRef]
49. Ramo, N.L.; Lee, J.H.T.; Troyer, K.L.; Kwon, B.K.; Puttlitz, C.M.; Cripton, P.; Streijger, F.; Shetye, S.S. Comparison of in vivo and ex vivo viscoelastic behavior of the spinal cord. *Acta Biomater.* **2017**, *68*, 78–89. [CrossRef]
50. Troyer, K.L.; Puttlitz, C.M. Nonlinear viscoelasticty plays an essential role in the functional behavior of spinal ligaments. *J. Biomech.* **2012**, *45*, 684–691. [CrossRef]
51. Troyer, K.L.; Estep, D.J.; Puttlitz, C.M. Viscoelastic effects during loading play an integral role in soft tissue mechanics. *Acta Biomater.* **2012**, *8*, 234–243. [CrossRef]
52. Stella, J.A.; Liao, J.; Sacks, M.S. Time-dependent biaxial mechanical behavior of the aortic heart valve leaflet. *J. Biomech.* **2007**, *40*, 3169–3177. [CrossRef]
53. Liao, J.; Yang, L.; Grashow, J.; Sacks, M.S. The Relation Between Collagen Fibril Kinematics and Mechanical Properties in the Mitral Valve Anterior Leaflet. *J. Biomech. Eng.* **2006**, *129*, 78–87. [CrossRef]
54. Little, R.; Wead, W. Diastolic viscoelastic properties of active and quiescent cardiac muscle. *Am. J. Physiol. Content* **1971**, *221*, 1120–1125. [CrossRef] [PubMed]
55. Tsaturyan, A.K.; Izacov, V.J.; Zhelamsky, S.V.; Bykov, B.L. Extracellular fluid filtration as the reason for the viscoelastic behaviour of the passive myocardium. *J. Biomech.* **1984**, *17*, 749–755. [CrossRef]
56. Miller, C.E.; Vanni, M.A.; Keller, B.B. Characterization of passive embryonic myocardium by quasi-linear viscoelasticity theory. *J. Biomech.* **1997**, *30*, 985–988. [CrossRef]
57. Miller, C.E.; Wong, C.L. Trabeculated embryonic myocardium shows rapid stress relaxation and non-quasi-linear viscoelastic behavior. *J. Biomech.* **2000**, *33*, 615–622. [CrossRef]
58. Loeffler, L.; Sagawa, K. A one dimensional viscoelastic model of cat heart muscle studied by small length perturbations during isometric contraction. *Circ. Res.* **1975**, *36*, 498–512. [CrossRef]
59. Faconti, L.; Bruno, R.M.; Ghiadoni, L.; Virdis, S.; Virdis, A. Ventricular and Vascular Stiffening in Aging and Hypertension. *Curr. Hypertens. Rev.* **2015**, *11*, 100–109. [CrossRef]
60. Borlaug, B.A.; Redfield, M.M.; Melenovsky, V.; Kane, G.C.; Karon, B.L.; Jacobsen, S.J.; Rodeheffer, R.J. Longitudinal changes in left ventricular stiffness: A community-based study. *Circ. Heart Fail.* **2013**, *6*, 944–952. [CrossRef]
61. Wang, Z.; Schreier, D.A.; Hacker, T.A.; Chesler, N.C. Progressive right ventricular functional and structural changes in a mouse model of pulmonary arterial hypertension. *Physiol. Rep.* **2013**, *1*, e00184. [CrossRef]
62. Wang, Z.; Chesler, N.C. Pulmonary vascular mechanics: Important contributors to the increased right ventricular afterload of pulmonary hypertension. *Exp. Physiol.* **2013**, *98*, 1267–1273. [CrossRef]
63. Wang, Z.; Chesler, N.C. Pulmonary Vascular Wall Stiffness: An Important Contributor to the Increased Right Ventricular Afterload with Pulmonary Hypertension. *Pulm. Circ.* **2011**, *1*, 212–223. [CrossRef]
64. Tabima, D.M.; Philip, J.L.; Chesler, N.C. Right ventricular-pulmonary vascular interactions. *Physiology* **2017**, *32*, 346–356. [CrossRef] [PubMed]
65. Ibrahim, E.-S.H. Myocardial tagging by Cardiovascular Magnetic Resonance: Evolution of techniques–pulse sequences, analysis algorithms, and applications. *J. Cardiovasc. Magn. Reson.* **2011**, *13*, 36. [CrossRef] [PubMed]
66. Seo, Y.; Ishizu, T.; Aonuma, K. Current status of 3-dimensional speckle tracking echocardiography: A review from our experiences. *J. Cardiovasc. Ultrasound* **2014**, *22*, 49–57. [CrossRef] [PubMed]

67. Voigt, J.-U.; Pedrizzetti, G.; Lysyansky, P.; Marwick, T.H.; Houle, H.; Baumann, R.; Pedri, S.; Ito, Y.; Abe, Y.; Metz, S.; et al. Definitions for a common standard for 2D speckle tracking echocardiography: Consensus document of the EACVI/ASE/Industry Task Force to standardize deformation imaging. *Eur. Hear. J. Cardiovasc. Imaging* **2014**, *16*, 1–11.
68. Bossone, E.; D'Andrea, A.; D'Alto, M.; Citro, R.; Argiento, P.; Ferrara, F.; Cittadini, A.; Rubenfire, M.; Naeije, R. Echocardiography in Pulmonary Arterial Hypertension: From Diagnosis to Prognosis. *J. Am. Soc. Echocardiogr.* **2013**, *26*, 1–14. [CrossRef]
69. Leng, S.; Jiang, M.; Zhao, X.-D.; Allen, J.C.; Kassab, G.S.; Ouyang, R.-Z.; Tan, J.-L.; He, B.; Tan, R.-S.; Zhong, L. Three-Dimensional Tricuspid Annular Motion Analysis from Cardiac Magnetic Resonance Feature-Tracking. *Ann. Biomed. Eng.* **2016**, *44*, 3522–3538. [CrossRef]
70. Zou, H.; Xi, C.; Zhao, X.; Koh, A.S.; Gao, F.; Su, Y.; Tan, R.-S.; Allen, J.; Lee, L.C.; Genet, M.; et al. Quantification of Biventricular Strains in Heart Failure With Preserved Ejection Fraction Patient Using Hyperelastic Warping Method. *Front. Physiol.* **2018**, *9*, 1295. [CrossRef]
71. Genet, M.; Stoeck, C.; von Deuster, C.; Chuan Lee, L.; Guccione, J.; Kozerke, S. Finite Element Digital Image Correlation for Cardiac Strain Analysis from 3D Whole-Heart Tagging. In Proceedings of the ISMRM 24rd Annual Meeting and Exhibition, Singapore, 7–13 May 2016.
72. Genet, M.; Stoeck, C.T.; von Deuster, C.; Lee, L.C.; Kozerke, S. Equilibrated warping: Finite element image registration with finite strain equilibrium gap regularization. *Med. Image Anal.* **2018**, *50*, 1–22. [CrossRef]
73. Phatak, N.S.; Maas, S.A.; Veress, A.I.; Pack, N.A.; Di Bella, E.V.R.; Weiss, J.A. Strain measurement in the left ventricle during systole with deformable image registration. *Med. Image Anal.* **2009**, *13*, 354–361. [CrossRef]
74. Finsberg, H.N.T.; Sundnes, J.S.; Xi, C.; Lee, L.C.; Zhao, X.; Tan, J.L.; Genet, M.; Zhong, L.; Wall, S.T. Computational quantification of patient specific changes in ventricular dynamics associated with pulmonary hypertension. *Am. J. Physiol. Heart Circ. Physiol.* **2019**, *317*, H1363–H1375. [CrossRef] [PubMed]
75. Khan, S.; Fakhouri, F.; Majeed, W.; Kolipaka, A. Cardiovascular magnetic resonance elastography: A review. *NMR Biomed.* **2018**, *31*, e3853. [CrossRef] [PubMed]
76. Arani, A.; Arunachalam, S.P.; Chang, I.C.Y.; Baffour, F.; Rossman, P.J.; Glaser, K.J.; Trzasko, J.D.; McGee, K.P.; Manduca, A.; Grogan, M.; et al. Cardiac MR elastography for quantitative assessment of elevated myocardial stiffness in cardiac amyloidosis. *J. Magn. Reson. Imaging* **2017**, *46*, 1361–1367. [CrossRef] [PubMed]
77. Fan, L.; Yao, J.; Yang, C.; Tang, D.; Xu, D. Infarcted Left Ventricles Have Stiffer Material Properties and Lower Stiffness Variation: Three-Dimensional Echo-Based Modeling to Quantify In Vivo Ventricle Material Properties. *J. Biomech. Eng.* **2015**, *137*, 81005. [CrossRef]
78. Mojsejenko, D.; McGarvey, J.R.; Dorsey, S.M.; Gorman, J.H., 3rd; Burdick, J.A.; Pilla, J.J.; Gorman, R.C.; Wenk, J.F. Estimating passive mechanical properties in a myocardial infarction using MRI and finite element simulations. *Biomech. Model. Mechanobiol.* **2015**, *14*, 633–647. [CrossRef]
79. Tang, D.; Yang, C.; Geva, T.; Del Nido, P.J. Patient-specific MRI-based 3D FSI RV/LV/patch models for pulmonary valve replacement surgery and patch optimization. *J. Biomech. Eng.* **2008**, *130*, 41010. [CrossRef]
80. Acosta, S.; Puelz, C.; Rivière, B.; Penny, D.J.; Brady, K.M.; Rusin, C.G. Cardiovascular mechanics in the early stages of pulmonary hypertension: A computational study. *Biomech. Model. Mechanobiol.* **2017**, *16*, 2093–2112. [CrossRef]
81. Avazmohammadi, R.; Mendiola, E.A.; Soares, J.S.; Li, D.S.; Chen, Z.; Merchant, S.; Hsu, E.W.; Vanderslice, P.; Dixon, R.A.F.; Sacks, M.S. A Computational Cardiac Model for the Adaptation to Pulmonary Arterial Hypertension in the Rat. *Ann. Biomed. Eng.* **2019**, *47*, 138–153. [CrossRef]
82. Xi, C.; Latnie, C.; Zhao, X.; Tan, J.L.; Wall, S.T.; Genet, M.; Zhong, L.; Lee, L.C. Patient-Specific Computational Analysis of Ventricular Mechanics in Pulmonary Arterial Hypertension. *J. Biomech. Eng.* **2016**, *138*, 111001. [CrossRef]
83. Humphrey, J.D. *Cardiovascular Solid Mechanics: Cells, Tissues, and Organs*; Springer: New York, NY, USA, 2002; ISBN 0387951687.
84. Fan, R.; Tang, D.; Yao, J.; Yang, C.; Xu, D. 3D Echo-Based Patient-Specific Computational Left Ventricle Models to Quantify Material Properties and Stress/Strain Differences between Ventricles with and without Infarct. *Comput. Model. Eng. Sci.* **2014**, *99*, 491–508.
85. Hassaballah, A.I.; Hassan, M.A.; Mardi, A.N.; Hamdi, M. An inverse finite element method for determining the tissue compressibility of human left ventricular wall during the cardiac cycle. *PLoS ONE* **2013**, *8*, e82703. [CrossRef] [PubMed]

86. Hess, O.M.; Grimm, J.; Krayenbuehl, H.P. Diastolic simple elastic and viscoelastic properties of the left ventricle in man. *Circulation* **1979**, *59*, 1178–1187. [CrossRef] [PubMed]
87. Hess, O.M.; Schneider, J.; Koch, R.; Bamert, C.; Grimm, J.; Krayenbuehl, H.P. Diastolic function and myocardial structure in patients with myocardial hypertrophy. Special reference to normalized viscoelastic data. *Circulation* **1981**, *63*, 360–371. [CrossRef] [PubMed]
88. Pouleur, H.; Karliner, J.S.; Lewinter, M.M.; Covell, J.W. Diastolic viscous properties of the intact canine left ventricle. *Circ. Res.* **1979**, *67*, 352–359. [CrossRef] [PubMed]
89. Ahmad, F.; Soe, S.; White, N.; Johnston, R.; Khan, I.; Liao, J.; Jones, M.; Prabhu, R.; Maconochie, I.; Theobald, P. Region-Specific Microstructure in the Neonatal Ventricles of a Porcine Model. *Ann. Biomed. Eng.* **2018**, *46*, 2162–2176. [CrossRef]
90. Dokos, S.; Smaill, B.H.; Young, A.A.; Legrice, I.J.; Legrice, I.J. Shear properties of passive ventricular myocardium. *Am. J. Physiol. Hear. Circ. Physiol.* **2002**, *283*, 2650–2659. [CrossRef]
91. Rankin, J.S.; Arentzen, C.E.; Mchale, P.A.; Ling, D.; Anderson, R.W. Diastolic anisotropic properties of the left ventricle in the conscious dog. *Circ. Res.* **1977**, *69*, 765–778.
92. Stroud, J.D.; Baicu, C.F.; Barnes, M.A.; Spinale, F.G.; Zile1, M.R. Viscoelastic properties of pressure overload hypertrophied myocardium: Effect of serine protease treatment. *Am. J. Physiol. Hear. Circ. Physiol.* **2002**, *315*, 1691–1702. [CrossRef]
93. Cansız, F.B.C.; Dal, H.; Kaliske, M. An orthotropic viscoelastic material model for passive myocardium: Theory and algorithmic treatment. *Comput. Methods Biomech. Biomed. Eng.* **2015**, *18*, 1160–1172. [CrossRef]
94. Shetye, S.S.; Troyer, K.L.; Streijger, F.; Lee, J.H.T.; Kwon, B.K.; Cripton, P.A.; Puttlitz, C.M. Nonlinear viscoelastic characterization of the porcine spinal cord. *Acta Biomater.* **2014**, *10*, 792–797. [CrossRef]
95. Troyer, K.L.; Shetye, S.S.; Puttlitz, C.M. Experimental Characterization and Finite Element Implementation of Soft Tissue Nonlinear Viscoelasticity. *J. Biomech. Eng.* **2012**, *134*, 114501. [CrossRef] [PubMed]
96. Wheatley, B.B.; Morrow, D.A.; Odegard, G.M.; Kaufman, K.R.; Haut Donahue, T.L. Skeletal muscle tensile strain dependence: Hyperviscoelastic nonlinearity. *J. Mech. Behav. Biomed. Mater.* **2016**, *53*, 445–454. [CrossRef] [PubMed]
97. Gültekin, O.; Sommer, G.; Holzapfel, G.A. An orthotropic viscoelastic model for the passive myocardium: Continuum basis and numerical treatment. *Comput. Methods Biomech. Biomed. Eng.* **2016**, *19*, 1647–1664. [CrossRef] [PubMed]
98. Huyghe, J.M.; van Campen, D.H.; Arts, T.; Heethaar, R.M. The constitutive behaviour of passive heart muscle tissue: A quasi-linear viscoelastic formulation. *J. Biomech.* **1991**, *24*, 841–849. [CrossRef]
99. Rosano, G.M.; Vitale, C. Metabolic Modulation of Cardiac Metabolism in Heart Failure. *Card. Fail. Rev.* **2018**, *4*, 99–103. [CrossRef]
100. Leonard, B.L.; Smaill, B.H.; LeGrice, I.J. Structural Remodeling and Mechanical Function in Heart Failure. *Microsc. Microanal.* **2012**, *18*, 50–67. [CrossRef]
101. Gupte, R.S.; Vijay, V.; Marks, B.; Levine, R.J.; Sabbah, H.N.; Wolin, M.S.; Recchia, F.A.; Gupte, S.A. Upregulation of Glucose-6-Phosphate Dehydrogenase and NAD(P)H Oxidase Activity Increases Oxidative Stress in Failing Human Heart. *J. Card. Fail.* **2007**, *13*, 497–506. [CrossRef]
102. Chen, B.; Frangogiannis, N.G. Macrophages in the Remodeling Failing Heart. *Circ. Res.* **2016**, *119*, 776–778. [CrossRef]
103. Helmes, M.; Trombitás, K.; Centner, T.; Kellermayer, M.; Labeit, S.; Linke, W.A.; Granzier, H. Mechanically driven contour-length adjustment in rat cardiac titin's unique N2B sequence. Titin is an adjustable spring. *Circ. Res.* **1999**, *84*, 1339–1352. [CrossRef]
104. Francisco, J.V. *Interstitial Fibrosis in Heart Failure*; Springer: New York, NY, USA, 2005; ISBN 978-0-387-22824-2.
105. Bishop, J.E.; Laurent, G.J. Collagen turnover and its regulation in the normal and hypertrophying heart. *Eur. Heart J.* **1995**, *16*, 38–44. [CrossRef]
106. Brower, G.L.; Gardner, J.D.; Forman, M.F.; Murray, D.B.; Voloshenyuk, T.; Levick, S.P.; Janicki, J.S. The relationship between myocardial extracellular matrix remodeling and ventricular function. *Eur. J. Cardio-Thoracic Surg.* **2006**, *30*, 604–610. [CrossRef] [PubMed]
107. Golob, M.J.; Wang, Z.; Prostrollo, A.J.; Hacker, T.A.; Chesler, N.C. Limiting collagen turnover via collagenase-resistance attenuates right ventricular dysfunction and fibrosis in pulmonary arterial hypertension. *Physiol. Rep.* **2016**, *4*, e12815. [CrossRef] [PubMed]

108. Plaksej, R.; Kosmala, W.; Frantz, S.; Herrmann, S.; Niemann, M.; Störk, S.; Wachter, R.; Angermann, C.E.; Ertl, G.; Bijnens, B.; et al. Relation of circulating markers of fibrosis and progression of left and right ventricular dysfunction in hypertensive patients with heart failure. *J. Hypertens.* **2009**, *27*, 2483–2491. [CrossRef] [PubMed]
109. Segura, A.M.; Frazier, O.H.; Buja, L.M. Fibrosis and heart failure. *Heart Fail. Rev.* **2014**, *19*, 173–185. [CrossRef]
110. Yu, L.; Ruifrok, W.P.T.; Meissner, M.; Bos, E.M.; Van Goor, H.; Sanjabi, B.; Van Der Harst, P.; Pitt, B.; Goldstein, I.J.; Koerts, J.A.; et al. Genetic and pharmacological inhibition of galectin-3 prevents cardiac remodeling by interfering with myocardial fibrogenesis. *Circ. Hear. Fail.* **2013**, *6*, 107–117. [CrossRef]
111. Begoña, L.; Ramón, Q.; Arantxa, G.; Javier, B.; Mariano, L.; Javier, D. Impact of Treatment on Myocardial Lysyl Oxidase Expression and Collagen Cross-Linking in Patients With Heart Failure. *Hypertension* **2009**, *53*, 236–242.
112. López, B.; González, A.; Querejeta, R.; Larman, M.; Díez, J. Alterations in the Pattern of Collagen Deposition May Contribute to the Deterioration of Systolic Function in Hypertensive Patients With Heart Failure. *J. Am. Coll. Cardiol.* **2006**, *48*, 89–96. [CrossRef]
113. Berk, B.C.; Fujiwara, K.; Lehoux, S. ECM remodeling in hypertensive heart disease. *J. Clin. Investiga.* **2007**, *117*, 568–575. [CrossRef]
114. Rossi, M.A. Pathologic fibrosis and connective tissue matrix in left ventricular hypertrophy due to chronic arterial hypertension in humans. *J. Hypertens.* **1998**, *16*, 1031–1041. [CrossRef]
115. Brower, G.L.; Janicki, J.S. Contribution of ventricular remodeling to pathogenesis of heart failure in rats. *Am. J. Physiol. Circ. Physiol.* **2001**, *280*, H674–H683. [CrossRef]
116. Ryan, T.D.; Rothstein, E.C.; Aban, I.; Tallaj, J.A.; Husain, A.; Lucchesi, P.A.; Dell'Italia, L.J. Left Ventricular Eccentric Remodeling and Matrix Loss Are Mediated by Bradykinin and Precede Cardiomyocyte Elongation in Rats With Volume Overload. *J. Am. Coll. Cardiol.* **2007**, *49*, 811–821. [CrossRef] [PubMed]
117. Zheng, J.; Chen, Y.; Pat, B.; Dell'italia, L.A.; Tillson, M.; Dillon, A.R.; Powell, P.C.; Shi, K.; Shah, N.; Denney, T.; et al. Microarray identifies extensive downregulation of noncollagen extracellular matrix and profibrotic growth factor genes in chronic isolated mitral regurgitation in the dog. *Circulation* **2009**, *119*, 2086–2095. [CrossRef] [PubMed]
118. Takashi, N.; Hiroyuki, T.; Hirofumi, T.; Masaru, T.; Keiko, S.; Toshiyuki, K.; Makoto, U.; Kyoko, I.-Y.; Tsutomu, I.; Akira, T. Regulation of Fibrillar Collagen Gene Expression and Protein Accumulation in Volume-Overloaded Cardiac Hypertrophy. *Circulation* **1997**, *95*, 2448–2454.
119. LeWinter, M.M.; Meyer, M. Mechanisms of diastolic dysfunction in heart failure with a preserved ejection fraction: If it's not one thing it's another. *Circ. Heart Fail.* **2013**, *6*, 1112–1115. [CrossRef]
120. Fernández-Golfín, C.; Pachón, M.; Corros, C.; Bustos, A.; Cabeza, B.; Ferreirós, J.; de Isla, L.P.; Macaya, C.; Zamorano, J. Left ventricular trabeculae: Quantification in different cardiac diseases and impact on left ventricular morphological and functional parameters assessed with cardiac magnetic resonance. *J. Cardiovasc. Med.* **2009**, *10*, 827–833. [CrossRef]
121. Van de Veerdonk, M.C.; Dusoswa, S.A.; Tim Marcus, J.; Bogaard, H.-J.; Spruijt, O.; Kind, T.; Westerhof, N.; Vonk-Noordegraaf, A. The importance of trabecular hypertrophy in right ventricular adaptation to chronic pressure overload. *Int. J. Cardiovasc. Imaging* **2014**, *30*, 357–365. [CrossRef]
122. Zile, M.R.; Baicu, C.F.; Gaasch, W.H. Diastolic Heart Failure—Abnormalities in Active Relaxation and Passive Stiffness of the Left Ventricle. *N. Engl. J. Med.* **2004**, *350*, 1953–1959. [CrossRef]
123. Prasad, A.; Hastings, J.L.; Shibata, S.; Popovic, Z.B.; Arbab-Zadeh, A.; Bhella, P.S.; Okazaki, K.; Fu, Q.; Berk, M.; Palmer, D.; et al. Characterization of static and dynamic left ventricular diastolic function in patients with heart failure with a preserved ejection fraction. *Circ. Heart Fail.* **2010**, *3*, 617–626. [CrossRef]
124. Røe, Å.T.; Aronsen, J.M.; Skårdal, K.; Hamdani, N.; Linke, W.A.; Danielsen, H.E.; Sejersted, O.M.; Sjaastad, I.; Louch, W.E. Increased passive stiffness promotes diastolic dysfunction despite improved Ca^{2+} handling during left ventricular concentric hypertrophy. *Cardiovasc. Res.* **2017**, *113*, 1161–1172. [CrossRef]
125. Sakata, Y.; Ohtani, T.; Takeda, Y.; Yamamoto, K.; Mano, T. Left Ventricular Stiffening as Therapeutic Target for Heart Failure With Preserved Ejection Fraction. *Circ. J.* **2013**, *77*, 886–892. [CrossRef]
126. Kraigher-Krainer, E.; Shah, A.M.; Gupta, D.K.; Santos, A.; Claggett, B.; Pieske, B.; Zile, M.R.; Voors, A.A.; Lefkowitz, M.P.; Packer, M.; et al. Impaired Systolic Function by Strain Imaging in Heart Failure With Preserved Ejection Fraction. *J. Am. Coll. Cardiol.* **2014**, *63*, 447–456. [CrossRef] [PubMed]

127. Opitz, C.F.; Hoeper, M.M.; Gibbs, J.S.R.; Kaemmerer, H.; Pepke-Zaba, J.; Coghlan, J.G.; Scelsi, L.; D'Alto, M.; Olsson, K.M.; Ulrich, S.; et al. Pre-Capillary, Combined, and Post-Capillary Pulmonary Hypertension: A Pathophysiological Continuum. *J. Am. Coll. Cardiol.* **2016**, *68*, 368–378. [CrossRef] [PubMed]
128. Puwanant, S.; Park, M.; Popović, Z.B.; Tang, W.H.W.; Farha, S.; George, D.; Sharp, J.; Puntawangkoon, J.; Loyd, J.E.; Erzurum, S.C.; et al. Ventricular geometry, strain, and rotational mechanics in pulmonary hypertension. *Circulation* **2010**, *121*, 259–266. [CrossRef] [PubMed]
129. Sachdev, A.; Villarraga, H.R.; Frantz, R.P.; McGoon, M.D.; Hsiao, J.-F.; Maalouf, J.F.; Ammash, N.M.; McCully, R.B.; Miller, F.A.; Pellikka, P.A.; et al. Right Ventricular Strain for Prediction of Survival in Patients With Pulmonary Arterial Hypertension. *Chest* **2011**, *139*, 1299–1309. [CrossRef] [PubMed]
130. Haeck, M.L.; Scherptong, R.W.; Marsan, N.A.; Holman, E.R.; Schalij, M.J.; Bax, J.J.; Vliegen, H.W.; Delgado, V. Prognostic Value of Right Ventricular Longitudinal Peak Systolic Strain in Patients With Pulmonary Hypertension. *Circ. Cardiovasc. Imaging* **2012**, *5*, 628–636. [CrossRef]
131. Finsberg, H.; Xi, C.; Tan, J.L.; Zhong, L.; Genet, M.; Sundnes, J.; Lee, L.C.; Wall, S.T. Efficient estimation of personalized biventricular mechanical function employing gradient-based optimization. *Int. J. Numer. Methods Biomed. Eng.* **2018**, *34*, e2982. [CrossRef]
132. Kennish, A.; Yellin, E.; Frater, R.W.; Frater, W. Dynamic stiffness profiles in the left ventricle. *J. Appl. Physiol.* **1975**, *39*, 665–671. [CrossRef]
133. LeWinter, M.M.; Engler, R.; Pavelec, R.S. Time-dependent shifts of the left ventricular diastolic filling relationship in conscious dogs. *Circ. Res.* **1979**, *45*, 641–653. [CrossRef]
134. Gaasch, W.H.; Zile, M.R. Left Ventricular Diastolic Dysfunction and Diastolic Heart Failure. *Ann. Rev. Med.* **2004**, *55*, 373–394. [CrossRef]
135. Burlew, B.S.; Weber, K.T. Cardiac Fibrosis as a Cause of Diastolic Dysfunction. *Herz* **2002**, *27*, 92–98. [CrossRef]
136. Weber, K.T.; Brilla, C.G. Pathological hypertrophy and cardiac interstitium. Fibrosis and renin-angiotensin-aldosterone system. *Circulation* **1991**, *83*, 1849–1865. [CrossRef] [PubMed]
137. Maron, B.J.; Bonow, R.O.; Cannon, R.O.; Leon, M.B.; Epstein, S.E. Hypertrophic Cardiomyopathy. *N. Engl. J. Med.* **1987**, *316*, 844–852. [CrossRef] [PubMed]
138. Borbély, A.; Van Der Velden, J.; Papp, Z.; Bronzwaer, J.G.; Edes, I.; Stienen, G.J.; Paulus, W.J. Cardiomyocyte Stiffness in Diastolic Heart Failure. *Circulation* **2005**, *111*, 774–781. [CrossRef] [PubMed]

bioengineering

MDPI

Article

Compressive Mechanical Properties of Porcine Brain: Experimentation and Modeling of the Tissue Hydration Effects

Raj K. Prabhu [1,2,*], Mark T. Begonia [1,2], Wilburn R. Whittington [1,3], Michael A. Murphy [1], Yuxiong Mao [1], Jun Liao [4], Lakiesha N. Williams [5], Mark F. Horstemeyer [6] and Jianping Sheng [7]

1. Center for Advanced Vehicular Systems, Mississippi State University, Mississippi State, MS 39795, USA; mbegonia@vt.edu (M.T.B.); whittington@me.msstate.edu (W.R.W.); mam526@cavs.msstate.edu (M.A.M.); maxmcn@gmail.com (Y.M.)
2. Department of Agricultural & Biological Engineering, Mississippi State University, Mississippi State, MS 39762, USA
3. Department of Mechanical Engineering, Mississippi State University, Mississippi State, MS 39762, USA
4. Department of Bioengineering, University of Texas Arlington, Arlington, TX 76010, USA; jun.liao@uta.edu
5. J. Crayton Pruitt Family Department of Biomedical Engineering, University of Florida, Gainesville, FL 32611, USA; lwilliams@bme.ufl.edu
6. School of Engineering, Liberty University, Lynchburg, VA 24515, USA; mhorstemeyer@liberty.edu
7. U.S. Army Tank Automotive Research, Development, and Engineering Center (TARDEC), Warren, MI 48397, USA; jianping.sheng.civ@mail.mil

* Correspondence: rprabhu@abe.msstate.edu

Received: 3 April 2019; Accepted: 2 May 2019; Published: 7 May 2019

Abstract: Designing protective systems for the human head—and, hence, the brain—requires understanding the brain's microstructural response to mechanical insults. We present the behavior of wet and dry porcine brain undergoing quasi-static and high strain rate mechanical deformations to unravel the effect of hydration on the brain's biomechanics. Here, native 'wet' brain samples contained ~80% (mass/mass) water content and 'dry' brain samples contained ~0% (mass/mass) water content. First, the wet brain incurred a large initial peak stress that was not exhibited by the dry brain. Second, stress levels for the dry brain were greater than the wet brain. Third, the dry brain stress–strain behavior was characteristic of ductile materials with a yield point and work hardening; however, the wet brain showed a typical concave inflection that is often manifested by polymers. Finally, finite element analysis (FEA) of the brain's high strain rate response for samples with various proportions of water and dry brain showed that water played a major role in the initial hardening trend. Therefore, hydration level plays a key role in brain tissue micromechanics, and the incorporation of this hydration effect on the brain's mechanical response in simulated injury scenarios or virtual human-centric protective headgear design is essential.

Keywords: porcine brain; mechanical behavior; hydration effects; Split-Hopkinson pressure bar; micromechanics; finite element analysis

1. Introduction

Traumatic brain injury (TBI), due to mechanical impact to the head, is a leading cause of death and life-long disability in the United States. Around 5.3 million Americans currently have long-term disabilities after sustaining a TBI [1]. In the United States, direct and indirect medical costs related to TBI amounted to an estimated $60 billion for the year 2000 alone [2]. Thus, TBI's profound impact on our society necessitates effective protective measures to curb consequent injuries and disabilities [3]. Ostensibly, to design protective equipment for the brain requires an understanding of its mechanical response during injurious loading conditions. Thus, microstructural investigation of the relationship between tissue hydration, cellular structure, and mechanical impact at various strain rates is critical to effective comprehension and modeling of underlying TBI mechanisms.

In the 1940s, pioneering work on the mechanical properties of brain parenchyma asserted that shear strain had a significant influence on brain trauma during an impact or at finite deformations [4,5]. Motivated by the shear strain theory, a number of subsequent studies on brain mechanical properties concentrated on shear experiments [6] and quantifying shear properties [7–10]. The strain rates for the experiments ranged from quasi-static to moderate strain rates (0.001–60 s^{-1}) and revealed a nonlinear stress–strain behavior. Hence, the brain was treated as a soft engineering material. A seminal study by Estes and McElhaney [11] on brain specimens under quasi-static compression showed a similar nonlinear response (with stiffening stress–strain behavior). In more recent studies, extensive shear tests were performed on human brains in the strain rate range of 0.1–90 s^{-1} [12]. Their results confirmed the earlier nonlinear stress–strain behavior.

Motivation to understand the role of axonal fibers (brain white matter) spurred a series of shear relaxation experiments on brain and brain stem materials, which showed strain-rate dependent behavior at quasi-static rates [13]. Furthermore, their studies noted the anisotropic nature of a brain due to the presence of axonal fibers, mainly in the quasi-static regime with the white matter being stiffer than grey matter [14]. White matter's greater stiffness was attributed to the fibrous texture of axons. Essentially, the difference in stiffness of white and grey matter contributed to the variation of stress–strain behaviors in the cerebrum, cerebellum, and brain stem especially at strains larger than 20% [15–17].

Similar nonlinear (hardening) mechanical behavior was also observed in tensile studies conducted at the turn of the 21th century [18,19]. The strain rates that were employed were in the quasi-static regime showing a high strain rate sensitivity. In another tensile response study, in-vitro testing on a cultured human brain showed subsequent swelling of neurons similar to that observed in the rat's brains after a TBI [20]. The axonal fibers demonstrated a delayed elastic response after an initial dynamic stretch injury. The evolution of this elastic response involved immediate undulations of axonal fibers after the dynamic stretch injury, which was followed by a slow reversion to its original shape (straight orientation) within an hour [21].

More recently, moderately-high-strain-rate compression tests showed a minor dependence on the heterogeneity of brain for strain rates greater than 40 s^{-1} [22]. Other noteworthy research established the premise for the onset of TBI in in-vitro brain tissue cultures in strains larger than 20% at strain rates greater than 40 s^{-1} [20,23,24]. Consequently, in the current study, strain rates greater than 50 s^{-1} were treated as high rate.

Initial hardening from the high strain rate behavior of soft biological materials was not evident in the quasi-static behavior where the stress monotonically increased as the deformation proceeded [18,19,22,25]. However, moderately high strain rate data (10–70 s^{-1}) of the human liver tissue were predominantly marked with an initial hardening trend that was followed by a softening and then further hardening at larger strains [26]. Some studies argued that the initial hardening was due to inertial effects and asserted that an annular geometry of the specimen averted the inertial effects and guaranteed a uniform stress-state [27,28]. However, in a recent study on the dynamic response of the brain, part of the inertial effects was shown to be intrinsic to the material and that non-uniform stress-states were realized in the material [29]. As a major component of the brain, water proved crucial in instigating the initial hardening [29,30]. To date, no researcher has analyzed the effect of the amount of water within the brain that has undergone high impact loads. Cheng and Bilston [31] did show that the quasi-static viscoelastic properties of brain white matter arose from the solid matrix with the interstitial fluid's migration in the white matter, providing the short-term elastic response, and modeled the white matter's response using a poroviscoelastic (PVE) model. However, the work of Cheng and Bilston [31] did not directly address the hydration effects at higher strain rates.

The contribution of our current study is the quantification of the difference in the mechanical properties of the wet and dry brain at quasi-static and high strain rates. We then present the micromechanics of brain's high rate deformation, based on a mixture theory of water and dehydrated brain, through finite element (FE) simulations. Sections 2.1 and 2.2 describe the materials and methods employed in the experiments. Section 2.3 describes the theory employed in the experiments, and Section 2.4 presents the statistical methodology employed for experimental data. Section 2.5 presents the theory for micromechanics and gives an overview of the Split-Hopkinson pressure bar (SHPB) FE simulations used to evaluate the micromechanics of water included in the brain and the dry brain in the specimen. Section 3 presents the results and the corresponding discussion. Finally, conclusions are summarized in Section 4.

2. Materials and Methods

The experimental procedures involved in this study used in vitro porcine samples, the protocol for which were approved by the Office of Regulatory Compliance and Safety (ORSC) at Mississippi State University (MSU), Mississippi State, MS, 39762. The IACUC approval number is 11-048.

2.1. Sample Preparation

Intact porcine heads from healthy males were collected from a local abattoir and transported to a necropsy laboratory (Mississippi State University College of Veterinary Medicine). Porcine brains were extracted from each skull and stored in a phosphate buffered saline (PBS) solution to minimize dehydration and degradation. Test specimens were prepared via scalpel incision through the corpus callosum to separate the two hemispheres. A stainless-steel cylindrical die was then used to dissect brain material through the sagittal plane of each hemisphere, producing cylindrical test specimens with the sulci and gyri characterizing the superior surface. Each test specimen consisted of gray and white matter. The average initial height of the test specimens was 15 mm while the average initial diameter was 30 mm. Brain extractions and dissection required approximately one hour for completion, and all compression experiments were conducted within three hours post-mortem [25]. For making 'dry brain' specimens, surgically extracted cylindrical porcine brain samples were lyophilized using a Freezone™ 1-liter benchtop freeze dryer (LABCONCO ®) for approximately 48 hours under 0.1 Pa and −50 °C. The lyophilizing process evaporated the water in the brain, giving the lyophilized parenchyma. Before testing, a 30 mm diameter die was used to dissect the lyophilized tissue specimens with an average thickness of 15 mm. Details of the samples obtained and porcine brains used along with testing conditions are given in Table 1.

Table 1. Details of samples, animal brain numbers, and testing conditions at each strain rate

Strain Rate (s^{-1})	Wet/Dry	Number of Samples	Number of Animals/Porcine Brains	Temperature (°C)	Pressure (MPa)
0.00625	Wet	6	3	20.85	0.1
0.025	Wet	7	4	20.85	0.1
0.1	Wet	16	7	20.85	0.1
50	Wet	6	3	20.85	0.1
250	Wet	4	2	20.85	0.1
450	Wet	5	2	20.85	0.1
550	Wet	7	3	20.85	0.1
750	Wet	4	2	20.85	0.1
0.00625	Dry	4	2	20.85	0.1
0.1	Dry	5	3	20.85	0.1
250	Dry	4	2	20.85	0.1
Total		68	33		

2.2. Testing Apparatuses

2.2.1. Mach-1™ for Quasi-static Testing of the Wet Brain

The MACH-1™ Micromechanical Testing System (BIOMOMENTUM, Quebec), Universal Motion Controller/Driver—Model ESP300 and load cell amplifier were used for the quasi-static compression experiments (Figure 1). The 1 kg (10 N) load cell had a resolution of 0.50 mg and was included with the Mach-1™ Micromechanical Testing System (BIOMOMENTUM, Quebec) to meet the sensitivity requirements for testing porcine brain tissue (~80% water content, referred to as wet brain) [32]. A circular platen with an estimated diameter of 50 mm was also selected to accommodate the smaller cross-sectional area of the test specimens [25]. In addition, a stainless-steel chamber was fabricated for housing test specimens immersed in 0.01 M PBS, prior to testing to minimize tissue dehydration. The samples were tested to their failure strains at applied strain rates of 0.00625, 0.025, and 0.1 s^{-1}.

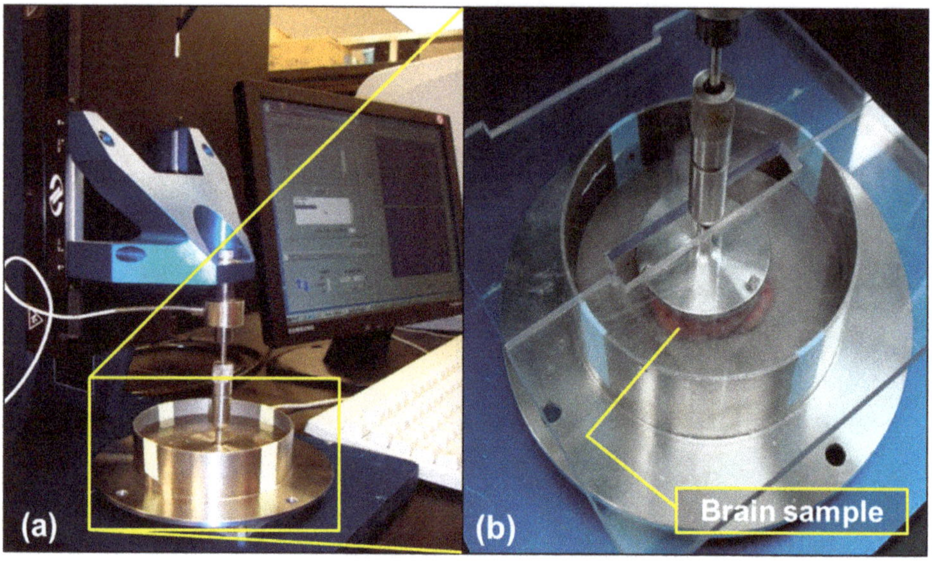

Figure 1. (**a**) Quasi-static compression test on the porcine brain using the Mach-1™ Micromechanical Testing System (BIOMOMENTUM, Quebec). (**b**) Wet porcine samples were immersed in 0.1 M neutral buffered PBS during quasi-static compression at strain rates of 0.00625, 0.025, and 0.1 s^{-1}.

2.2.2. Split-Hopkinson Pressure Bar (SHPB) for High Strain Rate Testing of Wet and Dry Brain Specimens

The Split-Hopkinson pressure bar (SHPB) comprises a striker bar, an incident bar, and a transmitted bar (Figure 2) [33–35] (see Appendix B). For both wet and dry brain, each specimen (30 mm diameter, 15 mm thickness) was placed between the incident and transmitted bars. The striker bar was then propelled at a specified velocity via a pneumatic device, hitting the incident bar and causing compression on the specimen lodged between the two aforementioned bars, which were instrumented with strain gauges to collect the corresponding data. The experimental setup consisted of DAQ modules, strain gauges, a laser speed meter, a pressure release valve, and polycarbonate bars (Figure 2). The strain gauge data was processed using David Viscoelastic software [29,36,37]. Using the SHPB apparatus, wet brain specimens were tested at 50, 250, 450, 550, and 750 s^{-1}, and dry brain specimens were tested at 250 s^{-1}.

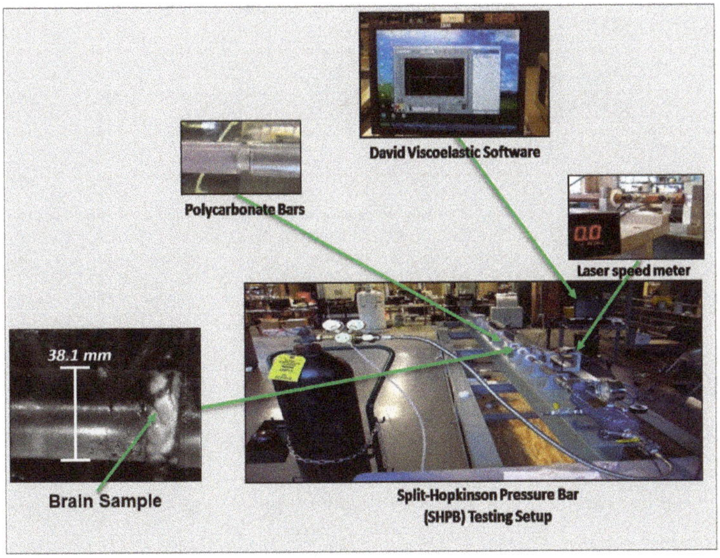

Figure 2. Overview of the Split-Hopkinson pressure bar (SHPB) used to conduct high-strain-rate tests on porcine brain specimens. Strain rates ranged from 50 to 750 s^{-1}.

2.2.3. Instron™5568 for Quasi-Static Testing of the Dry Brain

Quasi-static rate compression tests on the lyophilized porcine brain were performed using an Instron™ 5869 load frame (Figure 3). Due to the stiffer nature of the lyophilized dry brain, a larger load cell capacity (1 kN) was needed when compared to wet brain testing, and hence an Instron™ 5869 was used instead of the Mach-1™. Lyophilized brain specimens were cylindrical in geometry with a diameter of 15 mm (1 mm tolerance) and a height of 8 mm (2 mm tolerance), which was measured using digital calipers (Mitutoyo 500-752-10 CD-6" PSX). The strain was recorded using an Instron™ mechanical extensometer, which was used to control the strain rates of the experiments. Post-processing of the data was performed using Bluehill software (Instron™) at strain rates of 0.00625, 0.025, and 0.1 s^{-1}. Video imaging using LaVision™ software was also recorded on tests at strain rates of 0.00625, 0.025, and 0.1 s^{-1} to investigate the onset of barreling in samples and found that no barreling was observed up to 40% true strain (Figure 4).

Figure 3. Instron™ 5869 configuration used for compression testing of lyophilized (dry) porcine brain samples at strain rates ranging from 0.00625 to 0.1 s^{-1}.

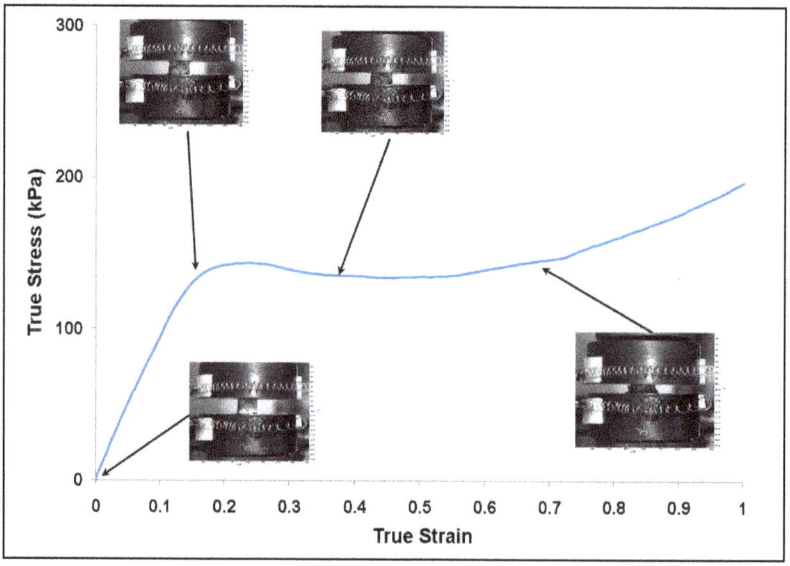

Figure 4. Plot of true stress–strain behavior for the lyophilized (dry) brain at a strain rate of 0.1 s^{-1}. The images show the deformation of the specimen during testing. The specimen started to barrel above 40% true stain. The water content in the specimen was 0% m/m.

2.3. Stress–Strain Experimental Data

Determining the true stress–strain behavior for the porcine brain at quasi-static rates were performed using standard procedures and formulations [38]. While testing wet and dry brain samples, using the Mach-I and Instron™ 5869 machines, respectively; precautions were taken to ensure a uniform stress state during testing. The LaVision™ video/software suite was used to ensure that the data measured were under a uniform cross-sectional area of the specimen and hence uniform stress state. For high strain rate tests using the SHPB, stress–strain calculations were made using standard wave theory equations. A detailed discussion on wave theory and its application to the SHPB is presented by Gary et al. [36] and Prabhu et al. [29,37].

During analysis, the tangent modulus was calculated based on the slope of the stress–strain response at 5% strain, the elastic–inelastic transition stress were determined based on the yield stress at high strain rates, and the quasi-static wet brain data was asserted to have elastic–inelastic behavior subsequent to 5% strain.

2.4. Statistical Analysis of the Experimental Data

Statistical analyses of three parameters, namely the tangent modulus, elastic–inelastic transition stress (σ_p or σ_t), and the true strain at σ_p or σ_t at various strain rates were conducted using the SigmaStat 3.0 software (SPSS, Chicago, IL). A one-way analysis of variance (ANOVA) method was used for statistical analysis on the three parameters, with a Holm-Sidak test being used for post-hoc comparisons. The mechanical difference at various strain rates, for a particular parameter, was considered to be statistically significant when $p < 0.05$.

2.5. Finite Element Simulation-Based Micromechanics of the Dry Brain and Water

FE simulations of the SHPB setup were similar to the experimental SHPB setup (Figure 5a). An FE simulation was initialized by allowing the striker bar to be set in motion. The speed of the striker corresponded to that of the SHPB experiment. As the striker bar came into contact with the incident bar, stress waves that arose from the striker bar propagated through the incident bar. Part of the wave traveled through the sample and part was reflected back into the incident bar. The applied striker bar speed and the resulting pressure became the boundary conditions for the FE simulations of the whole SHPB set-up with a reduced integration formulation and default hourglass controls. In all, the number of elements included in the SHPB FE model was 47,300, and the type of elements used were regular hexahedrons. The striker, incident and transmitted bars were treated as elastic materials. The Young's Modulus and Poisson's ratio were 2391.2 MPa and 0.36, respectively. Figure 5a gives an overview of the FE model that simulated the SHPB test. The FE simulation sample elements were randomly assigned materials properties of water and dry brain (Figure 5b) such that the effective mass of the water and dry brain elements in the specimen varied from 20% to 80% m/m. Table 2 gives the FE simulation cases from the combination of water and dry brain. Here, a micromechanics approach of calculating the total average stress of the specimen from the component level stresses, that is, of water and the dry brain was implemented.

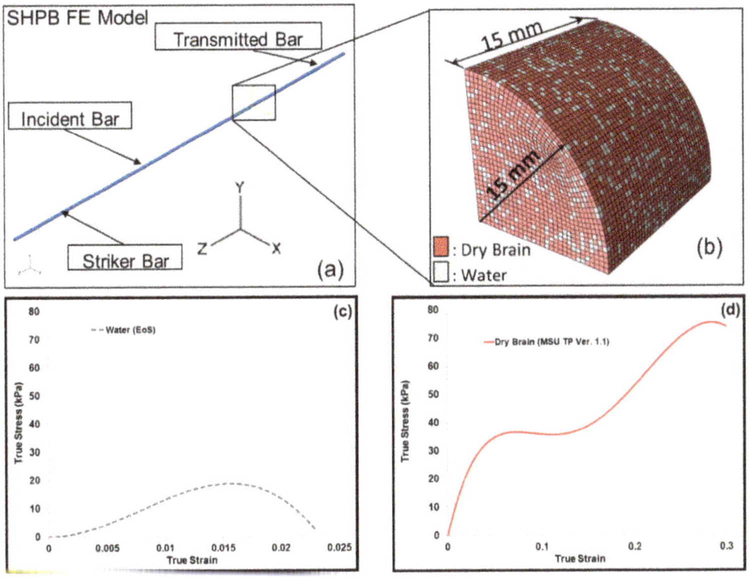

Figure 5. (**a**) A schematic of the finite element (FE) set up for Split-Hopkinson pressure bar (SHPB) tests and (**b**) FE simulation sample dimensions, with a sample having 20% (m/m) water and 80% (m/m) dry brain. The water content in the dry brain is ~0% (m/m). Local loading direction (negative z-direction) true stress–strain responses of (**c**) water and (**d**) dry brain. Mixture theory was applied to obtain the entire sample's mechanical response.

An Equation of State (EOS) was used to define the material property of water. The EOS assumed the Hugoniot form of Mie-Gruneisen's EOS [39]. The expression for the EOS used for water in the FE model is as follows in Equation (1)

$$U_s = c_0 + sU_p \tag{1}$$

where U_s is the wave velocity, U_p the particle velocity and c_0 and s are constants of the linear relationship of the wave velocity and particle velocity. A specific material model, MSU TP 1.1 [40], was used to capture the elastic and inelastic response of the dry brain material. A summary of the constitutive equations is shown in Table 3 (see Appendix A for additional details). The constitutive model (MSU TP Ver. 1.1) presented in this paper captures both the instantaneous and longer-term large deformation processes and could admit microstructural features within the internal state variables. With the microstructural features, we can use our internal state variable model so that eventually history effects could be captured and predicted. In the absence of the microstructural features, other constitutive models should be able to show the stress state under the high-rate loadings exhibited here since no varying history was induced. A one-dimensional version of the material model, MSU TP Ver. 1.1, was implemented in MATLAB [41] to obtain the material point simulator. The material point simulator was then optimized to calibrate the material model constants to the high strain rate experimental data of the dry brain at 250 s^{-1} (Figure A2). The values for the material constants for MSU TP 1.1 are shown in Table 4, Figure 5c,d give the local representative mechanical responses of water and dry brain, respectively. They denote the local elemental response for the FE model specimen consisting of a mixture of water and dry brain. Initially, a mesh refinement study for the FE model was also conducted to analyze the convergence of ABAQUS/Explicit solutions [42]. Figure 6. shows the simulation results of the incident and transmitted strain measurements at different mesh resolutions, respectively. Figure 6. also illustrates that meshes with 4703 and 12,432 elements did not converge, but the FE solutions with 47,300 elements and higher converged. FE simulations were then used in ABAQUS/EXPLICIT [42] to further analyze the different combinations of water and dry brain, similar to the way a specimen would undergo deformation during a SHPB experiment. The specimen geometry used for conducting FE simulations was cylindrical with a radius of 15 mm. The thickness of the cylindrical specimen was 15 mm.

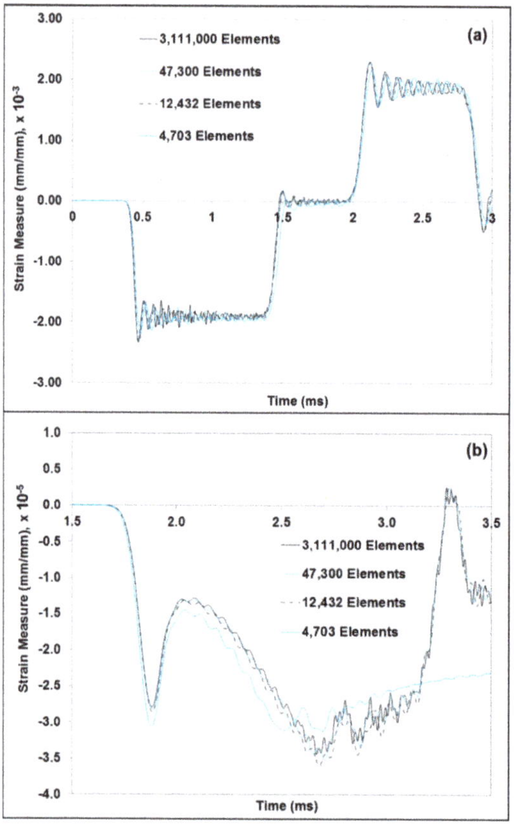

Figure 6. Comparison plots from the FEA mesh refinement study, which shows the strain measures for the (**a**) incident and reflected wave and (**b**) transmitted wave. The element type considered for this study was regular hexahedral. Four cases of mesh refinement, varying from 4703 to 3,111,000 elements, were considered.

Table 2. Overview of the finite element (FE) simulation cases for the micromechanics of dry brain and water content variation.

FE Simulation Case	Dry Brain % (m/m)	Water % (m/m)
1	80	20
2	60	40
3	40	60
4	20	80

Table 3. Summary of the model equations for MSU TP 1.1 (see Bouvard et al. [40]). See Appendix C for a summary of symbol descriptions.

Term/Function	Description
$\overline{\Psi} = \overline{\Psi}\left(\overline{C}^e, \overline{\xi}_1, \overline{\xi}_2,\right)$, where $\overline{C}^e$ is the elastic part of the right Cauchy-Green tensor, and $\overline{\xi}_1$, $\overline{\xi}_2$ and $\overline{E}^{\beta}$ are internal strain fields (internal state variables).	Free energy, $\overline{\Psi}$
$\sigma = J^{e-1}\tau = J^{e-1}F^e\overline{S}F^{eT}$, where J^{e-1} is the inverse of the determinant of F^e. F^e and F^{eT} are the elastic part of F and the transpose of elastic part of F. $\overline{S} = 2\frac{\partial\hat{\overline{\Psi}}}{\partial\overline{C}^e}$	Cauchy Stress, σ Second Piola-Kirchhoff Stress, $\overline{S}$
$\tau_1 = R^e\overline{M}_1R^{eT}$, where R^e and R^{eT} are the elastic part of the rotation tensor (R) and the transpose of the elastic part of R. $\overline{M} = 2\mu\overline{E}^e + \left(K - \frac{2}{3}\mu\right)\overline{Tr}\left(\overline{E}^e\right)\overline{I}$, where μ and K are the elastic shear and bulk moduli modeling the elastic behavior respectively. E^e is the elastic part of the Green-Lagrange strain tensor, and $\overline{I}$ is the identity matrix. $F = F^eF^p, F^e = R^eU^e, \overline{E}^e = \ln(U^e)$, where F^p is the plastic part of the deformation tensor, and U^e is the right stretch tensor.	Kirchhoff Stress (elasto-viscoplastic part, τ_1) Elastic Law (Mandel Stress, $\overline{M}$) Deformation Gradient
$\overline{\kappa}_1 = \frac{\partial\hat{\overline{\Psi}}}{\partial\overline{\xi}_1}, \overline{\kappa}_2 = \frac{\partial\hat{\overline{\Psi}}}{\partial\overline{\xi}_2}$, where $\overline{\kappa}_1$ and $\overline{\kappa}_2$ are stress-like thermodynamic conjugates of the $\overline{\xi}_1$ and $\overline{\xi}_2$ respectively. $\overline{\alpha} = \frac{\partial\hat{\overline{\Psi}}}{\partial\overline{E}^{\beta}}$, where $\overline{\alpha}$ is a stress-like thermodynamic conjugate of $\overline{E}^{\beta}$.	Stress-like internal state variables Stress-like internal state variable
$\dot{F}^p = \overline{D}^pF^p$, where $\overline{D}^p$ is the inelastic rate of deformation. $\overline{D}^p = \frac{1}{\sqrt{2}}\dot{\overline{\gamma}}^p\,\overline{N}^p$ with $\overline{N}^p = \frac{DEV(\overline{M}_1)}{\lVert DEV(\overline{M}_1)\rVert}$, where $\dot{\overline{\gamma}}^p$ is the viscous shear strain rate given by the following equation: $\dot{\overline{\gamma}}^p = \dot{\overline{\gamma}}_0^p\left[\sinh\left(\frac{\langle\overline{\tau}-(\overline{\kappa}_1+\overline{\kappa}_2+\alpha_p\overline{\pi})\rangle}{Y}\right)\right]^m$ with $\overline{\tau} = \frac{1}{\sqrt{2}}\lVert DEV(\overline{M}-\overline{\alpha})\rVert$ and $\overline{\pi} = -\frac{1}{3}\overline{Tr}(\overline{M})$, where $\overline{\tau} = (1/\sqrt{2})\lVert DEV(\overline{M}-\overline{\alpha})\rVert$ is an equivalent shear stress term and $\overline{\pi} = (1/3)\overline{Tr}(\overline{M})$ is the effective pressure term, $\dot{\overline{\gamma}}_0^p$ is a reference strain rate, m is a strain rate sensitivity parameter, and αp is a pressure sensitivity parameter and Y is the yield criterion. $\dot{\overline{\xi}}_1 = h_0\left(1 - \frac{\overline{\xi}_1}{\overline{\xi}^*}\right)\dot{\overline{\gamma}}^p, \dot{\overline{\xi}}^* = g_0\left(1 - \frac{\overline{\xi}^*}{\overline{\xi}^*_{sat}}\right)\dot{\overline{\gamma}}^p$, where ξ^* represents an evolving strain threshold or criterion that the macromolecular chains must overcome to slip. h_0 and g_0 are hardening moduli, and ξ^*_{sat} is the saturation value of ξ^*. $\dot{\overline{\xi}}_2 = h_1\left(\overline{\lambda}^p - 1\right)\left(1 - \frac{\overline{\xi}_2}{\overline{\xi}_{2sat}}\right)\dot{\overline{\gamma}}^p$ with $\overline{\lambda}^p = \frac{1}{\sqrt{3}}\sqrt{\overline{Tr}\left(\overline{B}^p\right)}$ and $\overline{B}^p = F^pF^{pT}$ where h_1 is the hardening modulus, and ξ_{2sat} is the saturation value of ξ_2. $\dot{\overline{\beta}} = R_{s_1}\left(\overline{D}^p\overline{\beta} + \overline{\beta}\,\overline{D}^p\right)$ and $\overline{\beta}(X, 0) = I$	Flow rule Equivalent plastic shear strain-rate Polymer chain resistance to plastic flow Polymer chain crystallization at large strain Evolution equation of stretch-like tensor $\overline{\beta}$
$\left\{\mu, K, \dot{\gamma}_0^p, m, Y\right\}\left\{\overline{\xi}_0, \overline{\xi}^*_{sat}, h_0, g_0, C_{\kappa_1}\right\}\left\{h_1, \overline{\xi}_{2sat}, C_{\kappa_2}\right\}$ $\left\{R_{S1}, \lambda_L, \mu_R\right\}\left\{\alpha_p, \overline{\xi}_{10}, \overline{\xi}_{20}\right\}$	Material constants

Table 4. Values of material constants for dry brain material using MSU TP Ver. 1.1 viscoplasticity model along with their definitions.

Model Constants	Constant Definition	Values
μ (MPa)	Shear Modulus	0.80
K (MPa)	Bulk Modulus	399.73
$\dot{\gamma}^{p}$ (s^{-1})	Reference Strain Rate	120,000
m	Strain Rate Sensitivity Parameter	0.90
Y_o (MPa)	Material Yield Parameter	9.00
α_p	Sensitivity Parameter	0
λ_L	Network Locking Stretch	2.00
μ_R	Rubbery Modulus	0.07
R_{s1}	Material Hardening Parameter	1.4
h_o	Hardening Modulus	0.41
$\xi^{o}{}_{1}$	Internal Strain-Like Parameter Initial Value	0.0045
$\xi^{*}{}_{sat}$	Internal Strain-Like Parameter Saturation Value	0.001
$\xi^{*}{}_{o}$	Energetic Strain Barrier	1.2
g_o	Hardening Modulus	0.3
$C\kappa_1$ (MPa)	Internal Stress-Like Parameter	0.41
h_1	Hardening Modulus	0
$e^{o}{}_{s2}$	Internal Strain-Like Parameter Initial Value	0
$e^{sat}{}_{s2}$	Internal Strain-Like Parameter Saturation Value	0.4
$C\kappa_2$ (MPa)	Internal Stress-Like Parameter	0

3. Results and Discussion

3.1. Experiment Response

Figure 7 shows plots of the experimental true stress–strain behavior of the wet porcine brain at both high and quasi-static strain rates under compression, which exhibited two distinct patterns of stress–strain behaviors. We also note that both the low- and high-rate testing methods showed significant strain-rate sensitivity.

The high-rate experimental data showed an initial hardening trend similar to the yield point in some high carbon steel alloys or thermoplastics [40], followed by softening and then further hardening at higher strains (Figure 7a). A similar high-rate phenomenon was also reported by Sparks and Dupaix [26], Prabhu [29], and Clemmer [30]. These initial hardening and softening trends were highly strain-rate dependent, thus illustrated by the initial hardening peak stress, σ_p. σ_p also marked the transition from elastic to inelastic deformations. We also observed that the strains corresponding to σ_p increased steadily as the strain rate increased (Figure 7a). Hence, σ_p occurred at different true strain values for strain rates of 50, 250, 450, 550, and 750 s^{-1}. Figure 8. represents a plot of σ_p versus different strain rates in the high-rate regime and shows that σ_p increased linearly as the strain rate increased. A one-way analysis of variance (ANOVA) method for statistical analyses on the tangent modulus, σ_p and the true strain at σ_p was conducted by Prabhu [29]. They showed that there was a significant difference for σ_p and the true strain at σ_p over the strain rates 50–750 s^{-1}, with *p* values <0.05. However, no statistical difference was observed for the tangent modulus.

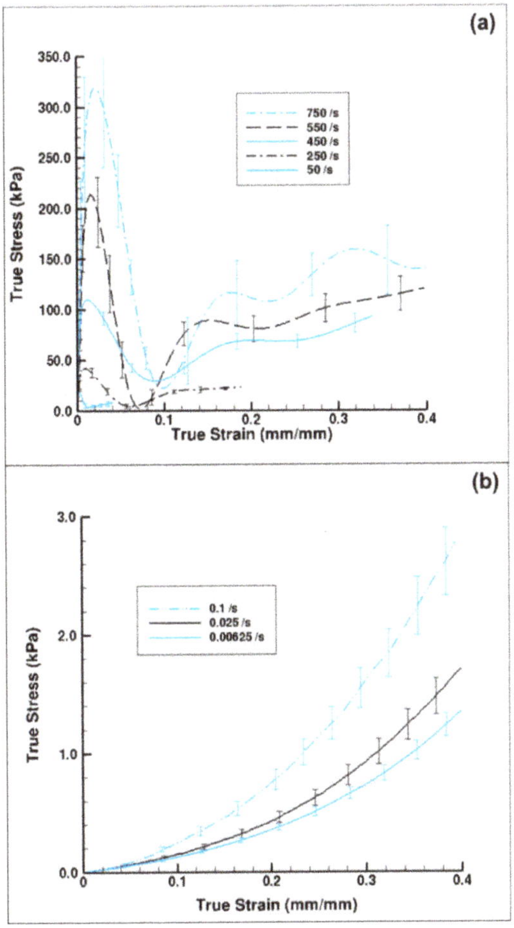

Figure 7. Comparison of (**a**) high strain rate (n = 4 (750 s^{-1}), 7 (750 s^{-1}), 5 (750 s^{-1}), 4 (750 s^{-1}), and 6 (750 s^{-1})) [29] and (**b**) quasi-static strain rate wet-brain material behavior (n = 16 (0.1 s^{-1}), 7 (0.025 s^{-1}), and 6 (0.00625 s^{-1})). The water content in the wet brain sample was ~80% m/m. Note the lower stress values for the quasi-static wet brain case and that the initial peak in stress observed under high-rate loading was not present in the quasi-static loading. The error bars represent standard error.

In contrast, the quasi-static wet brain tissue behavior followed a monotonic increase indicative of material hardening, which is a stress–strain relation more typically exhibited by soft tissues (Figure 7b), such as seen in muscle, brain, liver, and tendon tissues [23,38,43]. Unlike the high-rate behavior, at quasi-static rates, the material (wet brain) continues to harden after the initial elastic response with the mechanical behavior at the quasi-static regime being completely devoid of the initial hardening and softening trend noted in high-rate response. However, the elastic–inelastic transitions for both quasi-static and dynamic data occur at similar strain levels (Figure 7). At quasi-static rates, if one were to consider the elastic-viscoelastic transition stress, σ_t, at the true strain values signifying elastic-viscoelastic transition (analogous to the high strain rate σ_p marking elastic–inelastic transition), the variation of σ_t over quasi-static strain rates was observed to be nominal (Figure 9).

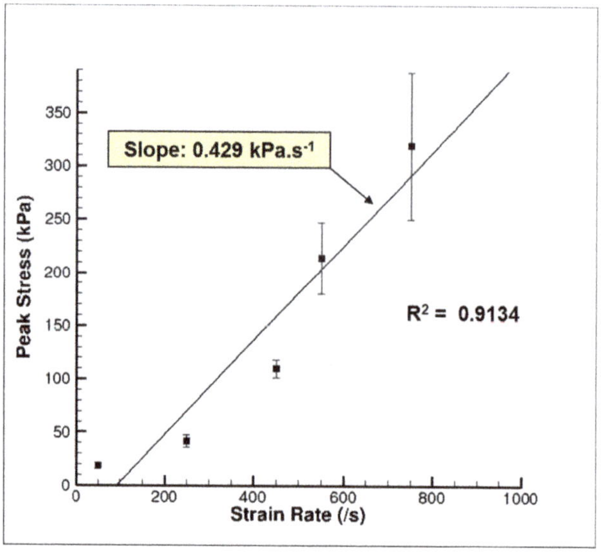

Figure 8. Plot of the initial hardening peak stress, σ_p, of the wet porcine brain from the high strain rate tests ranging from 50–750 s^{-1} [29]. The water content in the porcine brain was ~80% m/m. Twenty-six brain samples were analyzed. The error bars represent standard error.

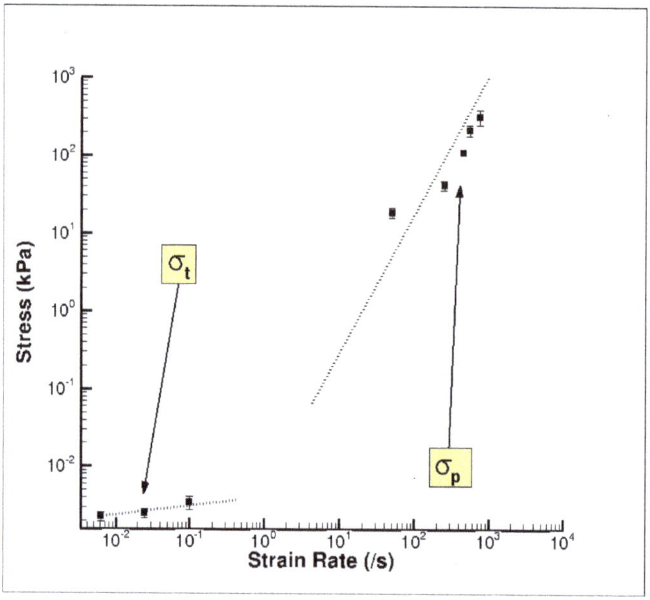

Figure 9. Comparison of the elastic–inelastic transition stress, σ_t, and the initial hardening peak stress, σ_p, of the wet porcine brain at varying strain rates using a log–log scale. The water content of the wet porcine brain was ~80% m/m. All 39 brain samples were analyzed. The error bars represent standard error.

Table 5 presents the variation of the mean σ_t, the strain at σ_t, and the tangent modulus at 0.01 true strain over quasi-static strain rates; along with p values. With $p < 0.05$, a statistical difference in tangent modulus was noted over the quasi-static strain rates, but no statistical difference was observed for σ_t and the true strain at σ_t. In other words, Figure 9 and Table 5 illustrate the rather soft variance of quasi-static σ_t values. While σ_t marks the elastic–inelastic transition leading to subsequent hardening of the material, σ_p marks the linear-nonlinear transition followed by a softening trend. A comparison of σ_t at quasi-static rates and σ_p at high strain rates indicates the presence of two different deformation mechanisms that are highly dependent on temporal rates (Figures 7 and 9; Table 5). The deformation mechanism at quasi-static rates was initially marked by an elastic regime that can be explained by the percolation of water through the specimen matrix [25,43]. The lower rates of the quasi-static regime are favorable to the slow migration of water through the various intercellular cavities in the matrix. At higher strains, the percolation of water is constrained due to the reducing specimen volume and the consequential tissue (cellular structure and matrix) compaction. The hardening trend observed at higher strains could be attributed to the resistance offered by the tissue as it compacts.

Table 5. Statistical comparison of the tangent modulus, elastic–inelastic transition stress (σ_t) and strain at σ_p/σ_t of the wet brain (~80% m/m water content) at quasi-static and high strain rates. $P < 0.05$: significantly different; $p > 0.05$: significantly indifferent.

Variable \ Strain Rate (s^{-1})	0.00625 s^{-1} n = 6	0.0250 s^{-1} n = 7	0.100 s^{-1} n = 16	*p*-Value
Tangent Modulus (kPa)	1.6822 ± 0.0047	1.7497 ± 0.0021	3.2378 ± 0.0371	*<0.05*
Transition Stress (kPa)	0.1046 ± 0.0046	0.1069 ± 0.0020	0.1400 ± 0.0377	*>0.05*
Strain at Peak Stress	0.075 ± 0.0040	0.075 ± 0.0060	0.072 ± 0.0050	*>0.05*

In comparison, the wet brain samples exhibited an initial mechanical response when tested at dynamic rates that are not observed in quasi-static tests (Figure 7). Song [27] suggested this initial mechanical response is solely due to radial inertial effects in soft materials, but Prabhu [29] asserted that at least a portion of this initial response is an intrinsic material property. If the observed initial mechanical response is due to the tissue being deformed faster than water can easily percolate out of the cells, some or all of the behavior would indeed be intrinsic to the material. In other words, the water present in the matrix and cellular structures offers inertial resistance to the sudden deformation, giving rise to a sharp initial mechanical response. The initial hardening trend is then followed by a softening trend is due to the rupturing of cells. Further compaction leads to the realignment of damage cellular and matrix components, producing a strain hardening effect at higher strains.

Figure 10 shows the experimental true stress–strain mechanical response of dry porcine brain under compression at the quasi-static strain rates of 0.00625 and 0.1 s^{-1} and the high strain rate of 250 s^{-1}. The wet brain behavior in Figure 7b was akin to most soft biological materials at quasi-static rates, which exhibit a toe region at the beginning, then an intermediate inelastic behavior, and finally a hardening at larger strains. However, the dry brain behavior in Figure 10 was more similar to the quasi-static behavior of metals, with an initial elastic region followed by a work hardening regime. Such contrasting trends in the material behaviors can be attributed to the presence of water in the wet brain (~80%) [25,29], which contributes to the strain rate sensitivity for both quasi-static and high rates. For the dry brain, Table 6 shows that an absence of water makes the tangent modulus, yield point (elastic–inelastic transition stress), and strain at yield point in the stress–strain behavior insensitive to the applied strain rate.

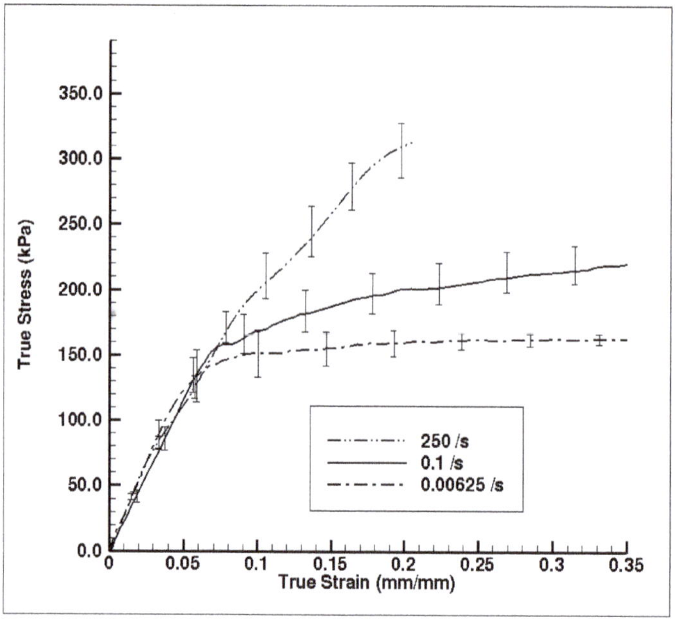

Figure 10. Comparison of experimental quasi-static and high strain rate dry brain (~0% water content m/m) stress–strain behavior (n = 4 (250 s^{-1}), 5 (0.1 s^{-1}), and 4 (0.00625 s^{-1})). Specimens possessed a diameter of 15 mm diameter and a thickness of 5 mm. The error bars represent standard error.

Table 6. Statistical Comparison of the tangent modulus, elastic–inelastic transition stress (σ_p/σ_t) and strain at σ_p/σ_t of the dry brain (~0% m/m water content) at quasi-static and high strain rates. $p < 0.05$: significantly different; $p > 0.05$: significantly indifferent.

Strain Rate (s^{-1}) / Variable	0.00625 s^{-1} n = 4	0.100 s^{-1} n = 5	250 s^{-1} n = 4	*p*-Value
Tangent Modulus (kPa)	2591.451 ± 424.4080	2282.160 ± 922.3810	2143.683 ± 620.0000	>0.05
Transition Stress (kPa)	108.602 ± 17.7860	138.313 ± 55.9020	166.389 ± 48.1420	>0.05
Strain at Transition Stress	0.0419 ± 0.0086	0.0606 ± 0.0100	0.0776 ± 0.0106	>0.05

The characteristic of the dry brain true stress–strain behavior is similar at the three applied strain rates that were employed (Figure 10); the stress–strain behavior was marked by an initial toe region with an elastic response up to a 'yield point' and then followed by a work hardening at larger strains. As noted from Figure 10, the variation in yield point did not change as the applied strain rate changed, although the work hardening increased (concave-down) as the applied strain rate increased. Table 6 presents the results of the statistical analyses on the dry brain tangent modulus, elastic–inelastic transition stress (σ_p/σ_t) or stress at yield point, and the true strain at σ_p/σ_t. The *p*-value for all three parameters was greater than 0.05 at quasi-static and high strain rates, implying statistical indifference. Hence, the tangent modulus, yield point (elastic–inelastic transition stress), and strain at yield point of the true stress–strain behavior for the dry brain was observed to be strain rate insensitive. The similarities between the quasi-static and dynamic rates for the dry brain in Figure 10 highlight the effect of water on the mechanical response of brain tissue. Since there is minimal difference between the quasi-static and dynamic mechanical results for dry brain prior to the yield point, much of the observed initial mechanical response is due to water contained in the sample. This finding reinforces the above supposition that the initial mechanical response observed for the wet brain when tested at dynamic rates is due to water being forced out of the tissue matrices and cellular structures faster than the water can easily percolate. Therefore, future testing must consider how much the initial mechanical response is due to radial inertial effects and how much is a true tissue mechanical response.

3.2. Simulation Response

As further investigations were performed on the tissue hydration effects at high strain rates using finite element analysis (FEA), the FE simulation results showed a substantial increase of the initial hardening trend as the strain rate increased for wet brain material (Figure 7a). Figure 11 shows the FE simulation of the various proportions of water (20%, 40%, 60%, and 80% m/m) and the dry brain (80%, 20%, 40%, and 60% m/m) along with the experimental high strain rate response of the wet (~80% water) and the dry brain (~0% water) at 250 s^{-1}. Examining the FE simulation results in Figure 11 shows that the trend of the FE simulation true stress–strain curves change as the m/m content of water and the dry brain were changed. The specimen with the lowest water content (20% m/m) had the highest stress, and the observed stress became lower as the m/m water content in the specimen was increased. In other words, the initial stress response in the high-rate response of the specimen was inversely proportional to the tissue hydration at high strain rates. This behavior is unlike the brain tissue response reported by Cheng and Bilston [31], where the creep rate of the brain white matter at quasi-static rates depended on the water movement out of the sample. At high strain rates, the water resistance to rapid movement is distinctly seen in the initial response of the FE specimens with larger m/m water content. Furthermore, as the m/m water content in the sample increased the initial hardening trend also increased. This again validates the assertion that cellular cavity water content plays a significant role in the inertial effects of the brain tissue. The greater the hydration of the brain tissue the higher the initial hardening trend is within the high strain rate mechanical response. Additionally, the experimental results shown in Figure 11 give the upper and lower bounds for the FE simulation results as they contain the lowest and highest amount of water.

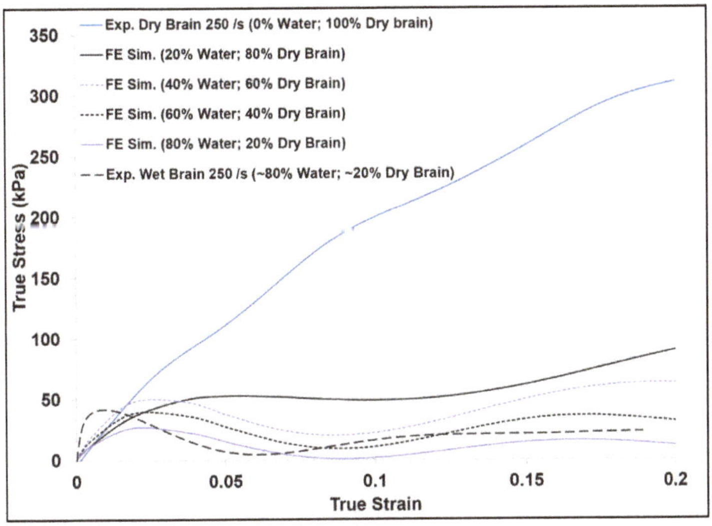

Figure 11. Comparison of finite element (FE) simulation and experimental true stress–strain (σ_{33}) behavior of specimens with various water and dry brain m/m contents at a strain rate of 250 s^{-1}. Here compressive stress and strain are taken as positive.

Results for the current experimental and computational investigation show that water plays an important role in the mechanical response of brain tissue at all strain rates. Furthermore, it can be inferred that water plays a crucial role in the strain rate sensitivity and deformation of the brain. Prior studies by Prabhu [29] and Clemmer [30] have shown that the presence of water in matrix and cells notably influences the stress-state at quasi-static and high strain rates. Weinberg and Ortiz [44] showed that the cavitation of water at high strain rates causes biological damage to surrounding tissue. This behavior needs to be included in material models and accounted for when simulating human brain injury scenarios.

This behavior can be implemented into brain tissue material models to better design protective gear for use in dynamic rate conditions by incorporating the effects of water into brain constitutive material models. Including this data has the potential to better capture the brain's mechanical response to trauma [45] and to generate higher fidelity finite element simulation responses. These simulations could benefit the design of protective safety equipment, such as helmet designs [46], because it would allow quickly iterating through designs and materials to optimize protective headgear and include how the brain would be affected.

Furthermore, these findings have implications in the way that brain tissue, and perhaps some other soft materials, need to be examined at dynamic rates with regards to the initial mechanical response. Distinguishing between inertial effects and inherent water-related material properties will provide a better understanding of the brain's micro and macro mechanics, which can then be potentially linked to physiological damage. This difference affects the way that brain tissue is considered and analyzed at dynamic rates as well as the way material models for brain should be implemented in macroscale simulation models. The current study provides a novel description for the effect of hydration level on the mechanical behavior of the brain. It also lays the experimental basis for a mixture theory-based material modeling of the brain. In extension, this change in the simulated material model will affect the way macroscale models of the brain will respond to dynamic deformations and any protective headgear design they are used for. Hence, quantifying the role of water at these strain rate ranges could help unlock a better understanding of the physics and pathophysiology of TBI. Lastly, this study is limited without histological quantification of water's role during deformation, a process that will constitute the scope of future research.

4. Conclusions

The goal of this research was to assess hydration effects on the mechanical behavior of porcine brain over a range of strain rates (quasi-static and high dynamic rates) and analyze the tissue hydration effects using FEA at high strain rates. Experimental results show a strong strain rate dependence for the wet brain (~80% m/m); however, the dry brain's tangent modulus, yield point (elastic–inelastic transition stress), and strain at yield point were strain rate insensitive (Figures 7 and 10). Two different phenomenological behaviors emerged at the different applied strain rates as well. At higher strain rates (~700 s^{-1}), the wet brain's behavior was marked by an initial hardening trend, followed by a strain-softening, and then strain-hardening (concave-up) as the deformation increased. In contrast at quasi-static strain rates (~0.01 s^{-1}), the wet brain's behavior was marked by an initial toe region and then by concave-up strain hardening similar to other soft tissue phenomenological behavior [47]. The tangent modulus, initial hardening peak stress (σ_p) and strain at σ_p of the wet brain high strain rate data showed a statistical significance (Figure 8 and Table 5) with inconsequential rate dependence of elastic–viscoelastic transition stress (σ_t) and strain at σ_t in the quasi-static range (Figure 9 and Table 5).

At quasi-static rates, the difference between wet (Figure 7b) and dry (Figure 10) brain material was significant in terms of the magnitude of the stresses and the characteristic stress–strain behavior. The dry brain material incurred concave-down hardening while the wet brain incurred concave-up hardening. Significant differences in σ_t and strain at σ_t over quasi-static strain rates point to distinct characteristics of the mechanical responses of wet and dry brain (Tables 5 and 6); however, no significant difference was observed in the tangent modulus, σ_t, and strain at σ_t of the dry brain at quasi-static and higher strain rates (Table 6). Micromechanical FEA using various proportions of water in the dry brain further showed that water played a major role on the initial hardening trend (Figure 11). In summary, the different mechanical properties and behavior trends could be attributed to water's dominant role at quasi-static and high rates. As such, the need to develop constitutive models with the effect of water is crucial to the understanding of the physics and pathophysiology of TBI.

Author Contributions: Conceptualization, R.K.P., J.L., M.F.H., and L.N.W.; Methodology, W.R.W., M.T.B., and Y.M.; Software, R.K.P., W.R.W., and Y.M.; Validation, R.K.P.; Formal analysis, R.K.P. and J.L.; Investigation, R.K.P., M.T.B., and W.R.W.; Resources, L.N.W. and M.F.H.; Data curation, R.K.P.; Writing—Original draft preparation, R.K.P.; Writing—review and editing, M.A.M., M.F.H., and L.N.W.; Visualization, R.K.P., M.A.M., and Y.M.; Supervision, R.K.P., L.N.W., and M.F.M.; Project administration, L.N.W., M.F.H., J.L., and J.S.; Funding acquisition, M.F.H. and L.N.W.

Funding: The authors would like to recognize the Center for Advanced Vehicular Systems (CAVS) and the Agricultural and Biological Engineering Department at Mississippi State University for supporting this work. This material is based upon work supported by the U.S. Army TACOM Life Cycle Command under Contract no. W56HZV-08-C-0236, through a subcontract with Mississippi State University, and was performed for the Simulation Based Reliability and Safety (SimBRS) research program. Also, this material is based upon work supported by the National Nuclear Security Administration, (Department of Energy) under award number DE-FC26-06NT42755.

Acknowledgments: The authors would like to thank David Adams and Erin E. Colebeck for their effort in this research.

Conflicts of Interest: The authors declare no conflict of interest.

Appendix A.

The material model, MSU TP Ver. 1.1, has been developed to account for the elastic and inelastic deformation of an amorphous polymer. The kinematics of the material model draws its inspiration from the works of Kröner [48] and Lee [49] in which a multiplicative decomposition of the deformation gradient $\mathbf{F}$ (into elastic and inelastic components) was proposed. This multiplicative decomposition renders itself to an elastoviscoplastic model. The expression for $\mathbf{F}$ is given as

$$\mathbf{F} = \mathbf{F}^{e}\mathbf{F}^{p}\ ,\ \mathrm{J} = \det\mathbf{F} = \ \mathrm{J}^{e}\mathrm{J}^{p}\ ,\ \mathrm{J}^{e} = \det \mathbf{F}^{e}\ ,\ \mathrm{J}^{p} = \det \mathbf{F}^{p} \tag{A1}$$

where J is the determinant of $\mathbf{F}$. Such a description is physically motivated by the mechanisms underlying the elasticity and inelasticity in amorphous polymers. $\mathbf{F}^{e}$ represents the elastic part due to 'reversible elastic mechanisms', such as bond stretching and mainly chain rotation inducing the different conformations of the intermolecular structure in the polymeric material. $\mathbf{F}^{p}$ represents the inelastic or plastic part due to 'irreversible mechanisms', such as permanent chain stretching and the dissipative mechanisms due to the relative slippage of molecular chains in polymers. An overview of the material model with representative elastic and inelastic parts is given in Figure A1.

The kinematic multiplicative decomposition (Equation (A1)) suggests that there exists an intermediate configuration between the undeformed $\mathbf{B}_{o}$ and the current $\mathbf{B}$ configuration, which is denoted here by $\overline{\mathbf{B}}$. All calculations of the inelastic flow and evolution equations were performed in this intermediate configuration $\overline{\mathbf{B}}$. The Clausius–Duhem inequality [50] is then prescribed in Equation (A2).

$$\boldsymbol{\tau} : \boldsymbol{d}_{e} + \overline{\boldsymbol{M}} : \overline{\boldsymbol{D}}_{\mathrm{p}} - \dot{\overline{\psi}} \geq 0 \tag{A2}$$

Where $\boldsymbol{\tau}$ is the Cauchy stress in the current configuration ($\mathbf{B}$), $\overline{\boldsymbol{D}}^{e}$ is the elastic rate of deformation tensor, $\overline{\boldsymbol{M}}$ is the Mandel stress residing in the intermediate configuration ($\overline{\mathbf{B}}$), $\overline{\boldsymbol{D}}^{p}$ is the plastic rate of deformation tensor, and $\dot{\overline{\psi}}$ is the rate of change of the strain energy density function. Inelastic flow was of the model was assumed to be incompressible, and was void of plastic rotational component, that is, $\overline{\boldsymbol{w}}^{p} = \boldsymbol{0}$ [50,51]. The flow rule for the inelastic deformation captures the polymer's viscous flow observed during the yielding of the material. The inelastic deformations normally arise from the localized slips and viscous flow processes of polymeric chains that result in permanent displacement. In order to capture the rate of change of such processes, the flow rule represented by Equation (A3).

$$\dot{\mathbf{F}}^{\mathbf{P}} = \overline{\mathbf{D}}^{\mathbf{P}}\mathbf{F}^{\mathbf{P}},\ \overline{\mathbf{D}}^{\mathbf{P}} = \left(1/\sqrt{2}\right).\dot{\overline{\gamma}}^{\mathrm{P}}\overline{\mathbf{N}}^{\mathbf{P}} \tag{A3}$$

where $\overline{\mathbf{N}}^{\mathbf{P}} = \frac{\overline{\mathbf{DEV}}\left(\overline{\mathbf{M}}-\overline{\alpha}\right)}{\|\overline{\mathbf{DEV}}\left(\overline{\mathbf{M}}-\overline{\alpha}\right)\|}$ is the direction of the viscous flow, $\overline{\mathbf{D}}^{\mathbf{P}}$ the inelastic rate of deformation, and $\dot{\overline{\gamma}}^{\mathrm{P}}$ is the viscous shear strain rate given by the equation

$$\dot{\overline{\gamma}}^{\mathrm{P}} = \dot{\overline{\gamma}}_0^{\mathrm{P}}\left[Sinh\left(\frac{\overline{\tau} - \left(\overline{\kappa}_1 + \overline{\kappa}_2 + \alpha_{\mathrm{p}}\overline{\pi}\right)}{\mathrm{Y}}\right)\right]^{\mathrm{m}} \tag{A4}$$

where $\overline{\tau} = \left(1/\sqrt{2}\right)\|\overline{\mathbf{DEV}}\left(\overline{\mathbf{M}} - \overline{\alpha}\right)\|$ is an equivalent shear stress term and $\overline{\pi} = (1/3)\overline{\mathbf{Tr}}\left(\overline{\mathbf{M}}\right)$ is the effective pressure term. In Equation (A4), m is the strain rate sensitivity parameter, Y is a material parameter, and α_p is the pressure sensitivity parameter. Equation (A4) has been derived from previous works on plastic flow rules by Fotheringham and Cherry [52], Anand [53], and Richeton [54,55]. The inverse hyperbolic sine function captures inelastic material response that leads to thermal dissipation, and m corresponds to the nature of the motion of polymeric chains.

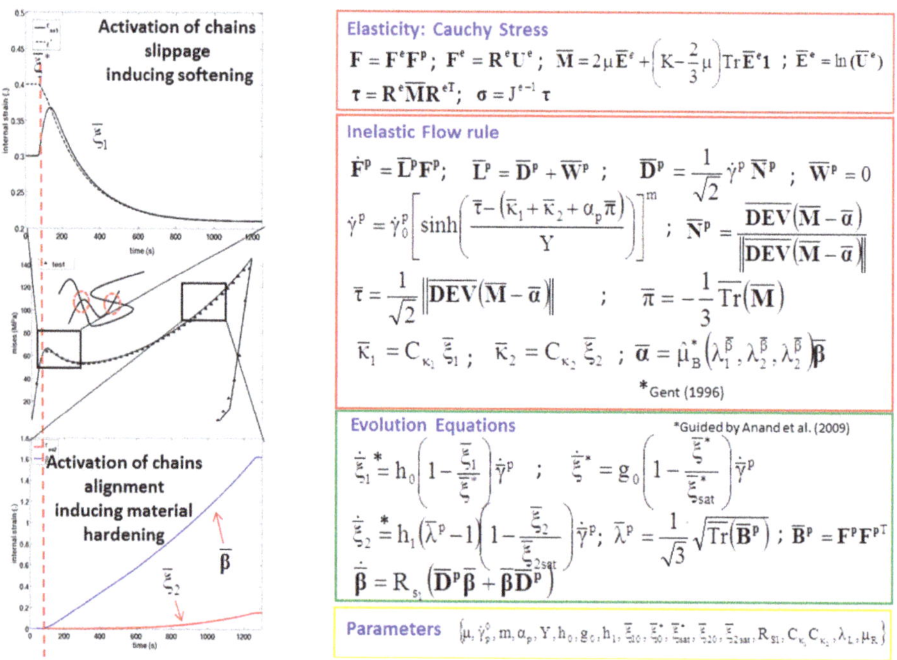

Figure A1. Overview of the material model, MSU TP Ver. 1.1, with salient features of the flow rule and evolution equations [40].

$\overline{\xi}_1$, $\overline{\xi}_2$, and $\overline{\mathbf{E}}^{\beta}$ are internal strain fields (internal state variables) induced due by the presence of defects such as bio-polymeric chain locking or entanglement of macromolecular chains. They are associated with the irreversible mechanisms related to the inelastic material behavior. While $\overline{\xi}_1$ captures the strain-like quantity arising due to polymeric chain locking and entanglement, $\overline{\xi}_2$ represents the internal strain-like quantity induced from the alignment of polymeric chains. Additionally, $\overline{\mathbf{E}}^{\beta}$ is a tensorial variable associated with the stretch-like tensor $\overline{\beta}$, which represents directional-dependent hardening induced at large deformations by the stretching of polymeric chains and fibers. Furthermore, $\overline{\kappa}_1$, $\overline{\kappa}_2$, and $\overline{\alpha}$ are stress-like thermodynamic conjugates of the $\overline{\xi}_1$, $\overline{\xi}_2$, and $\overline{\mathbf{E}}^{\beta}$ respectively. Additionally, $\overline{\xi}^{*}$ corresponds to the strain threshold (corresponding to an energetic barrier) after which the polymeric chains can slip. In the evolution equations of the internal state variables pertaining to inelastic deformation, h_o, h_1, and g_o are the hardening moduli for the internal strain-likes terms $\overline{\xi}_1$, $\overline{\xi}_2$, and $\overline{\xi}^{*}$ respectively. Here $\overline{\xi}_{2sat}$ and $\overline{\xi}^{*}_{sat}$ relate to the saturation values of $\overline{\xi}_2$ and $\overline{\xi}^{*}$. As mentioned above, $\overline{\mathbf{E}}^{\beta}$, which arises from the stretching of polymeric chains and fibers at large deformation was derived using statistical approach [56] for hyperelasticity. Here $\hat{\mu}_B$ represents the modulus of internal shear stress tensor $\overline{\alpha}$. The evolution of $\overline{\beta}$ follows a kinematic hardening-type rule for metals that was first proposed by Prantil et al. [57] and then adopted by Ames et al. [58] for polymers, and R_{s_1} is a constant specific to the material.

A one-dimensional version (material point simulator) of the above-mentioned material model was implemented in the MATLAB [41], and then the experimental data was used to calibrate the material point simulator using optimization functions available in MATLAB (Figure A2). Once the material point simulator and the experimental data had correlation, the derived material parameters were then implemented in ABAQUS/Explicit [42] through a three-dimensional user material subroutine to describe the material's mechanical response.

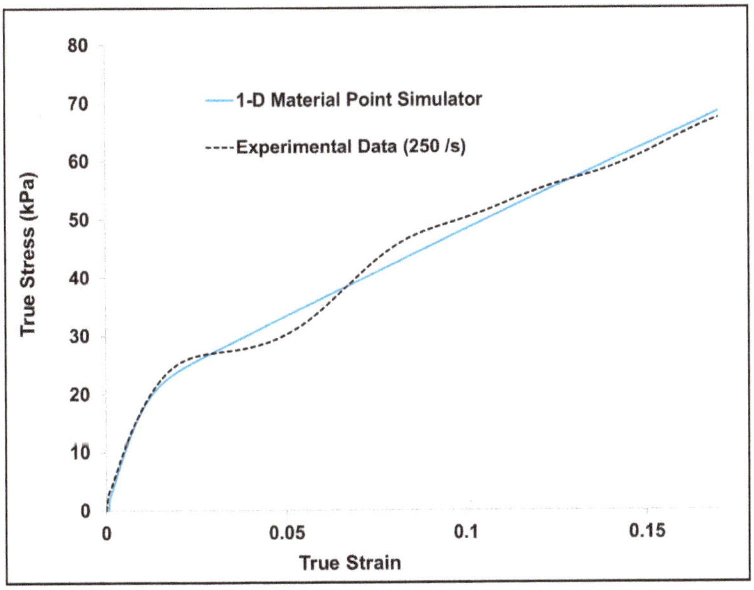

Figure A2. Comparison of experiment and one-dimensional (1-D) material point simulator σ_{33} for dry porcine brain sample compression at 250 s^{-1}. The material parameters for MSU TP Ver.1.1 material model were calibrated to the experimental data.

Appendix B.

Split-Hopkinson pressure bar (SHPB) is a high strain rate testing apparatus comprised of a series of bars, namely: the striker bar, incident bar, and transmitted bar. These bars transmit a single shock wave, provided by the impact of the striker bar, through a specimen sandwiched between the incident and transmitted bars. This shock wave is useful for gathering stress strain relations during high strain rate deformation. An illustration of the SHPB apparatus is shown in Figure A3.

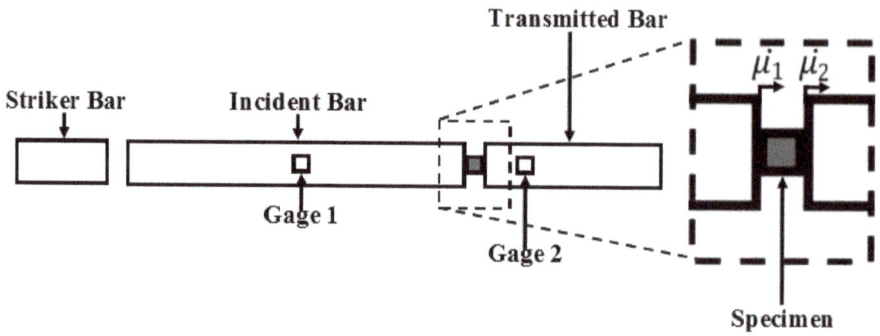

Figure A3. Illustration showing the location for the bars, strain gauges, and specimen in a typical Split-Hopkinson pressure bar (SHPB) configuration. Velocities for the incident and transmitted bars are denoted by $\dot{u}_1$and $\dot{u}_2$.

The theory of an SHPB is based on the elastic wave propagation in long cylindrical bars, and on the principle of superposition of waves. The stress wave traveling through the bar is assumed to propagate longitudinally with viscoelastic dispersion, due to our use of polycarbonate as the bar material. The initial wave, the incident wave, propagates down the incident bar and is recorded by a strain gauge attached to the incident bar, known as gauge 1. Due to the specimen having different mechanical impedance than the bars, a portion of the incident wave is transmitted through the specimen, while the remainder of the wave is reflected off of the specimen-bar interface. The transmitted portion, known as the transmitted wave, is captured by a strain gauge located on the transmitted bar. The reflected wave is captured by gauge 1.

The velocities of the bars near the specimen-bar interfaces, $\dot{u}$, at some time, t, are shown to be

$$\dot{u}_1(t) = -c_1 * (\varepsilon_i(t) - \varepsilon_r(t)) \tag{A5}$$

$$\dot{u}_2(t) = -c_2 * \varepsilon_t(t) \tag{A6}$$

where c_1 is the longitudinal wave speed, ε is the strain gauge record, and the subscripts i, r, and t are the incident, reflected, and transmitted waves, respectively. Subscripts 1 and 2 represent the incident and transmitted bars and are illustrated in Figure A3. By knowing the velocities of the bars at the specimen–bar interfaces, the strain rate, $\dot{\varepsilon}_s$, and strain ε_s, of the specimen can be found as

$$\dot{\varepsilon}_s(t) = \frac{\dot{u}_1(t) - \dot{u}_2(t)}{L_s} \tag{A7}$$

$$\varepsilon_s(t) = \int \dot{\varepsilon}_s(t)dt \tag{A8}$$

where L is the instantaneous length and the subscript s refers to the specimen. The forces, F, in the bars can then be found:

$$\mathrm{F_1(t) = A_1E_1 * (\varepsilon_i(t) + \varepsilon_r(t))} \tag{A9}$$

$$\mathrm{F_2(t) = A_2E_2\varepsilon_t(t)} \tag{A10}$$

where A and E are the cross-sectional area and elastic modulus, respectively. Knowledge of the forces on both sides of the specimen is important as this is the most widely used method of examining uniform loading in the specimen which is critical for a valid compression test. If the forces at each specimen side agree well, the specimen stress, σ_s, can then be found as

$$\sigma_s(t) = \frac{F_2(t)}{A_s}. \tag{A11}$$

Using only the transmitted bar force due to the reduced ringing in the bar signal. These calculations allow for the determination of the effect of the true strain rate on the true stress–strain behavior of materials.

Furthermore, in performing the quasi-static tests on the wet and dry brain, the following equations were used for calculating the true stress–strain response of the material. The engineering stress was given as

$$\sigma_{\mathrm{E}} = \frac{\mathrm{P}}{\mathrm{A}_0} \tag{A12}$$

where P is the force applied on the material, and A_0 the initial cross-sectional area of the sample. The engineering strain was defined as

$$\varepsilon_{\mathrm{E}} = \frac{\ell - \ell_0}{\ell_0} \tag{A13}$$

where l_0 is the initial height of the sample being tested, and l is the height of the sample at any given time. Using Equations (A12) and (A13), the true stress and strain were calculated using the following formulae:

$$\sigma_T = \sigma_E(1 + \varepsilon_E) \tag{A14}$$

$$\varepsilon_T = ln(1 + \varepsilon_E) \tag{A15}$$

Equations (A14) and (A15) were used for calculating the true stress–strain behavior of dry and wet brain samples at quasi-static rates.

Appendix C.

Table A1. Summary of symbol descriptions.

Symbol	Description
U_s	Wave velocity
U_p	Particle velocity
c_0, s	Wave velocity and particle velocity linear relationship constants
$\overline{\Gamma}$	Free energy
$\dot{\overline{\psi}}$	Rate of change of the strain energy density function
$\overline{\mathrm{C}}^{\mathrm{e}}$	Elastic part or the right Cauchy–Green tensor
$\overline{\xi}_1, \overline{\xi}_2, \overline{\mathbf{E}}^{\beta}$	Internal strain fields (internal state variables)
$\dot{\overline{\xi}}^*$	Evolving strain threshold
$\overrightarrow{\xi}_{\mathrm{sat}}, \overline{\xi}_{2\mathrm{sat}}$	Saturation value of $\dot{\overline{\xi}}^*$, $\overline{\xi}_2$
σ	Cauchy stress
$\mathbf{F}$	Deformation gradient
$\mathbf{F}^{\mathbf{e}}, \mathbf{F}^{\mathbf{e}\mathrm{T}}$	Elastic part of F, transpose of the elastic part of F
$\mathbf{F}^{\mathbf{p}}, \mathbf{F}^{\mathbf{p}\mathrm{T}}$	Plastic part of F, transpose of the plastic part of F
J, J^e, J^p	Determinant of F, determinant of $\mathbf{F}^{\mathbf{e}}$, Determinant of $\mathbf{F}^{\mathbf{p}}$
$\overline{\mathbf{D}}^e, \overline{\mathbf{D}}^{\mathbf{p}}$	Elastic rate of deformation, inelastic rate of deformation
$\overline{S}$	Second Piola–Kirchoff stress
$\overline{w}^p$	Plastic rotational deformation
$\mathbf{R}^{\mathbf{e}}, \mathbf{R}^{\mathbf{e}\mathrm{T}}$	Elastic portion of the rotation tensor, transpose of the elastic portion of the rotation tensor
$\mathbf{U}^{\mathbf{e}}$	Right stretch tensor

Table A1. *Cont.*

Symbol	Description
$\overline{\mathbf{M}}$	Mandel stress
τ_1	Kirchoff stress (elasto-viscoplastic part)
μ	Elastic shear moduli
$\overline{\mathbf{I}}$	Identity matrix
$\overline{\mathbf{E}}^e$	Elastic part of the Green–Lagrange strain
K	Bulk moduli
$\overline{\kappa}_1$, $\overline{\kappa}_2$	Stress-like thermodynamic conjugates to $\overline{\xi}_1$,$\overline{\xi}_2$
$\overline{\alpha}$	Stress-like thermodynamic conjugate to $\overline{\mathbf{E}}^{\beta}$
$\overline{\dot{\gamma}}^p$	Viscous shear strain rate
$\overline{\dot{\gamma}}_0^p$	Reference viscous shear strain rate
$\overline{\pi}$	Effective pressure
$\overline{\tau}$	Equivalent shear stress
αp	Stress-like thermodynamic conjugate of $\overline{\mathbf{E}}^{\beta}$
Y	Yield criterion
$\overline{\beta}$	Stretch tensor
$\dot{\overline{\beta}}$	Rate of stretch tensor
h_0, g_0, h_1	Hardening moduli
$\overline{\mathbf{B}}^p$	Intermediate plastic configuration
σ_t	Elastic-viscoelastic transition stress
σ_p	Elastic-plastic transition stress
$\overline{\mathbf{N}}^p$	Direction of viscous flow
$\dot{u}_1$	Hoppy bar velocities
t	Time
ε	Strain
$\dot{\varepsilon}_s$, ε_s	Hoppy bar sample strain rate, hoppy bar sample strain
L_s	Sample instantaneous length
A_1, A_2	Hoppy bar areas
F_1, F_2	Hoppy bar forces
E_1, E_2	Hoppy bar elastic moduli
ε_i, ε_t	Strain in the incident and transmitted hoppy bars
σ_s	Hoppy bar sample stress
A_s	Hoppy bar sample area
σ_E, ε_E	Engineering stress, engineering strain
A_0	Initial sample area
P	Force applied on the material
ℓ, ℓ_0	Current and reference lengths
σ_T, ε_T	True stress, true strain

References

1. Langlois, J.A.; Rutland-Brown, W.; Wald, M.M. The Epidemiology and Impact of Traumatic Brain Injury A Brief Overview. *J. Head Trauma Rehabilm* **2006**, *21*, 375–378. [CrossRef]
2. Finkelstein, E.A.; Corso, P.S.; Miller, T.R. *The Incidence and Economic Burden of Injuries in the United States*; Oxford University Press: New York, NY, USA, 2006; ISBN 9780195179484.
3. Thurman, D.J.; Alverson, C.; Dunn, K.A.; Guerrero, J.; Sniezek, J.E. Traumatic Brain Injury in the United States: A Public Health Perspective. *J. Head Trauma Rehabil.* **1999**, *14*, 602–615. [CrossRef] [PubMed]
4. Holbourn, A.H.S. MECHANICS OF HEAD INJURIES. *Lancet* **1943**, *242*, 438–441. [CrossRef]
5. Pudenz, R.H.; Shelden, C.H. The Lucite Calvarium—A Method for Direct Observation of the Brain. *J. Neurosurg.* **1946**, *3*, 487–505. [CrossRef] [PubMed]
6. Ommaya, A.K. Mechanical properties of tissues of the nervous system. *J. Biomech.* **1968**, *1*, 127–138. [CrossRef]
7. Fallenstein, G.T.; Hulce, V.D.; Melvin, J.W. Dynamic mechanical properties of human brain tissue. *J. Biomech.* **1969**, *2*, 217–226. [CrossRef]

8. Stalnaker, R.L. Mechanical Properties of the Head. West Virginia University, 1969. Available online: http://wbldb.lievers.net/10055045.html (accessed on 30 April 2019).
9. Shuck, L.Z.; Advani, S.H. Rheological Response of Human Brain Tissue in Shear. *J. Basic Eng.* **1972**, *94*, 905–911. [CrossRef]
10. McElhaney, J.H.; Melvin, J.W.; Roberts, V.L.; Portnoy, H.D. Dynamic Characteristics of the Tissues of the Head. In *Perspectives in Biomedical Engineering*; Kenedi, R.M., Ed.; Palgrave Macmillan UK: London, UK, 1973; pp. 215–222, ISBN 978-1-349-01604-4.
11. Estes, M.S.; McElhaney, J. Response of Brain Tissue of Compressive Loading. In Proceedings of the American Society of Mechanical Engineers Biomechanical and Human Factors Conference, Washington, DC, USA, 31 May–3 June 1970.
12. Donnelly, B.R.; Medige, L. Shear Properties of Human Brain Tissue. *Trans. ASME, J. Biomech. Eng.* **1997**, *119*, 423–432.
13. Arbogast, K.B.; Meaney, D.F.; Thibault, L.E. Biomechanical Characterization of the Constitutive Relationship for the Brainstem. In *SAE Technical Paper 952716, Proceeding of the 39th Stapp Car Crash Conference, 8–10 November*; Society of Automotive Engineers: Warrendale, PA, USA, 1995.
14. Arbogast, K.B.; Margulies, S.S. Regional Differences in Mechanical Properties of the Porcine Central Nervous System. *SAE Trans.* **1997**, *106*, 3807–3814.
15. Arbogast, K.B.; Margulies, S.S. Material characterization of the brainstem from oscillatory shear tests. *J. Biomech.* **1998**, *31*, 801–807. [CrossRef]
16. Prange, M.T.; Meaney, D.F.; Margulies, S.S. Directional properties of gray and white brain tissue undergoing large deformation. *Adv. Bioeng.* **1998**, *39*, 151–152.
17. Aimedieu, P.; Grebe, R.; Idy-Peretti, I. Study of brain white matter anisotropy. *Annu. Int. Conf. IEEE Eng. Med. Biol.* **2001**, *2*, 1009–1011.
18. Miller, K. How to test very soft biological tissues in extension? *J. Biomech.* **2001**, *34*, 651–657. [CrossRef]
19. Miller, K.; Chinzei, K. Mechanical properties of brain tissue in tension. *J. Biomech.* **2002**, *35*, 483–490. [CrossRef]
20. Bayly, P.V.; Black, E.E.; Pedersen, R.C.; Leister, E.P.; Genin, G.M. In vivo imaging of rapid deformation and strain in an animal model of traumatic brain injury. *J. Biomech.* **2006**, *39*, 1086–1095. [CrossRef] [PubMed]
21. Smith, D.H.; Wolf, J.A.; Lusardi, T.A.; Lee, V.M.-Y.; Meaney, D.F. High Tolerance and Delayed Elastic Response of Cultured Axons to Dynamic Stretch Injury. *J. Neurosci.* **1999**, *19*, 4263–4269. [CrossRef] [PubMed]
22. Tamura, A.; Hayashi, S.; Watanabe, I.; Nagayama, K.; Matsumoto, T. Mechanical Characterization of Brain Tissue in High-Rate Compression. *J. Biomech. Sci. Eng.* **2007**, *2*, 115–126. [CrossRef]
23. Bain, A.C.; Meaney, D.F. Tissue-level thresholds for axonal damage in an experimental model of central nervous system white matter injury. *J. Biomech. Eng.* **2000**, *122*, 615–622. [CrossRef]
24. Pfister, B.J.; Weihs, T.P.; Betenbaugh, M.; Bao, G. An in vitro uniaxial stretch model for axonal injury. *Ann. Biomed. Eng.* **2003**, *31*, 589–598. [CrossRef]
25. Begonia, M.T.; Prabhu, R.; Liao, J.; Horstemeyer, M.F.; Williams, L.N. The Influence of Strain Rate Dependency on the Structure–Property Relations of Porcine Brain. *Ann. Biomed. Eng.* **2010**, *38*, 3043–3057. [CrossRef]
26. Sparks, J.L.; Dupaix, R.B. Constitutive modeling of rate-dependent stress–strain behavior of human liver in blunt impact loading. *Ann. Biomed. Eng.* **2008**, *36*, 1883–1892. [CrossRef]
27. Song, B.; Chen, W.; Ge, Y.; Weerasooriya, T. Dynamic and quasi-static compressive response of porcine muscle. *J. Biomech.* **2007**, *40*, 2999–3005. [CrossRef]
28. Pervin, F.; Chen, W.W. Dynamic mechanical response of bovine gray matter and white matter brain tissues under compression. *J. Biomech.* **2009**, *42*, 731–735. [CrossRef]
29. Prabhu, R.; Horstemeyer, M.F.; Tucker, M.T.; Marin, E.B.; Bouvard, J.L.; Sherburn, J.A.; Liao, J.; Williams, L.N. Coupled experiment/finite element analysis on the mechanical response of porcine brain under high strain rates. *J. Mech. Behav. Biomed. Mater.* **2011**, *4*, 1067–1080. [CrossRef]
30. Clemmer, J.; Prabhu, R.; Chen, J.; Colebeck, E.; Priddy, L.B.; McCollum, M.; Brazile, B.; Whittington, W.; Wardlaw, J.L.; Rhee, H.; et al. Experimental Observation of High Strain Rate Responses of Porcine Brain, Liver, and Tendon. *J. Mech. Med. Biol.* **2016**, *16*, 1650032. [CrossRef]
31. Cheng, S.; Bilston, L.E. Unconfined compression of white matter. *J. Biomech.* **2007**, *40*, 117–124. [CrossRef]
32. Neeb, H.; Ermer, V.; Stocker, T.; Shah, N.J. Fast quantitative mapping of absolute water content with full brain coverage. *Neuroimage* **2008**, *42*, 1094–1109. [CrossRef]
33. Hopkinson, B. The Effects of Momentary Stresses in Metals. *Proc. R. Soc. London* **1904**, *74*, 498–506. [CrossRef]

34. Kolsky, H. An investigation of the mechanical properties of materials at very high rates of loading. *Proc. R. Sot. Lond. B62* **1949**, *676*, 676–700. [CrossRef]
35. Tucker, M.T.; Horstemeyer, M.F.; Whittington, W.R.; Solanki, K.N.; Gullett, P.M. The effect of varying strain rates and stress states on the plasticity, damage, and fracture of aluminum alloys. *Mech. Mater.* **2010**, *42*, 895–907. [CrossRef]
36. Gary, G.; Klepaczko, J.R.; Zhao, H. Generalization of split Hopkinson bar technique to use viscoelastic bars. *Int. J. Impact Eng.* **1995**, *16*, 529–530. [CrossRef]
37. Prabhu, R.; Whittington, W.R.; Patnaik, S.S.; Mao, Y.; Begonia, M.T.; Williams, L.N.; Liao, J.; Horstemeyer, M.F. A Coupled Experiment-finite Element Modeling Methodology for Assessing High Strain Rate Mechanical Response of Soft Biomaterials. *J. Vis. Exp.* **2015**, e51545. [CrossRef]
38. Miller, K. Most recent results in the biomechanics of the brain. *J. Biomech.* **2005**, *38*, 965. [CrossRef] [PubMed]
39. MacDonald, R.A.; MacDonald, W.M. Thermodynamic properties of fcc metals at high temperatures. *Phys. Rev. B* **1981**, *24*, 1715–1724. [CrossRef]
40. Bouvard, J.L.; Ward, D.K.; Hossain, D.; Marin, E.B.; Bammann, D.J.; Horstemeyer, M.F. A general inelastic internal state variable model for amorphous glassy polymers. *Acta Mech.* **2010**, *213*, 71–96. [CrossRef]
41. *MATLAB 2010*; The MathWorks Inc.: Natick, MA, USA, 2010.
42. *ABAQUS/Explicit User's Manual 2009*; Simulia Inc.: Providence, RI, USA, 2009.
43. Franceschini, G.; Bigoni, D.; Regitnig, P.; Holzapfel, G.A. Brain tissue deforms similarly to filled elastomers and follows consolidation theory. *J. Mech. Phys. Solids* **2006**, *54*, 2592–2620. [CrossRef]
44. Weinberg, K.; Ortiz, M. Shock wave induced damage in kidney tissue. *Comput. Mater. Sci.* **2005**, *32*, 588–593. [CrossRef]
45. Ratajczak, M.; Ptak, M.; Chybowski, L.; Gawdzińska, K.; Bedziński, R. Material and structural modeling aspects of brain tissue deformation under dynamic loads. *Materials (Basel)* **2019**, *12*, 271. [CrossRef] [PubMed]
46. Kaczyński, P.; Ptak, M.; Fernandes, F.A.O.; Chybowski, L.; Wilhelm, J.; de Sousa, R.J.A. Development and Testing of Advanced Cork Composite Sandwiches for Energy-Absorbing Structures. *Materials (Basel)* **2019**, *12*, 697. [CrossRef]
47. Yamada, H.; Evans, F.G. *Strength of Biological Materials*; Williams & Wilkins: Philadelphia, PA, USA, 1970.
48. Kroner, E. Allgemeine kontinuumstheorie der versetzungen und eigenspannungen. *Arch. Ration. Mech. Anal.* **1960**, *4*, 273–334. [CrossRef]
49. Lee, E.H. Elastic plastic deformation at finite strain. *ASME J. Appl. Mech.* **1969**, *36*. [CrossRef]
50. Gurtin, M.E.; Anand, L. The Decomposition F = FeFp, Material Symmetry, and Plastic Irrotationality for Solids that are Isotropic-Viscoplastic or Amorphous. *Int. J. Plast.* **2005**, *21*, 1686–1719. [CrossRef]
51. Boyce, M.C.; Weber, G.G.; Parks, D.M. On the kinematics of finite strain plasticity. *J. Mech. Phys. Solids* **1989**, *37*, 647–665. [CrossRef]
52. Fotheringham, D.G.; Cherry, B.W. Strain rate effects on the ratio of recoverable to non-recoverable strain in linear polyethylene. *J. Mater. Sci.* **1978**, *13*, 231–238. [CrossRef]
53. Anand, L. Single-crystal elasto-viscoplasticity: Application to texture evolution in polycrystalline metals at large strains. *Comput. Methods Appl. Mech. Eng.* **2004**, *193*, 5359–5383. [CrossRef]
54. Richeton, J.; Ahzi, S.; Vecchio, K.S.; Jiang, F.C.; Adharapurapu, R.R. Influence of temperature and strain rate on the mechanical behavior of three amorphous polymers: Characterization and modeling of the compressive yield stress. *Int. J. Solids Struct.* **2006**, *43*, 2318–2335. [CrossRef]
55. Richeton, J.; Ahzi, S.; Vecchio, K.S.; Jiang, F.C.; Makradi, A. Modeling and validation of the large deformation inelastic response of amorphous polymers over a wide range of temperatures and strain rates. *Int. J. Solids Struct.* **2007**, *44*, 7938–7954. [CrossRef]
56. Gent, A.N. A new constitutive relation for rubber. *Rubber Chem. Technol.* **1996**, *69*, 59–61. [CrossRef]
57. Prantil, V.C.; Jenkins, J.T.; Dawson, P.R. An analysis of texture and plastic spin for planar polycrystals. *J. Mech. Phys. Solids* **1993**, *41*, 1357–1382. [CrossRef]
58. Ames, N.M.; Srivastava, V.; Chester, S.A.; Anand, L. A thermo-mechanically coupled theory for large deformations of amorphous polymers. Part II: Applications. *Int. J. Plast.* **2009**, *25*, 1495–1539. [CrossRef]

bioengineering

MDPI

Article

Mechanical Response of Porcine Liver Tissue under High Strain Rate Compression

Joseph Chen [1,†], Sourav S. Patnaik [1,†], R. K. Prabhu [1], Lauren B. Priddy [1], Jean-Luc Bouvard [1], Esteban Marin [1], Mark F. Horstemeyer [1], Jun Liao [1,2,*] and Lakiesha N. Williams [1,3,*]

1 Department of Biological Engineering and Center for Advanced Vehicular Systems, Mississippi State University, Mississippi State, MS 39762, USA; chen.joseph@berkeley.edu (J.C.); sourav.patnaik@utsa.edu (S.S.P.); rprabhu@abe.msstate.edu (R.K.P.); lbpriddy@abe.msstate.edu (L.B.P.); jean-luc.bouvard@mines-paristech.fr (J.-L.B.); marineb@corning.com (E.M.); marineb@corning.com (M.F.H.)
2 Department of Bioengineering, University of Texas at Arlington, Arlington, TX 76010, USA
3 Department of Biomedical Engineering, University of Florida, Gainesville, FL 32611, USA
* Correspondence: jun.liao@uta.edu (J.L.); lwilliams@bme.ufl.edu (L.N.W.)
† Co-first authors.

Received: 15 February 2019; Accepted: 22 May 2019; Published: 30 May 2019

Abstract: In automobile accidents, abdominal injuries are often life-threatening yet not apparent at the time of initial injury. The liver is the most commonly injured abdominal organ from this type of trauma. In contrast to current safety tests involving crash dummies, a more detailed, efficient approach to predict the risk of human injuries is computational modelling and simulations. Further, the development of accurate computational human models requires knowledge of the mechanical properties of tissues in various stress states, especially in high-impact scenarios. In this study, a polymeric split-Hopkinson pressure bar (PSHPB) was utilized to apply various high strain rates to porcine liver tissue to investigate its material behavior during high strain rate compression. Liver tissues were subjected to high strain rate impacts at 350, 550, 1000, and 1550 s^{-1}. Tissue directional dependency was also explored by PSHPB testing along three orthogonal directions of liver at a strain rate of 350 s^{-1}. Histology of samples from each of the three directions was performed to examine the structural properties of porcine liver. Porcine liver tissue showed an inelastic and strain rate-sensitive response at high strain rates. The liver tissue was found lacking directional dependency, which could be explained by the isotropic microstructure observed after staining and imaging. Furthermore, finite element analysis (FEA) of the PSHPB tests revealed the stress profile inside liver tissue and served as a validation of PSHPB methodology. The present findings can assist in the development of more accurate computational models of liver tissue at high-rate impact conditions allowing for understanding of subfailure and failure mechanisms.

Keywords: soft tissue; liver; high-rate compression; polymeric split-Hopkinson pressure bar; finite element modeling

1. Introduction

The liver is the most frequently injured intra-abdominal organ because of its location within the abdomen and its fragile material properties [1]. In 2007, 1.7 million car accidents in the United States resulted in injury (National Highway Traffic Safety Administration) with the liver being one of the most commonly injured abdominal organs from motor vehicle accidents [2,3]. Efforts to determine the optimal safety measures for automobile-related accidents have largely relied on crash dummies, which have significant limitations in recapitulating injury impact to humans [4]. Since the 1970s, there have been no substantial changes in assessing injury. Injury assessment reference values (IARVs) proposed by General Motors for dummies in crash tests were determined via force and acceleration calculations and defined a tolerance level of 5% significant injury risk of various organs [5,6]. An improved, more cost-effective alternative to assess organ damage during car crash situations is the development of computational models that represent the human body and more accurately predict the risk of human tissue/organ injuries. Recent work in developing a geometrically correct "virtual human" has been performed with the goal of measuring bodily trauma in automobile accidents [7–10].

Besides an anatomically relevant mesh, the development of a biofidelic computational model requires knowledge of the mechanical properties of many human tissues and organs under different loading conditions, especially in high-impact situations. Quasi-static biomechanical characterizations of soft tissues have been performed since the 1970s to determine the mechanical properties of various tissue types; however, regarding the response of tissues that may be subjected to high-impact situations such as automobile accidents, sport injuries, and blunt trauma, these quasi-static tests are limited and cannot be extrapolated to high-rate applications. Mechanical testing therefore must be performed at higher strain rates to properly describe the tissue's response during blunt force impacts [11,12].

A standard protocol has not been well established for high strain rate mechanical testing on liver tissues. Sparks et al. customized a drop tower in which a weight was dropped onto a whole human liver organ, resulting in average strain rates up to 62 s^{-1} [13]. Others have used indentation instruments to generate strain rates up to 200 s^{-1} [14,15]. This study considered the split-Hopkinson pressure bar (SHPB) apparatus in an effort to establish a methodology of high strain rate testing of soft tissues. The SHPB apparatus has the ability to apply compressive stresses at high strain rates (100–10,000 s^{-1}) [11] and has been widely applied in testing of metals and inorganic polymers [16,17]. Elastic wave propagation in the SHPB system can be analyzed based on the principle of superposition of waves and the elastic wave propagation theory of classical mechanics. As a result, the stress, strain, and particle velocity can be estimated by analyzing the incident wave and the reflected wave at any cross-section [18].

When the SHPB is used for testing soft tissues, many issues must be considered to generate consistent and accurate data. The mechanical impedance of soft materials is extremely low compared to conventional metallic bars and, therefore, confound proper interpretation of the data. Incorporation of polymeric bars into the SHPB setup has allowed for testing of soft materials such as rubber and biological tissues, of which the acoustical impedance matches more closely with that of the softer polymeric bars. Unlike conventional metallic bars, polymeric bars do not impede wave translation and enable a smooth translation of energy generated by the impact of the incident bar and the soft specimen; this results in smoother, more noise-free curves [12]. A few groups have applied the SHPB apparatus for soft tissue biomechanics experimentation. For example, Song et al. tested porcine muscle along two perpendicular directions at dynamic strain rates up to 3700 s^{-1} using the SHPB apparatus. They found that both directions showed a nonlinear, strain rate-dependent behavior [12]. Similarly, Van Sligtenhorst et al. used a polymeric SHPB apparatus to obtain the mechanical response of bovine muscle at strain rates up to 2300 s^{-1} [11]. As they varied strain rates, they observed strain rate dependency. Recently, Pervin et al. used an aluminum SHPB apparatus to evaluate bovine liver tissue at strain rates ranging from 1000–3000 s^{-1}, and they also found the tissue to exhibit a nonlinear, strain rate-dependent response [19].

The objective of the present work is to investigate the tissue behavior of porcine liver at high-rate impacts using a custom-made PSHPB coupled with finite element analyses. The protocol for testing soft biological materials using PSHPB has been adopted from previous studies [18,20,21]; however, the protocol for procuring and preparing porcine liver samples was developed in-house. Both experimental results and computational simulations of liver tissue under high strain rate conditions will provide the framework to be incorporated into a human model. This model in the future will be implemented to optimize the automobile safety measures to reduce the risk of human injuries and death in high-impact situations.

2. Methodology

2.1. Sample Preparation

Porcine livers from healthy adult pigs were obtained from a local abattoir (age range: 6–9 months; weight range: 250–350 lbs.; sex: male). The specimens were stored in phosphate-buffered saline (PBS) (Sigma-Aldrich, St. Louis, MO, USA) at 4 °C soon after extraction and were transported to the laboratory. All testing was performed within 12 h of extraction. For PSHPB application, the tissue sample was carefully extracted to maintain a certain shape and size. A bar diameter larger than the sample size was important to ensure that most of the energy was transmitted through the sample [11]. Testing was performed on samples with aspect ratios ranging from 1:1 to 3:1, and it was determined that an aspect ratio of 3:1 produced consistent data (results not shown). Thus, a cylindrical die of 30 mm inner diameter was used to cut disc-shaped samples to approximately 27 mm in diameter and 9 mm thick for an aspect ratio of 3:1 (Figure 1). The aspect ratio and size corresponded well with previous studies [11,22]. In order to extract samples in different directions, a 9 mm slice was cut in the appropriate direction, laid on its side, and then punched out with the cylindrical die. The axis of the disc-shaped sample was aligned along one of the three orthogonal directions.

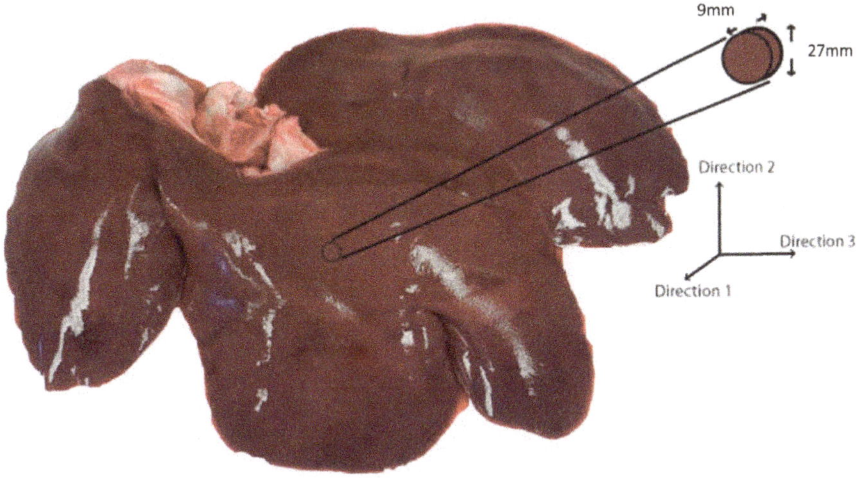

Figure 1. Three orthogonal directions (1, 2, and 3) based on porcine liver anatomy. Representative sample geometry and size.

2.2. High Strain Rate Testing Using a Polymeric Split-Hopkinson Pressure Bar (PSHPB)

High strain rate testing was performed as per previously established protocol [23–25]. The PSHPB, made of commercially extruded natural polycarbonate (PC 1000) rods, was composed of a striker bar, an incident bar, and a transmitted bar with lengths of 0.762, 2.438, and 1.219 m, respectively, and a diameter of 38.1 mm (Figure 2a). A cylindrical specimen was placed between the incident and transmitted bars, and the striker bar was propelled at a specified velocity by means of a pneumatic pressure system. The sample was compressed in the z-direction. The z-direction was normal to the 27 mm diameter surface of the liver specimen.

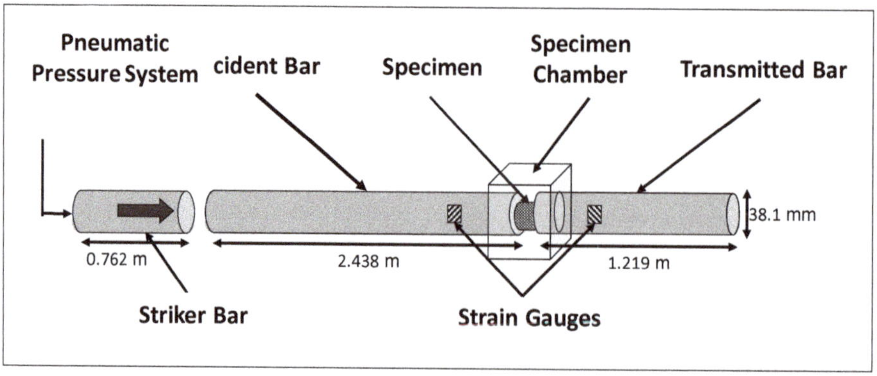

(**a**)

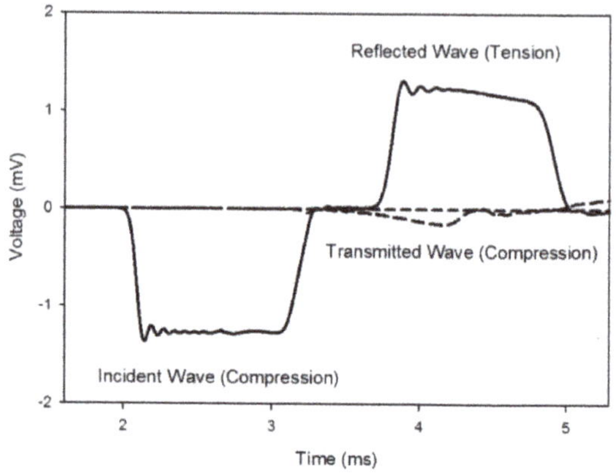

(**b**)

Figure 2. (**a**) Schematic of the polymeric split-Hopkinson pressure bar (PSHPB) apparatus. (**b**) Incident, reflected, and transmitted waves obtained from PSHPB testing on porcine liver tissue.

As the striker bar impacted the incident bar, a compressive wave (incident wave) was generated and propagated down the incident bar where it reached the specimen, which caused compression of the specimen. At this point, a portion of the wave was reflected back into the incident bar as a tensile wave (reflected wave). The remainder of the compressive wave (transmitted wave) was transmitted through the specimen and into the transmitted bar (Figure 2b). The incident, reflected, and transmitted waves were measured by two strain gauges—one gauge on both the incident and transmitted bars. The PSHPB experimental setup was based on the following assumptions: (i) the specimen undergoes uniform and uniaxial stress during deformation; (ii) the incident and transmitted bars are elastic; (iii) the edges of the bars in contact with the specimen remain flat and parallel; (iv) the incident, transmitted, and reflected waves undergo minimal dispersion as they travel along the bars; and (v) strains measured at the surface of the bars are indicative of those throughout the cross-section [26]. The theory behind the SHPB setup and the constitutive true stress–strain relationship of the sample deformation is briefly discussed in Appendix A. The experimental setup also included a laser speed meter for monitoring the incident bar speed and DAQ modules for data acquisition. Data were processed via DAVID Viscoelastic Software [18].

Cylindrical samples were extracted from three orthogonal directions based on porcine liver anatomy (Figure 1). For evaluating strain rate sensitivity, samples were extracted along Direction 1

(Figure 1), and strain rates of 350 ($n = 4$), 550 ($n = 4$), 1000 ($n = 4$), and 1550 s^{-1} ($n = 5$) were applied. The range of strain rates was chosen based on the deformation rate from impact at 55 km/h (more than 1000 s^{-1}) [27]. To evaluate the directional dependence (anisotropy) of tissue behavior, samples were dissected along three orthogonal directions (Directions 1, 2, and 3; $n = 4$ for each direction) and tested at a strain rate of 350 s^{-1}. For each test, a sample was glued between the incident and transmitted bars using cyanoacrylate glue (Cemedine, Japan) [28,29]. Liver tissue was kept moist with PBS throughout the testing procedure, and testing was carried out at room temperature (21–23 °C).

Statistical analyses of three parameters, namely the tangent modulus, peak stress to valley stress ratio, and ultimate stress to valley stress ratio, were conducted using the SigmaStat 3.0 software (SPSS, Chicago, IL, USA). A one-way analysis of variance (ANOVA) method was used for statistical analysis on the two parameters, and a Holm–Sidak test was used for post hoc comparisons. A paired Student's t-test was used to calculate the mechanical difference between the two parameters at different strain rates. For $p < 0.05$, the mechanical difference at various strain rates, for a particular parameter, was considered to be statistically significant.

2.3. Microstructural Analysis

To assess the microstructural characteristics of liver tissue along different orthogonal directions, samples were dissected along each orthogonal direction (1, 2, and 3) corresponding to the orientation of samples used for high strain rate testing. Liver samples were fixed in 10% neutral buffered formalin and dehydrated in a graded ETOH series. Samples were then embedded in Paraplast with CitriSolve as a transitional fluid, sectioned to a thickness of 7 μm, and subjected to hematoxylin and eosin (H&E) staining. In H&E staining, liver cell nuclei were stained black/purple, and extracellular matrix proteins were stained pink.

ImageAnalyzer v.2.2-0 software (CAVS, Mississippi State University, Starkville, MS, USA) was used for microstructural analyses of histological images from samples cut along each orthogonal direction [30]. The parameters obtained for each image during analysis included the following: object count, cell nuclear density, area fraction of cell nuclei, mean area of cell nuclei, and mean nearest neighbor distance (nnd). Total cell nuclei area was a measure of the total area of all cell nuclei, and area fraction was the ratio of total cell nuclei area to total image area. Mean area represented the average area of cell nuclei, and object count was the number of nuclei present in the image. Cell nuclear density equaled the object count divided by the total image area. Mean nnd was a measure of the average distance between neighboring nuclei.

2.4. Finite Element Modeling

Similar to Prabhu et al. [25], finite element (FE) simulations (ABAQUS/explicit solver version 6.9) of porcine liver high-rate tests were conducted to better understand the behavior of the liver tissue under high-rate compression. The bars in the FE simulations were modeled as an elastic polycarbonate material (Young's modulus of 2391 MPa and Poisson's ratio of 0.36) [25]. The hyperelastic and inelastic behaviors of liver tissue were fitted (using partial least square fitting in Matlab®) with a phenomenological internal state variable (ISV) material model developed by Bouvard et al. (Mississippi State University TP, version 1.0) [31,32]. The constitutive model (MSU TP version 1.1) used in this study captured both the instantaneous and long-term steady-state processes during deformation and could admit microstructural features within the internal state variables. With the microstructural features, the internal state variable model used for this study can eventually capture and predict history effects in tissues. In the absence of the microstructural features, other constitutive models should be able to show the nonuniformity of the stress state under the high-rate loadings exhibited here since no varying history was induced. Data from high strain rate PSHPB tests were utilized to calibrate the numerical model (referred here as MSU TP 1.1) using MATLAB (MathWorks Inc., Natick, MA, USA, 2010). Curve fitting (MSU TP 1.1) was performed for experimental data obtained at 205 s^{-1}, and strain rate dependency of the material model was validated with the stress–strain response of the tissue at 550 s^{-1}. Subsequent to the previous step, the model calibration was two-fold. The first step in calibration was performed such that the experimental and FEA strain gage data analyzed through DAVID Elastic (resulting in true stress–strain curves) were in good agreement. In the second step of calibration, the strain gage measurements from the SHPB experiment and FE simulation ere also correlated. Thus, correlation was performed for both strain measurements and the stress–strain of the specimen. It was noteworthy here that the material point simulator, being 1-D, gave a 1-D stress state for calibration, while the stress state in experiments and FE simulations were 3-D. The difference in the loading direction stress σ_{33} between the material point simulator, experiment, and FEA was due to the presence of a 3-D stress state in the experiment and FEA. So, the model calibration was performed through an iterative optimization scheme where model constants were varied appropriately until the experimental and specimen volume averaged FE simulation σ_{33} matched. The values for the material constants for MSU TP 1.1 are found in Table 1. Using calibrated data from PSHPB experiments, several FE simulations at strain rates of 350, 550, 1000, and 1550 s^{-1} were performed (ABAQUS/explicit solver version 6.9 [33]). The finite element model was composed of 22,010 hexahedral elements with the specimen containing 9200 elements. Mesh refinement was conducted to analyze the convergence of computational solutions. Boundary conditions included specified initial velocity for the striker bar. The finite element model simulated the experimental PSHPB setup corresponding to different strain rates, and a "contact" was defined between the surfaces of the specimen and polycarbonate rod.

Table 1. Values of material constants for liver material using the MSU TP 1.1 model.

Model/Material Constants.	Values
μ (MPa)	29
K (MPa)	12,492
γ_{vo} (s^{-1})	99,209.5
m	1.1
Y_o (MPa)	2
α_p	0
λ_L	7
μ_R	0.168627
R_{s1}	1.4
h_o	31
$\zeta^o{}_1$	0
$\zeta^*{}_{sat}$	0.01
$\zeta^*{}_o$	0.3
g_o	0.07
$C_{\kappa 1}$ (MPa)	0.4
h_1	0
$e^o{}_{s2}$	0
$e^{sat}{}_{s2}$	0.4
$C_{\kappa 2}$ (MPa)	0

3. Results

PSHPB experiments showed that liver tissue had a strain rate-sensitive behavior under high-rate compression (Figure 3a). Stresses were significantly higher as the strain rate increased from 350, 550, 1000, to 1550 s^{-1}. The resulting stress–strain behavior showed that the liver tissue exhibited an initial stiffening behavior, which was followed by softening. After softening, tissue hardening took place until yielding and ultimate failure. The nonmonotonic stress–strain behavior described above was apparent for all four strain rates (350, 550, 1000, and 1550 s^{-1}).

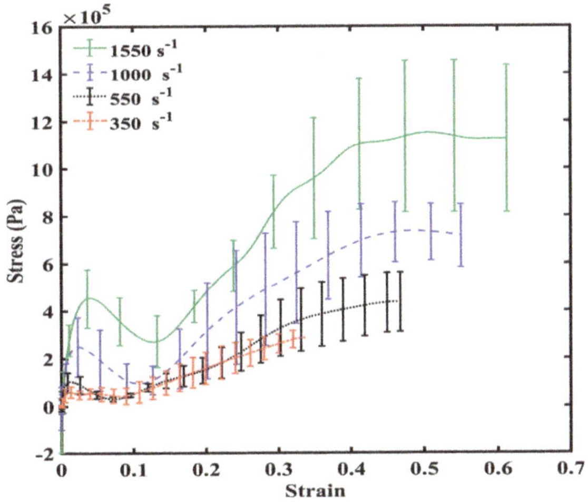

(**a**)

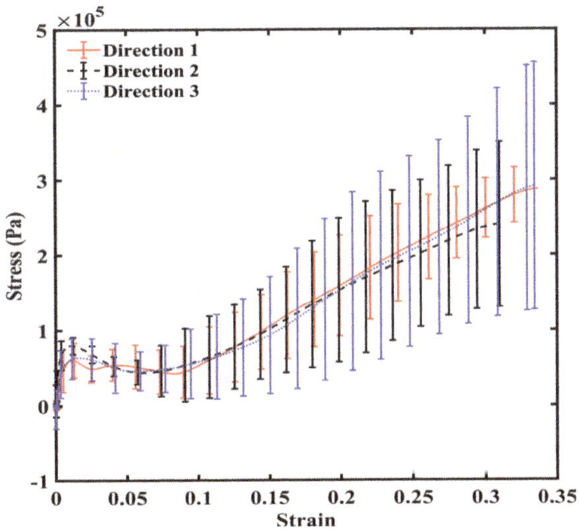

(**b**)

Figure 3. (**a**) True stress–strain response of porcine liver tissue at 350, 550, 1000, and 1550 s^{-1} in Direction 1. n = 4 for 350, 550, and 1000 s^{-1}; n = 5 for 1550 s^{-1}. Error bars indicate standard deviation. (**b**) Mechanical true stress–strain response of porcine liver tissue at 350 s^{-1} in Directions 1, 2, and 3 (n = 4) illustrating isotropy. Error bars indicate standard deviation.

To further examine the relationship between strain rate and liver tissue's mechanical response, data analyses of the stress–strain behaviors at 350, 550, 1000, to 1550 s^{-1} were performed by normalizing both the initial peak stress and the ultimate stress to the valley stress (lowest stress value following initial peak). Both the ratio of peak stress/valley stress and the ratio of ultimate stress/valley stress decreased with an increase of the strain rate (Table 2). Increasing strain rate from 550 to 1000 s^{-1} and from 1000 to 1550 s^{-1} yielded significant differences in the peak to valley stress ratios (ANOVA $p < 0.05$).

Table 2. Ratio of peak stress/valley stress and ratio of ultimate stress/valley stress shows an overall decreasing trend along with the increase of the strain rate. (n = 4).

Strain Rate (s^{-1})	Mean Peak Stress/Valley Stress	Mean Ultimate Stress/Valley Stress
350	5.37 ± 4.59	21.92 ± 16.39
550	3.55 ± 1.56	13.36 ± 6.09
1000	3.00 ± 0.84	12.09 ± 3.56
1550	1.42 ± 0.24	12.48 ± 2.12

The stress–strain behaviors of the liver tissues extracted from three orthogonal directions exhibited no significant differences at 350 s^{-1} (Figure 3b). Overall at high strain rates, porcine liver tissue demonstrated nonlinear, inelastic, strain rate-sensitive mechanical responses; all responses were characterized by an initial peak and subsequent hardening until yielding and failure. The isotropic mechanical behavior was verified by H&E staining of the samples, which showed black/purple cell nuclei and pink extracellular matrix of hepatocytes. This study revealed identical ultrastructures along the three orthogonal directions (Figure 4, Table 3). Image analysis of these stained images revealed no differences regarding each of the three directions in terms of cell nuclear density, area fraction of cell nuclei, mean area of cell nuclei, and mean nnd ($p > 0.05$; data not shown here). Figure 5 gives the strain-time responses (along the loading direction) for the experiment and FE simulation. As observed, there was very good correlation between the experiment and the FE model. The comparison of incident, reflected, and transmitted waves between the PSHPB test and the FE simulation showed that the dispersions of the stress waves were assessed appropriately in the FE model. This implied that the elastic assumption for the dispersion of the waves in the polymeric bars of the FE model was appropriate for modelling the high strain rate phenomenon. The high-frequency fluctuations observed in the FE model strain measurements (Figure 5) could be attributed to the "frictionless" nature of the FE model. While the experiment accompanied friction arising from PSHPB clamp contacts that dampened the fluctuation, the FE model neglected such contact from frictional effects as minimal. It can be observed from Figure 5 that the assumption of frictionless PSHB clamp contacts had not compromised the trend and agreement of the FE model strain data with the experimental data.

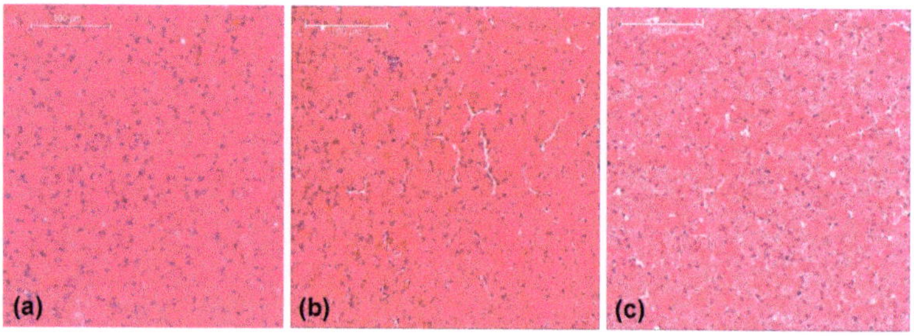

Figure 4. Liver histology revealing the tissue's homogeneity and isotropy. (**a**) Direction 1, (**b**) Direction 2, and (**c**) Direction 3. Liver tissues were fixed with 10% formalin at the load-free condition.

Table 3. Image analysis results from Figure 4a–c revealing the tissue's homogeneity and isotropy.

	Direction 1	Direction 2	Direction 3
Objects	797	795	803
Cell nuclear density (/mm^2)	4.08×10^3	5.91×10^3	4.11×10^3
Area fraction of cell nuclei	9.79%	13.4%	8.06%
Mean area of cell nuclei (μm^2)	23.98 ± 15.58	22.6 ± 15.71	19.59 ± 7.73
Mean nearest neighbor distance (nnd) (μm)	9.62 ± 3.14	8.44 ± 2.40	9.29 ± 3.12

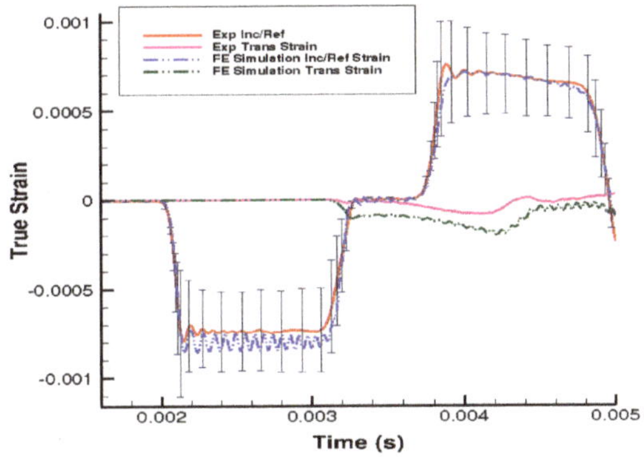

Figure 5. Comparison of the strain in the loading direction from the experimental data and finite element (FE) simulation results at 550 s^{-1}.

The stress state in the cylindrical liver sample was revealed by the FE model of the PSHPB test at a strain rate of 550 s^{-1} (Figure 6). The loading direction stress (σ_{33}) contour plots in the specimen are illustrated in Figure 6. The contour plots of σ_{33} (axial stress) and σ_{Mises} were found to vary dramatically during the initial hardening trend. The ε_{33} and von Mises contours of the sample revealed a nonuniform stress state throughout testing (Figures 6 and 7). The stresses varied over time as shown in Figure 8.

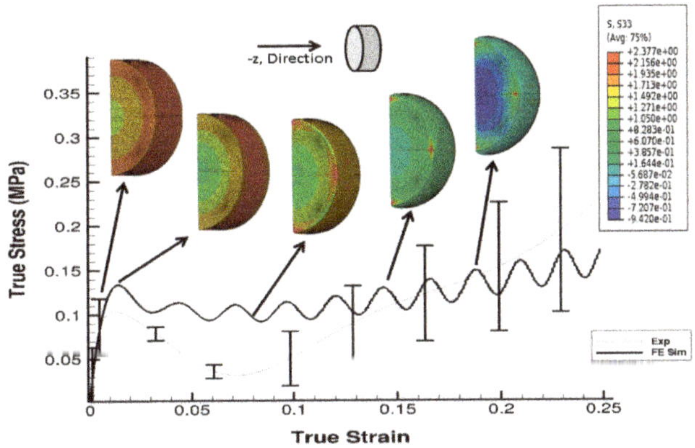

Figure 6. Contour plot (σ_{33}) with comparison of σ_{33} from the experiment and the FE simulation at 550 s^{-1}. Sample is a solid cylindrical disk, and σ_{33} data from FE simulation were processed in DAVID Viscoelastic Software.

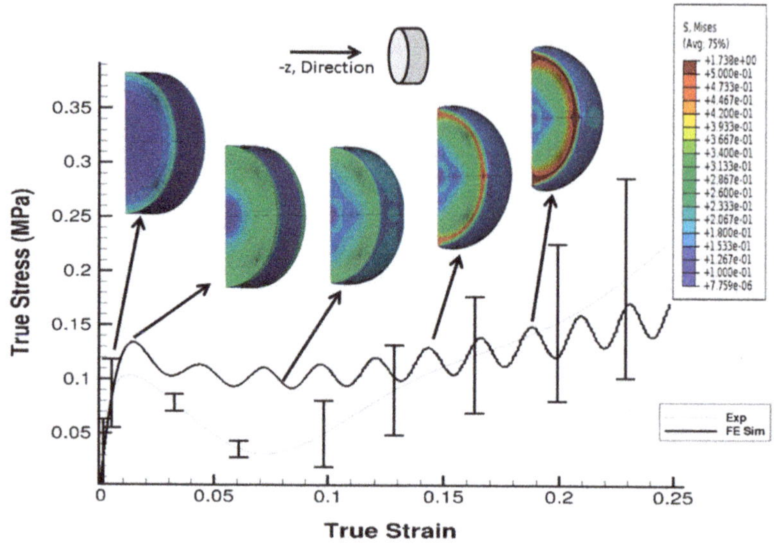

Figure 7. Contour plot (von Mises data) with comparison of σ_{33} from the experiment and FE simulation at 550 s^{-1}. Sample is a solid cylindrical disk, and σ_{Mises} data from the FE simulation were processed in DAVID Viscoelastic Software.

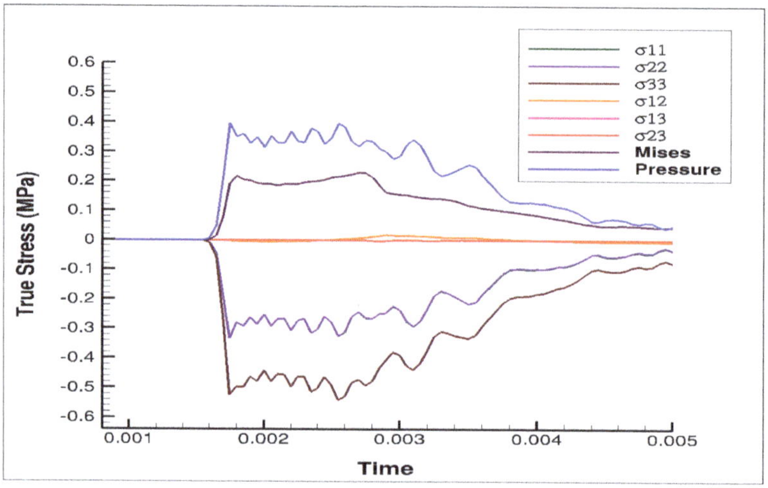

Figure 8. Plot of pressure, von Mises, and stress from different orthogonal directions (σ_{11}, σ_{22}, σ_{33}, σ_{12}, σ_{13}, and σ_{23}) over time, obtained from the FE simulation at 550 s^{-1}. Compressive stresses are negative.

4. Discussion

Hopkinson bar testing on soft tissues is a relatively new effort with few reports within the recent decade [11,12,22]. To obtain valid and accurate stress–strain data and further establish Hopkinson bar testing as the conventional high strain rate test for soft tissues, many variables were evaluated in this study. For example, specimen aspect ratio was an important factor for consideration so as to avoid unequal stress distribution in the sample and nonequilibrated input/output forces from an overly wide sample. Van Sligtenhorst et al. suggested an optimal aspect ratio of approximately 2:1 for bovine muscle tissue samples to produce equal stress distributions through the cross-section [11]. For choosing the appropriate aspect ratio, one must take into consideration two trends: (i) increasing the specimen aspect ratio can cause an increase in radial inertial effects; however, (ii) decreasing the specimen aspect ratio below a certain level could lead to nonuniform deformation along the longitudinal axis of the sample [11]. Van Sligtenhorst et al. and Song et al. showed the effects of specimen aspect ratio on the accuracy of PSHPB testing, and as a result, samples for the present study were prepared with the previously stated geometric criteria in the experimental design. By generating consistent, repeatable data that reflected the intrinsic mechanical response of liver tissue, the optimal aspect ratio for liver tissue was determined to be 3:1.

The obtained true stress-true strain curves showed a nonmonotonic characteristic overall (Figure 3a,b), which was similar to recently reported data from drop tower compression testing of human liver tissues at strains rates up to 62 s^{-1} [13]. Results from both studies indicated a loading path that initiated with a sharp stiffening response, followed by softening, subsequent hardening, and then yielding until ultimate failure. It is notable that this initial stiffening does not appear in stress–strain plots obtained in the regime of low strain rates (<10 s^{-1}), which often exhibit a monotonic, concave-upwards stress–strain relationship [28,34]. Song et al. hypothesized that the stiffening was purely a result of inertia [12]; however, this may or may not be the complete conclusion. Similarly, Sparks et al. cited inertia as the main factor in initial stiffening but included dynamic changes in specimen geometry during loading as a factor in stiffening [13]. Additionally, our study showed that the initial hardening trend observed in the high strain rate response of soft biological materials may completely be due to inertial effects, but part of it could also be attributed to the intrinsic behavior of the material [23]. Further, Prabhu et al. also asserted that the high water content (70–80%) of soft biological materials contributed to part of the initial hardening observed in high strain rate responses of soft biomaterials [23].

As observed in Figures 6 and 7, the high-frequency oscillatory behavior of the FE simulation stress–strain response was higher than that of the experimental response. The experiment, which represented more of a real-world scenario, encountered the lower levels of such high-frequency oscillations. In real life experimental systems, energy dissipation mechanisms act as a damping source, but in an FE model, which is an undamped system simulating a real world damped experimental system, proportional damping is commonly used to account for dissipative mechanisms (Appendix B).

It is interesting to note that high strain rate testing of different tissues results in different degrees of initial stiffening. The results obtained by Prabhu et al. for brain tissue and by Clemmer et al. for liver tissue at high rate compression demonstrated a higher initial hardening peak when compared to liver data reported in this study [23,24]. The above observation leads to a hypothesis correlating the initial stiffening with concentration of cellular content/water content [23]. Specifically, tissues with higher cellular content have a higher initial stiffening peak than those of a more fibrous nature (e.g., brain > liver > tendon). One of the future aims of this research is to characterize various soft tissues in an effort to confirm that the initial stiffening effect is actually an accurate representation of tissue behavior under high strain rate testing.

To our knowledge, no studies involving high strain rate Hopkinson pressure bar testing of porcine liver tissue exist. Moderate strain rate testing (20–62 s^{-1}) on human liver tissue was performed by Sparks et al. (2008), and they showed similar trends in stress–strain plots, despite the difference in methodology, in which a drop tower technique was used on intact human livers [13]. Though these strain rates were considered as fairly high in the report, they were relatively low compared to the strain rates obtained in the present study. Through repetition of testing using the PSHPB apparatus, various input velocities of the striker bar resulted in consistent strain rates in the porcine liver tissue (Table 4). Using the PSHPB method, striker bar speeds of approximately 6.5–17 mph corresponded to strain rates of 350–1550 s^{-1} in the liver tissue. For accurate replication of car crash scenarios, speed is a critical factor, and the impact speeds employed in the present study are more representative of speeds at which blunt trauma situations, such as those resulting from automobile accidents, occur.

Table 4. Correlation between striker bar speed and resultant strain rate of porcine liver tissue in high-rate tests.

Velocity of Striker Bar (mph)	Strain Rate of Liver Tissue (s^{-1})
6.487	350
9.843	550
13.645	1000
17.001	1550

The anisotropic mechanical response of liver tissue was also addressed in this study. The isotropy (or anisotropy) of liver tissue at high strain rates has not been well accepted in the present literature; therefore, evaluating this material property for modelling purposes was necessary. Previous work on high strain rate testing of bovine liver along two perpendicular directions determined that bovine liver behaved isotropically [19]. The present study extends this work to include three orthogonal directions in porcine liver tissue, along which high strain rate tests and histological analysis were performed. High strain rate mechanical testing clearly showed that no difference existed among stress–strain behaviors from testing along three orthogonal directions. Histological analysis of liver tissue, which examined characteristics as cell nuclear area, cell count, and mean distance between neighboring cells, revealed microstructural similarities among samples oriented along Directions 1, 2, and 3. These findings suggest the liver is an isotropic medium. Although exact mechanisms for the isotropic response are not yet clear, future studies into the contributions of the extracellular matrix may be insightful. The heterogeneous response of the liver tissue behavior during plastic deformation in an experimental setup is specimen-, microstructure-, or location-dependent, which is difficult to integrate in a simulated homogenous finite element model. In the PSHPB experimental setting, porcine liver specimens were glued to the setup, whereas the finite element model incorporated only minimal contact between the specimen and the polycarbonate rod. Even though utilization of glue during biomechanical testing of porcine liver tissue is not uncommon [34], it is difficult to quantitatively delineate the mechanical role of glue in PSHPB; hence, it was assumed to be constrained-body contact that was applied in the FE studies. Furthermore, high water content is essential for viscoelastic responses of soft tissues [1]. Incorporation of this mixture theory-based viscoelastic response in the material model can potentially capture the "softening" and further "hardening" responses; however, this component is beyond the scope of the current 1D material point simulator and will be addressed in future studies.

5. Conclusions

The use of a PSHPB apparatus for high strain rate testing of porcine liver tissue reveals the inelasticity, isotropy, and strain rate sensitivity of liver tissue. In conclusion, (i) the liver tissue response at high-rate compression was characterized by an initial hardening peak, followed by softening, and then by strain hardening to failure; (ii) the liver mechanical stress–strain behavior increased as the applied strain rate increased; and (iii) isotropic high-rate material behavior was observed along all three orthogonal directions and was confirmed by the liver histological microstructure.

In addition to these three conclusions, some other important points are worth mentioning. The wave propagation predicted by the finite element PSHPB simulation was consistent with the experimental results, thus substantiating the present results of the PSHPB. However, the computational simulation of the PSHPB process also showed that a uniform stress state was not fully achieved in the cylindrical sample. This limitation implies that future work is warranted to perfect the PSHPB technique in soft tissue high-rate characterization.

High strain rate tests conducted using an SHPB apparatus show that the porcine liver is strain rate-dependent (Figure 3a). The anisotropy of the material at a high rate is marginal but marked with high variation in the sample-to-sample mechanical behavior (Figure 3b). The material response is marked by an initial hardening effect, followed by a softening trend, and then further hardening at larger strains (Figure 3a,b). Simulations of the PSHPB test in ABAQUS/explicit solver (version 6.9) showed that the axial stress σ_{33} was primarily concentrated in the central region of the specimen (Figures 6 and 7). Specimen volume-averaged FE simulation results indicated that a homogeneous stress state was not maintained during specimen deformation (Figures 6 and 7). The range of stresses exhibited by the specimen in the FEA can be observed in Figure 8.

This novel approach using polymeric bars for high-rate impact of porcine liver tissue serves as a benchmark for future high strain rate testing of soft tissues. Experimental data coupled with the finite element model can be implemented in large-scale, computational models of the human body for simulation of high strain rate scenarios, such as automobile accidents, for validating the efficacy of various safety features.

Author Contributions: Data curation, J.C., S.S.P. and R.K.P.; Formal analysis, J.C., S.S.P., J.-L.B., E.M. and M.F.H.; Funding acquisition, J.L. and L.N.W.; Methodology, L.B.P.; Project administration, J.L. and L.N.W.; Writing—original draft, J.C. and S.S.P.; Writing—review & editing, S.S.P, J.L. and L.N.W.

Funding: This research was funded by Department of Energy, under award number [DE-FC26-06NT42755].

Acknowledgments: This work was supported by the National Nuclear Security Administration, Department of Energy, under award number [DE-FC26-06NT42755]. J.L. is supported in part by 1R01EB022018-01. The authors would like to thank the Center for Advanced Vehicular Systems Southern Regional Center for Lightweight materials, the MSU Electron Microscope Center, and the MSU Department of Agricultural and Biological Engineering. The authors would like to thank Mac McCollum and Wilburn Whittington for their effort in this research.

Conflicts of Interest: The authors declare no conflict of interest.

Appendix A. Working Principle of Split Hopkinson Pressure Bar (PSHPB)

The theory for PSHPB is based on classical mechanics of elastic wave propagation in the bars and on the principle of superposition of waves. In elastic wave propagation theory, stress, strain, and particle velocity caused from a single pressure wave (here compressive) are proportional to each other. Hence, knowledge of a single pressure wave at any cross-section of the bars enables us to calculate the wave nature at any other cross-section. Using the knowledge of incident wave and reflected wave at any cross section (and through the principle of superposition), stress, strain, and particle velocity can be calculated. Here the stress, the strain, and the particle velocity are simply the sum of those related to the incident wave and reflected wave, which are in opposite directions [18]. The specific wave energy (W_s) absorbed by the specimen is given by

$$W_s = W_i - (W_o + W_r). \tag{A1}$$

This is the equation for energy balance. Here W_i is the energy of the incident wave, W_o is the energy of the transmitted wave, and W_r is the energy of the reflected wave. Stress state homogeneity and balance of input and output are assumed. The bar's response, being elastic in nature, can be used to calculate the specimen's inelastic response through the energy balance of the stress waves. The expression for true stress–strain for a PSHPB experiment is given as:

$$\sigma_t(t) = \frac{F(t)}{S_s(0)}(1 - 2\nu\varepsilon_n(t)). \tag{A2}$$

With true strain defined as

$$\varepsilon_t(t) = -ln\left(\frac{1}{1-\varepsilon_n(t)}\right), \tag{A3}$$

with

$$\varepsilon_n(t) = \int_0^t \dot{\varepsilon}_n(t)dt, \tag{A4}$$

$$\dot{\varepsilon}_n(t) = \left(\frac{V_i(t) - V_o(t)}{l_s(0)}\right), \tag{A5}$$

where $\varepsilon_t(t)$ is the true strain (A2), $\varepsilon_n(t)$ is the nominal strain (A3), $\dot{\varepsilon}_n(t)$ is the nominal strain rate (A4), $l_s(0)$ is the initial length of the sample (undeformed), $F(t)$ is the force, V is the velocity, $S_s(0)$ is the initial cross-sectional area of the sample, and v is Poisson's ratio (which is equal to 0.5 for incompressible materials). Subscript i and o represent variables on the incident and transmitted bars, respectively. The above formulae are applicable to compression with positive strains.

Appendix B. Mechanism of Rayleigh Viscous Damping

Real world dissipation mechanisms, such as friction, vibration, etc., in a high strain rate experiment are hard to quantify and are simulated using an FE model. However, a generalized damping mechanism, called Rayleigh or classical damping, can be introduced to simulate frictional and vibrational dissipations. Rayleigh damping was introduced as a specific type of viscous damping, which was calculated as a linear combination of mass matrix (M) and stiffness matrix (K). The equation for viscous damping matrix is as follows:

$$C = \alpha_R M + \beta_R K, \tag{A6}$$

where α_R and β_R are the coefficients of M and K, respectively, and are real scalars (A6).

The observed difference in the high-frequency oscillatory waves of the FE simulation in comparison to the experimental data (Figure 5) arose from the fact that frictional and vibrational dissipations were simulated in the FE model through viscous damping. Viscous damping is a numerical damping mechanism to account for experimental dissipations.

References

1. Feliciano, D.V. Surgery for liver trauma. *Surg. Clin. North Am.* **1989**, *69*, 273–284. [CrossRef]
2. Rouhana, S.W.; Foster, M.E. Lateral impact—An analysis of the statistics in the NCSS. In Proceedings of the 29th Stapp Car Crash Conference, Society of Automotive Engineers, Warrendale, PA, USA, 9–11 October 1985.
3. Elhagediab, A.M.; Rouhana, R.B. Patterns of abdominal injury in frontal automotive crashes. In Proceedings of the 16th International ESV Conference Proceedings, NHTSA, Windsor, ON, Canada, 31 May–4 June 1998.
4. O'Neill, B. Preventing passenger vehicle occupant injuries by vehicle design—A historical perspective from IIHS. *Traffic Inj. Prev.* **2009**, *10*, 113–126. [CrossRef]
5. Mertz, H.J. *NHTSA Docket 74-14. Notice 32. Enclosure 2 of Attachment 1 of Part III of General Motors Submission USG2284*; USG: Chicago, IL, USA, 1984.
6. Mertz, H.J.; Prasad, P.; Irwin, A.I. Injury risk curves for children and adults in frontal and rear collisions. In *Forty-First Stapp Car Crash Conference*; SAE 973318; SAE: Warrendale, PA, USA, 1997.
7. Deng, Y.-C.; Kong, W.; Ho, H. Development of a finite element human thorax model for impact injury studies. In *SAE Technical Paper Series*; 1999-01-0715; SAE: Warrendale, PA, USA, 1999.
8. Haug, E. Biomechanical models in vehicle accident simulation. In Proceedings of the 1996 NATO Advanced Study Institute on Crashworthiness of Transportation Systems: Structural Impact and Occupant Protection, Tróia, Portugal, 7–19 July 1996.

9. Iwamoto, M.; Kisanuki, Y.; Watanabe, I.; Furusu, K.; Miki, K. Development of a finite element model of the total human model for safety (THUMS) and application to injury protection. In Proceedings of the 2002 International Conference on the Biomechanics of Impact, Munich, Germany, 18–20 September 2002.
10. Kimpara, H.; Iwamoto, M.; Miki, K.; Lee, J.B.; Begeman, P.; Yang, K.H.; King, A.I. Biomechanical properties of the male and female chest subjected to frontal and lateral impacts. In Proceedings of the 2003 International IRCOBI Conference on the Biomechanics of Impact, Lisbon, Portugal, 25–26 September 2003.
11. Van Sligtenhorst, C.; Cronin, D.S.; Brodland, G.W. High strain rate compressive properties of bovine muscle tissue determined using a split Hopkinson bar apparatus. *J. Biomech.* **2006**, *39*, 1852–1858. [CrossRef]
12. Song, B.; Chen, W.; Ge, Y.; Weerasooriya, T. Dynamic and quasi-static compressive response of porcine muscle. *J. Biomech.* **2007**, *40*, 2999–3005. [CrossRef] [PubMed]
13. Sparks, J.; Dupaix, R. Constitutive modeling of rate-dependent stress–strain behavior of human liver in blunt impact loading. *Ann. Biomed. Eng.* **2008**, *36*, 1883–1892. [CrossRef]
14. Ottensmeyer, M.P.; Kerdok, A.E.; Howe, R.D.; Dawson, S.L. The Effects of testing environment on the viscoelastic properties of soft tissues. In *Medical Simulation*; Cotin, S., Metaxas, D., Eds.; Springer: Berlin/Heidelberg, Germany, 2004; pp. 9–18.
15. Kerdok, A.E.; Ottensmeyer, M.P.; Howe, R.D. Effects of perfusion on the viscoelastic characteristics of liver. *J. Biomech.* **2006**, *39*, 2221–2231. [CrossRef]
16. Bouvard, J.L.; Francis, D.K.; Tschopp, M.A.; Marin, E.B.; Bammann, D.J.; Horstemeyer, M.F. An internal state variable material model for predicting the time, thermomechanical, and stress state dependence of amorphous glassy polymers under large deformation. *Int. J. Plast.* **2013**, *42*, 168–193. [CrossRef]
17. Whittington, W.R.; Oppedal, A.L.; Turnage, S.; Hammi, Y.; Rhee, H.; Allison, P.G.; Crane, C.K.; Horstemeyer, M.F. Capturing the effect of temperature, strain rate, and stress state on the plasticity and fracture of rolled homogeneous armor (RHA) steel. *Mater. Sci. Eng. A* **2014**, *594*, 82–88. [CrossRef]
18. Zhao, H.; Gary, G.; Klepaczko, J.R. On the use of a viscoelastic split Hopkinson pressure bar. *Int. J. Impact Eng.* **1997**, *19*, 319–330. [CrossRef]
19. Pervin, F.; Chen, W.W.; Weerasooriya, T. Dynamic compressive response of bovine liver tissues. *J. Mech. Behav. Biomed. Mater.* **2011**, *4*, 76–84. [CrossRef]
20. Liu, Q.; Subhash, G. Characterization of viscoelastic properties of polymer bar using iterative deconvolution in the time domain. *Mech. Mater.* **2006**, *38*, 1105–1117. [CrossRef]
21. Kwon, J.; Subhash, G. Compressive strain rate sensitivity of ballistic gelatin. *J. Biomech.* **2010**, *43*, 420–425. [CrossRef]
22. Song, B.; Chen, W. Dynamic stress equilibration in split Hopkinson pressure bar tests on soft materials. *Exp. Mech.* **2004**, *44*, 300–312. [CrossRef]
23. Prabhu, R.; Horstemeyer, M.F.; Tucker, M.T.; Marin, E.B.; Bouvard, J.L.; Sherburn, J.A.; Liao, J.; Williams, L.N. Coupled experiment/finite element analysis on the mechanical response of porcine brain under high strain rates. *J. Mech. Behav. Biomed. Mater.* **2011**, *4*, 1067–1080. [CrossRef]
24. Clemmer, J.; Prabhu, R.; Chen, J.; Colebeck, E.; Priddy, L.B.; Mccollum, M.; Brazile, B.; Whittington, W.; Wardlaw, J.L.; Rhee, H.; et al. Experimental observation of high strain rate responses of porcine brain, liver, and tendon. *J. Mech. Med. Biol.* **2016**, *16*, 1650032. [CrossRef]
25. Prabhu, R.; Whittington, W.R.; Patnaik, S.S.; Mao, Y.; Begonia, M.T.; Williams, L.N.; Liao, J.; Horstemeyer, M.F. A coupled experiment-finite element modeling methodology for assessing high strain rate mechanical response of soft biomaterials. *J. Vis. Exp.* **2015**, *99*, e51545. [CrossRef]
26. Subhash, C.; Ravichandran, G. Split-Hopkinson pressure bar testing of ceramics. In *ASM Handbook Vol 8, Mechanical Testing and Evaluation*; ASM Int.: Materials Park, OH, USA, 2000; pp. 427–428.
27. Uenishi, A.; Yoshida, H.; Kuriyama, Y.; Takahashi, M. Material characterization at high strain rates for optimizing car body structures for crash events. *Nippon Steel Tech. Rep.* **2003**, *88*, 21–24.
28. Roan, E.; Vemaganti, K. The nonlinear material properties of liver tissue determined from no-slip uniaxial compression experiments. *J. Biomech. Eng.* **2007**, *129*, 450–456. [CrossRef]
29. Saraf, H.; Ramesh, K.T.; Lennon, A.M.; Merkle, A.C.; Roberts, J.C. Mechanical properties of soft human tissues under dynamic loading. *J. Biomech.* **2007**, *40*, 1960–1967. [CrossRef]
30. Chen, J.; Brazile, B.; Prabhu, R.; Patnaik, S.S.; Bertucci, R.; Rhee, H.; Horstemeyer, M.F.; Hong, Y.; Williams, L.N.; Liao, J. Quantitative analysis of tissue damage evolution in porcine liver with interrupted mechanical testing under tension, compression, and shear. *J. Biomech. Eng.* **2018**, *140*, 071010. [CrossRef]

31. Bouvard, J.L.; Brown, H.R.; Marin, E.B.; Wang, P.; Horstemeyer, M.F. Mechanical testing and material modeling of thermoplastics: Polycarbonate, polypropylene and acrylonitrile-butadiene-styrene. In *MRS Fall Symposium W 2009*; MRS: Cambridge, UK, 2009.
32. Bouvard, J.L.; Ward, D.K.; Hossain, D.; Marin, E.B.; Bammann, D.J.; Horstemeyer, M.F. A general inelastic internal state variable model for amorphous glassy polymers. *Acta Mech.* **2010**, *213*, 71–96. [CrossRef]
33. Smith, M. *ABAQUS/Explicit User's Manual, Version 6.9*; Hibbit, Karlsson, and Sorenson, Inc.: Providence, RI, USA, 2009.
34. Sakuma, I.; Nishimura, Y.; Chui, C.K.; Kobayashi, E.; Inada, H.; Chen, X.; Hisada, T. In vitro measurement of mechanical properties of liver tissue under compression and elongation using a new test piece holding method with surgical glue. In *Surgery Simulation and Soft Tissue Modeling*; Ayache, N., Delingette, H., Eds.; Springer: Berlin/Heidelberg, Germany, 2003; pp. 284–292.

bioengineering

MDPI

Article

Tissue Level Mechanical Properties and Extracellular Matrix Investigation of the Bovine Jugular Venous Valve Tissue

Adam A. Benson and Hsiao-Ying Shadow Huang *

Mechanical and Aerospace Engineering Department, Analytical Instrumentation Facility, North Carolina State University, R3158 Engineering Building 3, Campus Box 7910, 911 Oval Drive, Raleigh, NC 27695, USA; aabenson@ncsu.edu

* Correspondence: hshuang@ncsu.edu; Tel.: +1-919-513-0798

Received: 27 March 2019; Accepted: 10 May 2019; Published: 14 May 2019

Abstract: Jugular venous valve incompetence has no long-term remedy and symptoms of transient global amnesia and/or intracranial hypertension continue to discomfort patients. During this study, we interrogate the synergy of the collagen and elastin microstructure that compose the bi-layer extracellular matrix (ECM) of the jugular venous valve. In this study, we investigate the jugular venous valve and relate it to tissue-level mechanical properties, fibril orientation and fibril composition to improve fundamental knowledge of the jugular venous valves toward the development of bioprosthetic venous valve replacements. Steps include: (1) multi loading biaxial mechanical tests; (2) isolation of the elastin microstructure; (3) imaging of the elastin microstructure; and (4) imaging of the collagen microstructure, including an experimental analysis of crimp. Results from this study show that, during a 3:1 loading ratio (circumferential direction: 900 mN and radial direction: 300 mN), elastin may have the ability to contribute to the circumferential mechanical properties at low strains, for example, shifting the inflection point toward lower strains in comparison to other loading ratios. After isolating the elastin microstructure, light microscopy revealed that the overall elastin orients in the radial direction while forming a crosslinked mesh. Collagen fibers were found undulated, aligning in parallel with neighboring fibers and orienting in the circumferential direction with an interquartile range of −10.38° to 7.58° from the circumferential axis (n = 20). Collagen crimp wavelength and amplitude was found to be 38.46 ± 8.06 μm and 4.51 ± 1.65 μm, respectively (n = 87). Analyzing collagen crimp shows that crimp permits about 12% true strain circumferentially, while straightening of the overall fibers accounts for more. To the best of the authors' knowledge, this is the first study of the jugular venous valve linking the composition and orientation of the ECM to its mechanical properties and this study will aid in forming a structure-based constitutive model.

Keywords: collagen crimp; elastin; microstructures; force-controlled mechanical testing

---

1. Introduction

Venous valves are semi-lunar cusps that prevent retrograde blood flow in the venous system. For example, jugular vein valve insufficiency is a hypothesized etiology, given that many patients report Valsalva-associated maneuvers prior to a transient global amnesia event [1–3]. It is also hypothesized that increased intra-abdominal pressure is transmitted into the intracranial venous system, causing intracranial hypertension; jugular valve insufficiency may facilitate pressure transmission [4]. Moreover, chronic venous insufficiency (CVI) occurs when the saphenous venous valves in the vein are damaged or malfunctioning, leading to blood pooling at the distal end of the legs, causing swelling in the legs. Current treatments for CVI, such as venous valve replacement, have been developed and jugular venous valves were generally used in need of surgical replacement: replacing incompetent saphenous venous valves in the legs to prevent varicose veins, edema, poor circulation, valvular incompetence and all other symptoms of CVI. For example, Medtronic's Contegra pulmonary valved conduit [5] and transcatheter Melody pulmonary valve [6] are based on glutaraldehyde-fixed trileaflet bovine jugular venous valves. Many current bioprosthetic replacements use fixated xenografts, since the better endotheliazed autologous vein segments take extreme surgical precision and are often unavailable [7]. However, fixated tissues tend to have warped mechanical properties and can affect valvular hemodynamic performance, which was shown in a study of the mitral valve [8]. Therefore, this study highlights the characteristics that should be desired when creating tissue-engineered substitutes, including mechanical properties, fibril orientation and fibril composition.

Venous valves are arranged in different valvular patterns depending on the location throughout the venous system. The two most common valvular arrangements are tri-cuspid and bi-cuspid but uni-cuspid, quadri-cuspid and quinque-cuspid have been recorded [9,10]; however, uni-cuspid arrangements have been questioned as damaged bi-cupsid valves [9]. There are two enface sides of the venous valve—the parietal side and the luminal side. During the "opened position," the parietal side faces the venous valve pocket and the luminal side faces away from the adhered venous wall. The parietal side is composed mostly of collagen and the luminal side has an overlaying elastic laminae, which is an extension of the venous wall [9]. Both sides are covered by an endothelium layer. Specifically, there are three types of venous valves: parietal valves, free parietal valves and ostial valves. Ostial valves are located at the entrance of a small vein into a larger one. Parietal valves are located at a junction where two veins of equal diameter merge into one vein. At these junctions, there is a valve present at each branch. Parietal valves are located distally to the heart of the bovine jugular vein, for example, where the maxillary and linguofacial veins converge. The third type, free parietal valves, are valves not at a junction [9].

Throughout different regions of the body, there are various types of collagen found in the extracellular matrix (ECM). Collagen fibers are made up of bundles of collagen fibrils and are usually found with an undulated pattern. The size and geometry of collagen fibers vary depending on the tissue. Different forms of collagen fibers include cord- or tape-shaped geometry [11]. Since collagen fibers give the extracellular matrix structural integrity, they are immensely relevant to this study. In the venous valve, the collagen microstructure is much thicker than the elastin microstructure and is located on the parietal side [9]. Previously, Huang and Lu [10,12] has used biochemical analysis of the venous valve tissue to investigate soluble collagen concentration and developed histological images using Masson's trichrome stain. However, the study did not provide any implicit values regarding collagen crimp length, only a mere approximation. Knowing that crimp length is especially important because collagen fibers become increasingly stiffer when uncrimped. Elastin is defined by its ability to stretch up to 150% of its original length without attaining any permanent damage [13,14]. Physiologically, the luminal side is advantageous for the elastin location, because the luminal side experiences maximum stretch during anterograde flow. Elastin provides compliancy and recoil, allowing the tissue to undergo ongoing mechanical stress. Therefore, it is suggested that the elastin microstructure under tensile load has the ability to retract and aid the valve in closing during retrograde flow.

Increased elastin has been observed in diseased venous valve leaflets [15], as has been collagen disorganization [16] and decreased collagen expression [17]. Given the increased pressure loads and concomitant tissue stresses in CVI, maladaptive venous valve tissue remodeling may parallel that of the pulmonary autograft in the Ross procedure [18,19] or that of the saphenous vein when utilized as a high pressure conduit in coronary bypass surgery [20]. Given how critical the elastin-rich ventricularis layer is to the function of heart valve leaflets, increases in venous valve leaflet elastin shown by Mouton et al. with progression of CVI may have important implications for the leaflet mechanical properties. Without knowing tissue-level mechanical properties and the ECM composition of healthy and diseased venous valve leaflets, it would be challenging to design replacement venous valves capable of long-term durability and normal physiologic function.

Previous mechanical tests of the venous valve included both a uniaxial mechanical test by Ackroyd et al. [21] and a biaxial test by Huang et al. [10,12,22]. The biaxial mechanical test better represents the physiological loading of the venous valve tissue and is, therefore, a better model of its mechanical properties. Huang and Lu [12] and Huang and Kaul [22] have previously reported the stress-strain curves, tangent moduli and constitutive models for the bovine jugular venous valve but the study was only limited to displacement-controlled loading condition. In this paper, the biaxial mechanical test applies three different force-controlled loading ratios, which allow the first attempt to characterize venous valve fiber rotation during loading. The mechanical properties could then be elucidated and compared to the investigated ECM's collagen and elastin microstructures of the jugular venous valve.

The study by Saphir and Lev revealed images of longitudinal cross-sections of the jugular venous valve and Crissman provided the isolated elastin microstructure but to date, there is no study directly investigating the interplay of mechanical properties and the ECM in jugular venous valve tissue [23–25]. To the best of the authors' knowledge, this paper is the first study to present microstructures that investigate the fiber orientation of both collagen and elastin, directly relating fiber orientation to jugular venous valves' mechanical properties studied by a bi-axial mechanical test. To better understand the mechanical effects between elastin and collagen fibers, especially at the low-strain region, we incorporated sodium hydroxide digestion to identify elastin microstructure alone in the jugular venous valve. In this context, the mechanical properties and microstructure of the jugular venous valves is of interest. This paper is sought to outline the contribution of both the collagen and elastin microstructure to the physiological function of jugular venous valve tissue in hopes of redefining the properties that must be mimicked in bioprosthetic replacement.

2. Materials and Methods

This study only focused on external jugular vein free parietal valves to keep all findings consistent. Additionally, only mature cows (Holstein breed, female, 10+ years old, ~1250 lbs weight) were used for the experiments since the age of the cow was shown to have noticeable effects on the collagen and elastin content [23]. Jugular venous valves were studied because their greater size aided mechanical testing and handling during NaOH (Fisher Chemical, Pittsburgh, PA, USA) treatments to isolate elastin. Jugular veins were removed from the mature cow and shipped to the lab overnight in a temperature-controlled environment. Therefore, the Institutional Animal Care and Use Committee (IACUC) review and approval is not required since tissues harvested from an animal that was euthanized for reasons other than our proposed study (culled). All connective tissue was cut off the exterior of the vessels. The veins were slowly turned inside out using forceps [12]. Venous valves were carefully dissected from the venous wall while submerged in Hank's Balanced Salt Solution (HBSS). Since only free parietal valves were dissected, valves located at branches were not used for the experiments. After the dissection, venous valves were stored in HBSS at 1.11 °C (34 °F) until needed for further testing, which occurred no later than 72 h after dissection. A representative jugular venous valve tissue was showed in Figure 1, where a regular high-resolution scanner and a LSM-710 confocal microscope were used. Through a combination of z-stack and tile scanning, the three-dimensional cell nucleus distribution (by staining whole mount samples with DAPI (1:50 dilution) for 2 h and rinsing exhaustively with HBSS) and collagen fiber structures of the tissue across the entire surface area and through the thickness, were imaged based on collagen auto-fluorescence; the venous tissue was excited at 488 nm, with emissions collected from 490–590 nm.

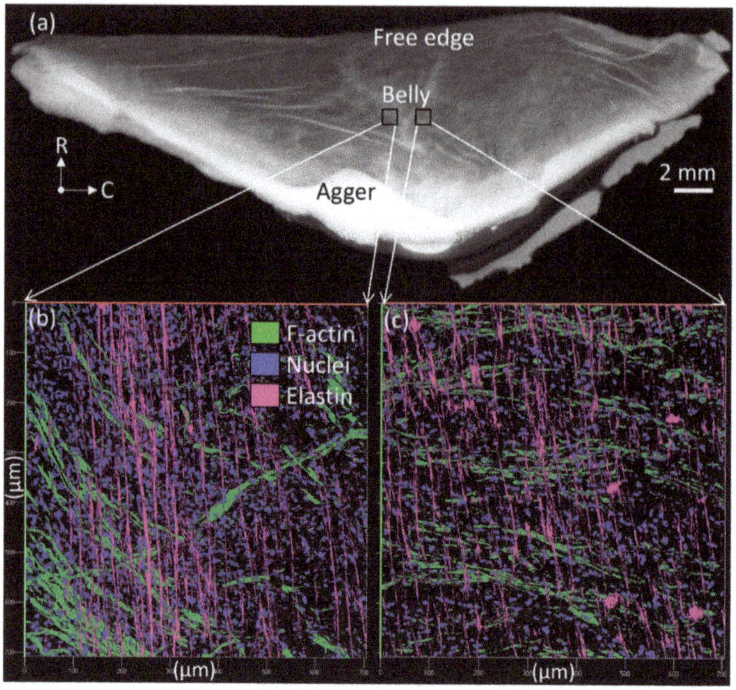

Figure 1. Jugular venous valve tissue viewed (**a**) from a regular high-resolution scanner and (**b**,**c**) in a LSM-710 confocal microscope (200×), where radial (R) and circumferential (C) directions were denoted in (**a**).

2.1. Light Microscopy of the Collagen Microstructure

Immediately following dissection, tissue was gently washed in HBSS. The sample was mounted on a microscope slide with HBSS and oriented so that the parietal side faced the eyepiece. The collagen microstructure was imaged with planar objectives on the inverted microscope (VWR Vista Vision, West Chester, PA, USA). The ImageJ (National Institutes of Health, MD, USA) ROI tool manager was used to quantify collagen orientation and crimp length and was related to the mechanical testing results. Specifically, the geometric analysis of the collagen crimp at relaxed conditions were associated with and used to investigate the circumferential direction's inflection point on the stress-strain curve.

2.2. Isolation of the Elastin

Elastins are known to stretch to 150–200% local strain but it is the composition and architecture of the multi-component ECM which dictates the final outcome. In this study, we used NaOH digestion to identify elastin microstructure alone in the jugular venous valve. In brief, NaOH digestion—more frequently called hot alkali extraction—can degrade collagen due to its harsh conditions and it is recommended not to exceed 50–60 min at high temperatures [26,27]. The hot alkali extraction was made famous by Lowry and later modified by Lansing and has been a prevalent extraction method ever since. The method involves gelatinizing collagen and using weak alkali to remove other proteins from the elastin residue [27].

Dissected venous valve tissue was removed from HBSS and carefully placed into a glass beaker filled with 200 mL of 0.1 N NaOH solution heated to 75 °C by a LHS-720 Series Digital Hot Plate (Omega, Stamford, CT, USA). Temperature was monitored with an E4 Compact Thermal Imaging Camera (Flir, Wilsonville, OR, USA). The tissue was digested for 45 min following a published procedure of the aortic valve [28]. After the digestion procedure, the tissue was washed in HBSS and biochemically analyzed using a soluble collagen assay kit (Sircol; Accurate Chemical and Scientific Corp., Westbury, NY, USA). The sample's wet weight mass and dry weight mass were measured via an analytical balance (VWR, West Chester, PA, USA). Next, the sample was vortexed in a 1-mL solution of acetic acid (0.5 M; Sigma-Aldrich, St. Louis, MO, USA) and pepsin (1 mg/mL Pepsin A (P-7000); Sigma Aldrich, St. Louis, MO, USA) in distilled water for 120 h. Collagen assay dye reagent of 0.1 mL was added to each sample and vortexed (VWR, West Chester, PA, USA) for binding. The sample was centrifuged with a Mini Spin (Eppendorf, Hamburg, Germany) at 13,400 RPM for 10 min. Excess dye and collagen extraction solution was removed, leaving the remaining solid mass of collagen content. One mL of 0.5 N NaOH was then added to the remaining collagen and vortexed. Solutions were placed in cuvettes and absorbance was measured by a Genesis 20 Spectrometer (Thermo Fisher Scientific, Waltham, MA, USA) at 550 nm [29,30]. Final collagen concentrations were determined by measuring the absorbance at 550 nm by a spectrometer versus collagen mass (both wet and dried) based on a collagen standard solution (Sircol Collagen Assay Collagen Standard 0.5 mg/mL). The proceeding process was repeated at 0.1 N NaOH 75 °C digestions for 45-min, 60-min and 75-min durations to determine an optimal duration for removing collagen content.

The 75-min duration was deemed the best process for removing almost all collagen content in jugular venous valve tissue, despite the hot alkali method suggested not exceeding 50–60 min [26]. To established a better comparison of initial dry weight of the sample to the dry weight of the sample after digestion, fresh samples (n = 7) were lyophilized in a Free Zone 2.5 (Labconco, Kansas City, MO, USA) and weighed on a VWR analytical balance. After the 75-min heated NaOH digestion, the samples were gently washed in distilled water, lyophilized and weighed on an analytical balance again. Washing had to be very gentle because the remaining elastin substrate was delicate and sticky. Lastly, the process was followed with another Sircol soluble collagen assay to evaluate the remaining collagen concentration data.

2.3. Light Microscopy of the Elastin Microstructure

In this work, both light microscopy and scanning electron microcopy (SEM) were used to image the elastin microstructure. The 75-min 75 °C 0.1 N NaOH digestion was used on every sample before imaging. After digestion, the sample was gently washed in HBSS and moved to a microscope slide mounted with HBSS. An inverted microscope (VistaVision, VWR, West Chester, PA, USA) with planar objectives was used to image the elastin microstructure of the venous valve tissue. Removal of the collagen microstructure caused the elastin microstructure to unstretch. Image stitching of the same location at different focuses was used on some images since the corrugated samples had multiple focus depths.

2.4. Scanning Electron Microscopy of the Elastin Microstructure

Following the light microscopy, SEM was used to visually investigate damage yielded by the NaOH treatment used to isolate the elastin microstructure. For the purpose of comparison, undigested samples were also imaged. Samples were placed directly into a critical point drying holder for filters (samples had to be kept flat) and then into 3% glutaraldehyde in 0.1M NaPO4 buffer, pH 7.3 at 4 °C for one week. Samples were rinsed in three 30-min changes of cold 0.1 M NaPO4 buffer, pH 7.3 (by moving the holder from one jar to another), followed by 30-min changes in cold 30% and 50% ethanol (EtOH). The sample holder was moved to a jar of 70% EtOH and held overnight at 4 °C. Completed dehydration with a 30-min change of cold 95% EtOH was followed by a 60-min change to cold 100% EtOH, warming to room temperature and two 60-min changes of 100% EtOH. Samples were critical point dried with a Samdri-795 critical point dryer (Tousimis, Rockville, MD, USA) for ten min at critical point in liquid CO_2. Samples were held in the desiccator overnight. All samples were sputter-coated (Hummer 6.2 Sputter System, Anatech USA, Union City, CA, USA); the non-digested samples were flat and coated with 25 Å from two sides plus 25 Å on top. The digested samples, which were convoluted, were coated with 25 Å from four sides plus 25 Å on top. Samples were returned to the desiccator until viewed. Samples were viewed in a JEOL JSM-5900LV SEM at 15 kV.

2.5. Force Control Mechanical Testing

Immediately following dissection, tissue was gently washed in HBSS. The mechanical test was conducted on a biaxial tester—the BioTester 5000 (CellScale, Waterloo, Ontario, Canada). Four rakes with five tungsten tines each were used for the boundary conditions of the venous valve tissue when testing [12,22,29,31]. The rakes were spaced in the radial and circumferential direction by 4500 μm during resting conditions. During testing, the venous valve tissue was submerged in HBSS and heated to a constant 37 °C. During all testing, only the belly region of the venous valve tissue was tested, removing the excess tissue that overlaid the tungsten rakes. The tissue was mounted so that the radial direction of the belly region aligned on the y-axis and the circumferential direction aligned on the x-axis of the BioTester [12,22,29,31]. Before testing, a 10 mN preload was loaded in both the radial and circumferential directions. If this consistent preload was not applied, the samples could not be equally compared since they began at different initial strain. The following loading ratios were implemented into the mechanical test: (1) The 1:1 ratio was tested to 900 mN. The tissue was stretched with the same force in both the circumferential and radial directions; (2) The 1:3 ratio was loaded with 300 mN of force in the circumferential direction and 900 mN of force in the radial direction; and (3) In the 3:1 ratio, the circumferential direction was loaded with 900 mN and the radial direction was loaded with 300 mN. The extracted data were averaged across seven samples (n = 7) for each loading condition.

2.6. Statistical Analysis

Statistical analyses were conducted using JMP (SAS, Cary, NC, USA). One-way ANOVA was used to test the significance of data collected for the wavelength and amplitude of collagen crimps. Statistical significance was tested at $p < 0.05$ to determine if it was appropriate to use data to form a collagen crimp model for strain analysis.

3. Results and Discussion

3.1. Anatomical Findings during Dissection

Out of the 55 total bovine jugular veins (~30 cm long vein segments), there were 159 free parietal valves. One-hundred eleven valves (111) were bi-cuspid, 46 of the valves were tri-cuspid and 2 of the valves were quadri-cuspid. In addition, it was noticed that not entirely absent in venous valve leaflets is the glycosaminoglycan-rich spongiosa layer characteristic of semilunar heart valve leaflets. In the heart valve leaflets, the central spongiosa layer has been hypothesized to function as a mechanical dampener [28,32].

In the jugular vein, 28.93% of venous valves were tri-cuspid, whereas in the saphenous vein, all of the venous valves dissected are bi-cuspid [10]. The saphenous vein's bi-cuspid geometry is always oriented so that its coaptation is colinear to its longitudinal direction of the venous wall's elliptical cross section [9]. The superficial and deep veins in the legs always have elliptical cross sections due to the compressive force between the subcutaneous fascia and/or muscle [33]. Edwards concluded that the alignment is most likely mandatory for a tight closure of the venous valve [33]. In other words, if the compressive force was oriented perpendicular to the coaptation of the venous valve, the bi-cuspid valve would have an ineffective closure. Comparatively, the jugular venous valve has been found to both have a common occurrence of bicuspid and tricuspid valves. This is in agreement with a previous study which discovered that bovine proximal external jugular vein segments tend to be bi-cuspid and distal external jugular vein segments tend to be tricuspid [12]. The common appearance of tri-cuspid valves may be in response to the vein in vivo being shaped distally with circular cross sections as opposed to elliptical, as in the proximal external jugular vein and saphenous vein. The smaller loads on the exterior of the vessel are justified by smaller muscles located in the neck compared to the chest and leg. The lateral pectoral groove most likely applies more compression proximally to the venous walls than the jugular groove applies distally to the venous walls. The lateral pectoral groove is squeezed by both the brachiocephalicus and descending pectoral muscles [34]. Perhaps when veins are less loaded and have circular cross sections, tri-cuspid valves perform better and remain competent longer than bi-cuspid valves. This would explain the cuspid distinction between the venous valves located in the distal external jugular vein segments and the proximal external jugular vein segments and saphenous veins.

Particularly, it is an interesting investigation because traditional percutaneous venous valve designs often have stents with circular cross sections but they are designed with bi-cuspid geometry [7,14]. Further providing justification, stent diameters are often oversized by 10% and the rigidity of the stent in some cases will dictate the shape of the host vein [7,14]. Due to the venous valves physiological existence, it may be true that the synergy of venous wall cross-sectional shape and cusp geometry play a role in overall venous valve competence. If the hypothesis is correct that tri-cuspid valves occur in the jugular vein due to the vessel's segments with circular cross sections, it could be that tri-cuspid valves would outperform bi-cuspid valves in the circular cross-sectioned stents that are often used in the lower legs.

3.2. Light Microscopy of the Collagen Microstructure

Collagen fibers are found in nearly all tissues and maintain their overall structure [35]. The collagen fiber network resides on the parietal side of the jugular venous valve just below the endothelium. Relaxed collagen fibers in the jugular venous valve have a wavy pattern and align in parallel with each other, orientating in the circumferential direction. An image of the collagen microstructure is shown in Figure 2a. Orientation of collagen fibers was measured through the center of the undulated collagen fiber segments (~150–200 μm length (Figure 2a)). It was found that most fibers align within a 10° range from the circumferential axis with an interquartile range of −10.38° to 7.58° (n = 20) and all fibers have an immediate parallel relation to neighboring fibers. The wavy pattern is a protection against direct tension on the collagen microstructure, similar to that of the aortic valve [36]. During loading, the collagen fibers must uncrimp before being loaded under tension. Since collagen ruptures at low strains (10–20%) and begins to yield at lower strains (1–2%), the crimps protect collagen fibers from overextending, undergoing plastic deformation and rupturing [37,38].

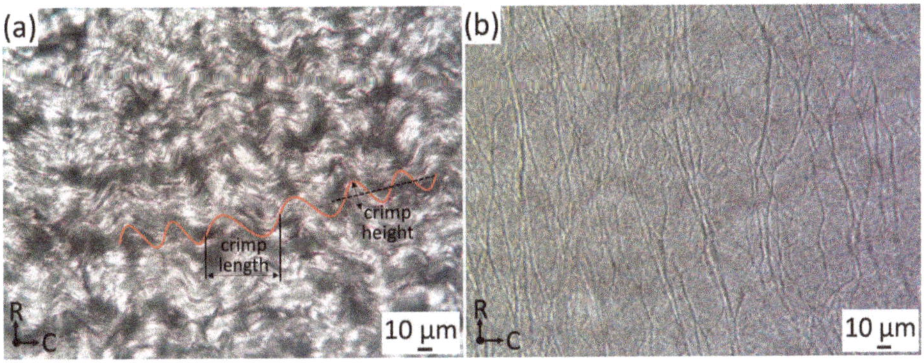

Figure 2. (**a**) Collagen crimp light microscopy images focused on the parietal side of jugular venous valve tissue (400×), where C = circumferential and R = radial directions. (**b**) Light microscopy of isolated elastin microstructure (luminal side) (400×). Please note that it was a projected image since several different focus levels were included in the image.

By analyzing the collagen fiber as a sinusoidal wave, we can make an approximation of how much strain it takes to uncrimp the collagen fibers. The average wavelength and amplitude was measured at

38.46 ± 8.06 μm and 4.51 ± 1.65 μm (n = 87), which was used to form the sinusoidal function. Therefore, the sinusoidal function was shown below:

$$y = 4.51 \sin\left(\frac{2\pi}{38.46}x\right) \tag{1}$$

The distance of a linearly stretched sinusoidal wave can be calculated using the arc length of a curve formula shown in Equation (2).

$$r = \int_a^b \sqrt{1 + \left(\frac{dx}{dy}\right)^2}\, dx \tag{2}$$

$$\int_{0\ \mu m}^{38.46\ \mu m} \sqrt{1 + \left(\left(\frac{2\pi}{38.46\ \mu m}\right)(4.51\mu m)\cos\left(\frac{2\pi}{38.46}x\right)\right)^2}\, dx = 43.24\ \mu m \tag{3}$$

After the elongated sinusoidal wave had been calculated, it was possible to find the percent true strain from uncrimping in the circumferential direction.

$$\ln\left(\frac{43.24\ \mu m}{38.46\ \mu m}\right) = 11.71\% \tag{4}$$

3.3. Isolation of the Elastin Microstructure

To the best of the authors' knowledge, this study is the first to use a variation of the hot alkali method to isolate the elastin microstructure in jugular venous valve tissue. The hot alkali method involves incubating a sample in a diluted NaOH solution, usually for 45 min, but some variations include extended time [39]. Remains of collagen were investigated with a Sircol Collagen Assay kit after the 0.1 N NaOH 75 °C heat treatment for the following timed tests: 45, 60 and 75 min. 45- and 60-min treatments still showed evidence of remaining collagen concentrations. The 75-min tests showed that in six out of eight samples, soluble collagen was completely removed from the venous valve's substrate. The remaining two samples had very low collagen concentrations compared to previous collagen assays conducted on non-digested tissue (361–616 mg/g dry weight) and were determined viable for imaging of the elastin microstructure [12].

After digestion, the samples had negligible or no amounts of collagen and were very delicate with only 11.63 ± 2.64% $\left(\frac{digested\ dry\ weight}{non-digested\ dry\ weight}\right)$ remaining. Comparisons of before and after digestion showed that the sample had noticeable amounts of shrinkage. It indicated that in relaxed conditions, the elastin microstructure was held in a preloaded position. In addition, the digested sample was very sticky, which was characteristic of purified tropoelastin [26]. However, the digestion was not perfect. Two out of the eight samples still had low amounts of soluble collagen concentrations after biochemical analysis. In defense of the digestion procedure, the remaining amounts of soluble collagen concentration, if any, were slim to none considering such high soluble collagen concentrations before digestion (proximal: 361 mg/g dry weight, middle: 439 mg/g dry weight and distal: 616 mg/g dry weight) [12].

3.4. Light Microscopy of the Elastin Microstructure

After isolating the elastin, the specimen was carefully transferred to a microscope slide, mounted with HBSS and imaged via light microscopy with a planar objective. Figure 2b showed the elastin microstructure located on the luminal side of the belly region of venous valve tissue and the large cusp fibers of the elastin microstructure having a radially crosslinked alignment was observed. We have observed that the light microscopy was difficult to focus because the remaining elastin microstructure had different regions of depth due to the sample shrinking and warping. Figure 2b needed to be stitched at several different focus levels to include the whole image.

3.5. Scanning Electron Microscopy of the Elastin Microstructure

Light microscopy was well-suited for qualitatively describing the macro scale of the elastin microstructure because of its fast preparation compared to SEM but for further magnification, SEM was advantageous for image quality. Therefore, SEM was used to test the quality of the isolated elastin digestion. Figure 3a showed the luminal side of the venous valve tissue before 75 °C NaOH digestion. Figure 3b showed the luminal side of venous valve tissue after 75 °C NaOH digestion. After ethanol dehydration and critical point drying, the samples were imaged in both non-digested and digested states. The non-digested samples clearly showed the endothelial cells' long axis alignment in the radial direction of the cusp shown in Figure 3b. The digested SEM images showed that not all of the basal lamina was digested but enough of it was removed to view parts of the elastin microstructure located in the belly region. The elastin microstructure showed no significant orientation at 5000× (Figure 3b) and also showed that certain fibers were damaged during the hot alkali digestion. These damaged fibers were disconnected and appear untaut, differing from the rest of the elastin fibers.

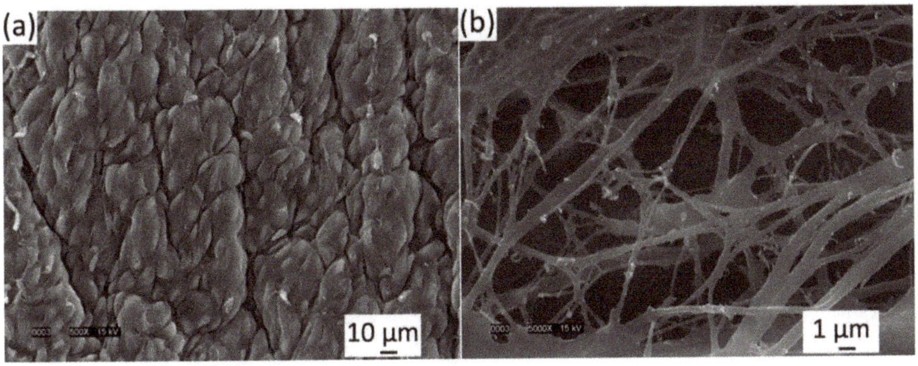

Figure 3. (**a**) Luminal side of venous valve tissue (**a**) before 75 °C NaOH digestion and (**b**) after 75 °C NaOH digestion viewed in a JEOL JSM-5900LV scanning electron microscope (SEM) at 15 kV. Elastin fibers are imaged clearly but damage can be seen.

As seen in Figure 3b and other studies, SEM showed that not all elastic fibers orient radially [24]. Light microscopy (400×) showed the overall macro-scaled radial crosslinked orientation (Figure 2b), whereas SEM (5000–10,000×) imaged smaller anastomosing elastic fibers orienting in all directions. These results must be approached cautiously, because it must be noted that the SEM critical point drying could alter the elastin's state, unlike light microscopy which yields more accurate imaging. Moreover, using SEM, it was discovered that freeze drying can fracture and twist the sample, where critical point drying was shown to shrink the sample [24]. The light microscopy imaging was not as clear or magnified as SEM, although the elastin microstructure was thought to be closer to its fresh state since the drying methods are avoided.

The 75-min hot alkali digestion used was effective enough for high magnification images of the elastin microstructure using lower magnification images using light microscopy and SEM. High SEM magnification showed that glycoproteins of the basal lamina were still intact, covering up parts of the elastin microstructure. Along with not completely isolating the elastin microstructure, the high magnification of SEM (10,000×) showed considerable amounts of damage to small anastomosing elastic fibers. Current methods used to isolate the elastin microstructure would not be advised for the future mechanical testing of the elastin microstructure, because the damaged elastin as well as the unwanted remains of collagen and the basal lamina would minimize the significance of the mechanical test's results. For future mechanical testing of the elastin microstructure, a more in depth NaOH digestion procedure or a collagenase digestion may be less invasive [30,40] of the elastin and allow for adequate mechanical testing of the isolated elastin microstructure.

3.6. *Biaxial Mechanical Testing*

A biaxial planar force-control mechanical test was used to investigate tissue-level fiber orientation and the rotation of the jugular venous valve. By implementing different loading ratios of circumferential: radial = 3:1, 1:1 and 1:3, the venous valve's fiber orientation and rotation were elucidated from the extracted stress-strain curves (Figure 4) and the curves demonstrated the pronounced mechanical anisotropy of tissues and the effects of transverse loading (in-plane coupling). Stiffness values of the linear regions of the 1:1 ratio for the circumferential and radial directions were 29.34 ± 1.72 MPa and 11.38 ± 0.84 MPa respectively (black curves in Figure 4; Table 1). The linear region's slope was evaluated between 75 N/m and 100 N/m of traction for both directions, using an average venous valve thickness, 40 μm. These values were similar to previous findings using a displacement-controlled mechanical test [12]. The averaged circumferential direction began to stiffen (enter the heal region) at 24.6% true strain and became linear approximately at 40%. The venous valve, which has anisotropic mechanical properties, had a radial direction considerably less stiff than the circumferential direction, not exiting the toe region until approximately 42% true strain. The 1:3 ratio (red curves in Figure 4) was loaded with 300 mN of force in the circumferential direction and 900 mN of force in the radial direction. In the circumferential direction, the heel region began at lower strain values than in the 1:1 ratio (~22% true strain). In the radial direction, the heel region began at greater strain values than the 1:1 ratio (~46% true strain). This created a leftward and rightward shift for the circumferential and radial stress strain curves compared to the 1:1 ratio. However, the stiffness values of the linear region were very comparable between 1:1 and 1:3 ratios. The linear stiffness values for the circumferential and radial direction were 33.15 ± 3.29 MPa and 11.16 ± 0.98 MPa, respectively (Table 1). In the 3:1 ratio, the circumferential direction was loaded with 900 mN and the radial direction was loaded with 300 mN (blue curves in Figure 4). The circumferential and radial directions both had leftward shifts on the stress strain graphs. The heel regions of both the circumferential and radial directions began stiffening before the 1:1 ratios corresponding directions (~19%, ~32%). The linear stiffness values for the circumferential and radial direction were 30.19 ± 2.78 MPa and 8.41 ± 1.54 MPa, respectively (Table 1).

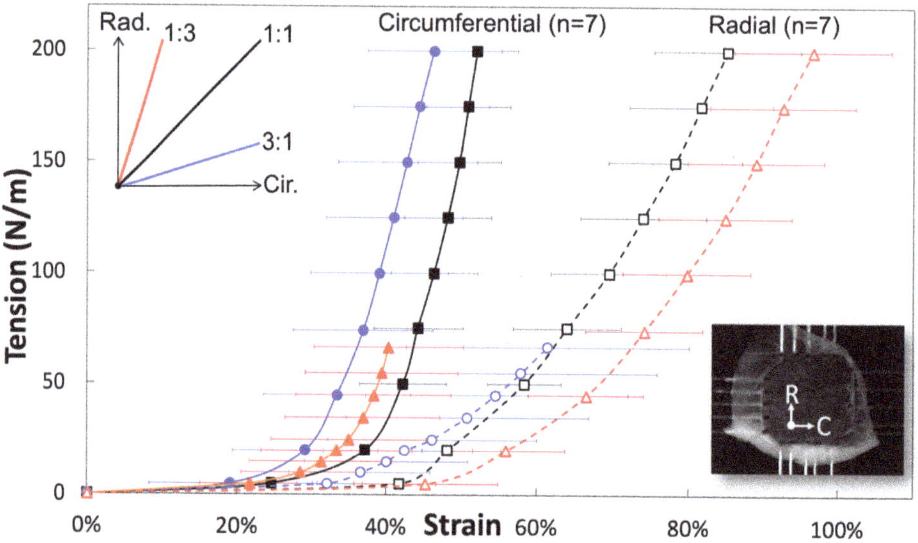

Figure 4. Representative circumferential and radial stress-strain curves (mean ± SEM) from fresh jugular venous valve leaflets, demonstrating the pronounced mechanical anisotropy of tissues and the effects of transverse loading (in-plane coupling).

Table 1. Modulus of elasticity of venous valvular tissue under three different loading ratios (mean ± SEM).

Modulus	1:1	1:3	3:1
Circumferential	29.34 ± 1.72 MPa	33.15 ± 3.29 MPa	30.19 ± 2.78 MPa
Radial	11.38 ± 0.84 MPa	11.16 ± 0.98 MPa	8.41 ± 1.54 MPa

Stiffness of the linear regions of all ratios closely resemble each other, except the radial 3:1 ratio could be due to rotated elastin fibers. These stiffness values showed that the linear region of the circumferential direction was close to three times stiffer than that of the radial direction in all loading ratios, except for the 3:1 ratio (due to its manipulated radial fiber components). The apparent differences between the ratios was the strain value at which the heel region begins. The 1:1 ratio showed that the jugular venous valve tissue's mechanical properties were planar anisotropic. The circumferential properties were substantially stiffer and discovered to be caused by the strict collagen alignment in the circumferential direction. The 1:3 ratio showed a leftward and rightward shift of the circumferential and radial directions on the stress-strain curves, respectively. This was caused by the reaction of the ECM to the loading condition and Poisson's ratio. Light microscopy images showed that the elastin fibers are crosslinked and oriented mostly radially (Figure 2b). During the 1:3 ratio loading, the elastin fibers straighten out in the radial direction and the circumferentially parallel-oriented collagen fibers remain oriented circumferentially. One significant difference in the loading ratios was overall stress; the 1:1 ratio has much more overall stress induced on the tissue than the 1:3 ratio, because both biaxial directions are loaded with 900 mN. In the 1:3 ratio, the circumferential direction was only loaded with 300 mN, while the radial direction was loaded with 900 mN. During 1:3 ratio testing, when the circumferential direction was at low strains, the radial direction was already at much higher strains. The overall stress of the radial direction being higher, while the circumferential direction was still loaded at low strains, made the stress-strain curve's circumferential direction shift leftward (or the heel region stiffen at lower strains). When the radial direction was at maximum force, the circumferential direction's strain was much less than in the 1:1 ratio. This results in the radial direction's heel region appearing to be less stiff because of Poisson's ratio. The overall difference of stress induced on the tissues from each loading ratio explained the leftward shift of the 1:3 ratio circumferential direction and the rightward shift of the 1:3 ratio's radial direction when compared to the 1:1 ratio.

The 3:1 ratio did not follow what was expected if only Poisson's ratio was considered. With higher forces in the circumferential direction and lower forces in the radial direction, the authors expected that the radial direction's heel region would stiffen and the circumferential direction would appear less stiff. This would be true if the re-alignment of the ECM had not affected its mechanical properties during loading. The radial direction's heel region stiffened at lower strains as expected but its linear region was not of the same stiffness as the other ratios and the circumferential direction's heel region stiffened unexpectedly at lower strains. The re-alignment of the ECM can explain this stress-strain curve shift and the more compliant radial direction. In the circumferential direction, the elastin contributes to low strain properties. While the loading ratio deformed the sample orienting elastin fibers circumferentially, the heel region began to stiffen at lower strain values. This created a leftward shift of the 3:1 ratio. Since less elastin was aligned radially at higher strain values, the linear region of the radial direction appeared less stiff (8.41 ± 1.54 MPa). This was noticeably less stiff then the similar 1:1 and 1:3 radial directions' linear region's stiffness (11.38 and 11.16 ± 0.98 MPa). The linear region's stiffness of the circumferential direction was unaffected by the elastin's re-orientation because at higher loads, uncrimped collagen completely takes over its mechanical properties in the circumferential direction. The circumferential directions' linear region's stiffness for the 1:1, 3:1 and 1:3 ratio are 29.34 ± 1.72 MPa, 30.19 ± 2.78 MPa and 33.15 ± 3.29 MPa, respectively (Table 1).

After all venous valve tissue was tested much past permanent slipping of collagen fibrils, mechanically tested close to rupture values. If collagen fibrils constantly slipped during loading, venous valves would be plastically deformed and soon destroyed. Despite these high loading values, the test was important in proving that the elastin layer contributes to both the circumferential and radial directions at lower strains. During the 3:1 loading ratio, the contribution of elastin in the circumferential direction was exaggerated and resulted in the stiffening of the curve at lower strain conditions. The force control test was also important for explaining that the collagen layer completely takes over the circumferential mechanical properties at higher loads when uncrimped. When elastin oriented partially circumferentially, the 3:1 ratio's circumferential linear region did not experience stiffened values. Lastly, the test was important in showing that at higher loads the radial direction's mechanical properties were still dominated by its elastin layer. With its elastin fibers oriented partially circumferentially, the 3:1 ratio's radial linear region appeared less stiff. The circumferential direction's inflection point of the 1:1 ratio occurs at ~24.6% true strain. It was hypothesized that straightening of the fibers accounts for the rest of the strain until the inflection point. Therefore, the authors hypothesized that straightening of the fibers account for ~12.89% of true strain. These results seem justified because collagen fibers did appear wavy in addition to the crimps. Additionally, the results seem physiologically similar to the aortic valve, where crimping accounts for 23% strain and straightening of the fibers accounts for 17% strain [38].

While collagen was uncrimping at low strains it was believed that the already pre-loaded elastin crosslinked mesh accounted for a majority of the circumferential mechanical properties. This assumption was reinforced by the force control testing. We believed elastin to be preloaded, because during the elastin isolation process the venous valve tissue consistently formed a corrugated radial direction and, when fixed, appeared stretched at the end of the digestion. Eventually, the strain reached the point where collagen was uncrimped. Once collagen was uncrimped, it immediately stiffened and the results of the stress-strain curve and elastin were no longer relevant in the stress-strain results in the circumferential direction. However, it was shown to still play a role in the radial direction's linear region by the force-control data. This mechanical testing should be followed with an isolated elastin microstructure mechanical test of the jugular venous valve tissue. Before this test is possible, a method must be perfected to isolate the elastin microstructure without damaging it. A collagenase digestion [30,40] may be a better option than the heated NaOH digestion for future work.

The bi-layer anatomy of the venous valve allows the radial direction to be much more compliant than the circumferential direction. The parietal side has strict collagen alignment circumferentially and the luminal side has its large elastic fibers crosslinked orienting radially. Collagen orienting circumferentially across the venous valve may have the ability to guide the valve from the closed to open position, extending the compliant radial direction. In this case, individual collagen fibers running parallel causes the venous valve, when opening, to fold between collagen fibers. This fold is believed to be guided by the crypts. Crypts are located along the parietal side of the venous valve [9]. Similarly, in the aortic valve, collagen fibers stabilize motion during mid-systole [41]. The collagen fibers, during the venous valve's closed position, are expected to carry the load of the above pressure column of blood. For example, cows spend the majority of their day (7–12 h) grazing with their head in the down position causing a pressure column from gravity [42]. In humans, pressure columns directed against the pockets of the jugular venous valve do not happen in daily activities due to our upright nature but do happen occasionally due to pressure wave impulses caused by coughing, straining or exterior compression of a venous segment [9]. In this case, the venous valve cannot completely prevent reflux but it will still partially mute it [9]. During pressure waves, venous valves are very important, because previous research has questioned if competent jugular venous valves obstruct reflux to the brain, preventing neurological disorders [43]. The strength of the venous valve tissue comes from collagen and causes it to have a breaking strength twice of the associated venous wall [21]. When loaded with reflux flow, collagen will uncrimp and mute the pressure wave. During the 1:1 loading ratio, the circumferential direction did not begin to acquire considerable force until the inflection point at 27.5% true strain. We hypothesize this to be the point at which collagen fibers are uncrimped and straightened. After the inflection point, the stiffness of the circumferential direction rapidly increased until the linear region, characteristic of yielding collagen fibers.

4. Conclusions

Overall, the goal of this study was to conclusively identify the mechanical properties of the venous valve's ECM with three different loading conditions. The mechanical test characterized fiber rotation and we related this information to the imaged collagen and elastin microstructures. The information from the imaged collagen and elastin microstructures, along with its mechanical properties, were used to understand how the bi-layer ECM contributes to the venous valves physiological function. In our study, light microscopy offers the first images of the venous valve's isolated elastin microstructure unaffected by these drying methods. For the first time, venous valve tissue's anisotropic behavior has been explained by the venous valve's bi-layer ECM. A heated 0.1 N NaOH digestion isolated the elastin microstructure developing accurate light microscopy imaging of venous valve tissue without invasive drying procedures, such as with previous SEM. Additionally, light microscopy of the collagen microstructure allowed first time characterization of crimp effects on venous valve mechanical properties, indicating crimp allows about 12% strain. Collagen aligning circumferentially and elastin orienting radially, causes the circumferential direction to be stiffer and the radial direction to be more compliant. Force-control testing also provided proof that the elastin's crosslinked mesh accounts for circumferential mechanical properties at low strains and affects the radial direction's tangent modulus of elasticity at higher strains. This new knowledge of venous valve tissue-level microstructures is important for advances in basic venous physiology and, additionally, will be important for novel approaches for preventing venous valve incompetence.

Author Contributions: A.A.B. and H.-Y.S.H. designed the experiments; A.A.B. acquired and analyzed the data; A.A.B. and H.-Y.S.H. interpreted the data and wrote the manuscript. Both authors have read and approved the final submitted manuscript.

Funding: This research was supported by the NSF CBET1553430

Conflicts of Interest: The authors declare no conflict of interest.

References

1. Spiegel, D.R.; Smith, J.; Wade, R.R.; Cherukuru, N.; Ursani, A.; Dobruskina, Y.; Crist, T.; Busch, R.F.; Dhanani, R.M.; Dreyer, N. Transient global amnesia: Current perspectives. *Neuropsychiatr. Dis. Treat.* **2017**, *13*, 2691–2703. [CrossRef] [PubMed]
2. Bartsch, T.; Alfke, K.; Stingele, R.; Rohr, A.; Freitag-Wolf, S.; Jansen, O.; Deuschl, G. Selective affection of hippocampal CA-1 neurons in patients with transient global amnesia without long-term sequelae. *Brain* **2006**, *129*, 2874–2884. [CrossRef] [PubMed]
3. Spiegel, D.R.; McCroskey, A.L.; Deyerle, B.A. A Case of Transient Global Amnesia: A Review and How It May Shed Further Insight into the Neurobiology of Delusions. *Innov. Clin. Neurosci.* **2016**, *13*, 32–41. [PubMed]
4. Nedelmann, M.; Kaps, M.; Mueller-Forell, W. Venous obstruction and jugular valve insufficiency in idiopathic intracranial hypertension. *J. Neurol.* **2009**, *256*, 964–969. [CrossRef]
5. Corno, A.F.; Hurni, M.; Griffin, H.; Jeanrenaud, X.; von Segesser, L.K. Glutaraldehyde-fixed bovine jugular vein as a substitute for the pulmonary valve in the Ross operation. *J. Thorac. Cardiovasc. Surg.* **2001**, *122*, 493–494. [CrossRef]
6. McElhinney, D.B.; Hennesen, J.T. The Melody (R) valve and Ensemble (R) delivery system for transcatheter pulmonary valve replacement. *Ann. N.Y. Acad. Sci.* **2013**, *1291*, 77–85. [CrossRef]
7. de Borst, G.J.; Moll, F.L. Percutaneous venous valve designs for treatment of deep venous insufficiency. *J. Endovasc. Ther.* **2012**, *19*, 291–302. [CrossRef]
8. Jensen, M.O.; Lemmon, J.D.; Gessaghi, V.C.; Conrad, C.P.; Levine, R.A.; Yoganathan, A.P. Harvested porcine mitral xenograft fixation: Impact on fluid dynamic performance. *J. Heart Valve Dis.* **2001**, *10*, 111–124.

9. Gottlob, R.; May, R. *Venous Valves: Morphology, Function, Radiology, Surgery*; Springer: Wien, Austria; New York, NY, USA, 1986.
10. Lu, J.; Huang, H.S. Biaxial mechanical behavior of bovine saphenous venous valve leaflets. *J. Mech. Behav. Biomed. Mater.* **2018**, *77*, 594–599. [CrossRef]
11. Ushiki, T. Collagen fibers, reticular fibers and elastic fibers. A comprehensive understanding from a morphological viewpoint. *Arch. Histol. Cytol.* **2002**, *65*, 109–126. [CrossRef]
12. Huang, H.S.; Lu, J. Biaxial mechanical properties of bovine jugular venous valve leaflet tissues. *Biomech. Model. Mechanobiol.* **2017**, *16*, 1911–1923. [CrossRef] [PubMed]
13. Muiznieks, L.D.; Keeley, F.W. Molecular assembly and mechanical properties of the extracellular matrix: A fibrous protein perspective. *Biochim. Biophys. Acta* **2013**, *1832*, 866–875. [CrossRef]
14. de Wolf, M.A.; de Graaf, R.; Kurstjens, R.L.; Penninx, S.; Jalaie, H.; Wittens, C.H. Short-Term Clinical Experience with a Dedicated Venous Nitinol Stent: Initial Results with the Sinus-Venous Stent. *Eur. J. Vasc. Endovasc. Surg.* **2015**, *50*, 518–526. [CrossRef]
15. Mouton, W.G.; Habegger, A.K.; Haenni, B.; Tschanz, S.; Baumgartner, I.; Ochs, M. Valve disease in chronic venous disorders: A quantitative ultrastructural analysis by transmission electron microscopy and stereology. *Swiss Med. Wkly.* **2013**, *143*, w13755. [CrossRef] [PubMed]
16. Budd, T.W.; Meenaghan, M.A.; Wirth, J.; Taheri, S.A. Histopathology of veins and venous valves of patients with venous insufficiency syndrome: Ultrastructure. *J. Med.* **1990**, *21*, 181–199.
17. Markovic, J.N.; Shortell, C.K. Genomics of varicose veins and chronic venous insufficiency. *Semin. Vasc. Surg.* **2013**, *26*, 2–13. [CrossRef]
18. Schoof, P.H.; Takkenberg, J.J.; van Suylen, R.J.; Zondervan, P.E.; Hazekamp, M.G.; Dion, R.A.; Bogers, A.J. Degeneration of the pulmonary autograft: An explant study. *J. Thorac. Cardiovasc. Surg.* **2006**, *132*, 1426–1432. [CrossRef] [PubMed]
19. Matthews, P.B.; Jhun, C.S.; Yaung, S.; Azadani, A.N.; Guccione, J.M.; Ge, L.; Tseng, E.E. Finite element modeling of the pulmonary autograft at systemic pressure before remodeling. *J. Heart Valve Dis.* **2011**, *20*, 45–52.
20. Ozturk, N.; Sucu, N.; Comelekoglu, U.; Yilmaz, B.C.; Aytacoglu, B.N.; Vezir, O. Pressure applied during surgery alters the biomechanical properties of human saphenous vein graft. *Heart Vessels* **2013**, *28*, 237–245. [CrossRef] [PubMed]
21. Ackroyd, J.S.; Pattison, M.; Browse, N.L. A study of the mechanical properties of fresh and preserved human femoral vein wall and valve cusps. *Br. J. Surg.* **1985**, *72*, 117–119. [CrossRef]
22. Kaul, N.; Huang, H.S. Constitutive modeling of jugular vein-derived venous valve leaflet tissues. *J. Mech. Behav. Biomed. Mater.* **2017**, *75*, 50–57. [CrossRef] [PubMed]
23. Saphir, O.; Lev, M. The venous valve in the aged. *Am. Heart J.* **1952**, *44*, 843–850. [CrossRef]
24. Crissman, R.S.; Pakulski, L.A. A rapid digestive technique to expose networks of vascular elastic fibers for SEM observation. *Stain Technol.* **1984**, *59*, 171–180. [CrossRef]
25. Crissman, R.S. Comparison of 2 Digestive Techniques for Preparation of Vascular Elastic Networks for Sem Observation. *J. Electron Microsc. Tech.* **1987**, *6*, 335–348. [CrossRef]
26. Mecham, R.P. Methods in elastic tissue biology: Elastin isolation and purification. *Methods* **2008**, *45*, 32–41. [CrossRef]
27. Jackson, D.; Cleary, G. The Determination of Collagen and Elastin. In *Methods of Biochemical Analysis*; John Wiley & Sons Inc.: New York, NY, USA, 1967; pp. 25–67.
28. Tseng, H.; Grande-Allen, K.J. Elastic fibers in the aortic valve spongiosa: A fresh perspective on its structure and role in overall tissue function. *Acta Biomater.* **2011**, *7*, 2101–2108. [CrossRef]
29. Huang, H.-Y.S.; Balhouse, B.N.; Huang, S. Application of simple biomechanical and biochemical tests to heart valve leaflets: Implications for heart valve characterization and tissue engineering. *Proc. Inst. Mech. Eng. Part H J. Eng. Med.* **2012**, *226*, 868–876. [CrossRef]
30. Huang, S.; Huang, H.-Y.S. Biaxial Stress Relaxation of Semilunar Heart Valve Leaflets during Simulated Collagen Catabolism: Effects of Collagenase Concentration and Equibiaxial Strain-State. *Proc. Inst. Mech. Eng. Part H J. Eng. Med.* **2015**, *229*, 721–731. [CrossRef] [PubMed]
31. Huang, H.-Y.S.; Huang, S.; Frazier, C.P.; Prim, P.; Harrysson, O. Directional Mechanical Property of Porcine Skin Tissues. *J. Mech. Med. Biol.* **2014**, *14*, 14500699. [CrossRef]

32. Buchanan, R.M.; Sacks, M.S. Interlayer micromechanics of the aortic heart valve leaflet. *Biomech. Model. Mechanobiol.* **2014**, *13*, 813–826. [CrossRef]
33. Edwards, E. The orientation of venous valves in relation to body surfaces. *Anat. Rec.* **1936**, *64*, 369–385. [CrossRef]
34. Budras, K.-D.; Habel, R. Chapter 5: Vertebral Column, Thoracic Skeleton and Neck. In *Bovine Anatomy*; Budras, K.-D., Ed.; Die Deutsche Bibliothek: Hannover, Germany, 2003; pp. 56–61.
35. Gelse, K.; Poschl, E.; Aigner, T. Collagens—Structure, function and biosynthesis. *Adv. Drug Deliv. Rev.* **2003**, *55*, 1531–1546. [CrossRef] [PubMed]
36. Wiltz, D.; Arevalos, A.; Balaoing, L.; Blancas, A.; Sapp, M.; Zhang, X.; Grande-Allen, J. Extracellular Matrix Organization, Structure and Function. In *Calcific Aortic Valve Disease*; Elena, A., Ed.; InTech: Rijeka, Croatia, 2013; pp. 3–30.
37. Chen, Q.; Thouas, G. Histology and Tissue Properties II Connective Tissues. In *Biomaterials A Basic Introduction*; CRC Press: Boca Raton, FL, USA, 2015; pp. 609–632.
38. Scott, M.; Vesely, I. Aortic valve cusp microstructure: The role of elastin. *Ann. Thorac. Surg.* **1995**, *60*, S391–S394. [CrossRef]
39. Robert, L.; Hornebeck, W. *Elastin and Elastases*; CRC Press: Boca Raton, FL, USA, 1989.
40. Barbour, K.; Huang, H.-Y.S. Strain Effects on Collagen Proteolysis in Heart Valve Tissues. *Mech. Time-Depend. Mater.* **2019**, *23*, 1–16. [CrossRef]
41. De Hart, J.; Peters, G.W.; Schreurs, P.J.; Baaijens, F.P. Collagen fibers reduce stresses and stabilize motion of aortic valve leaflets during systole. *J. Biomech.* **2004**, *37*, 303–311. [CrossRef]
42. Vallentine, J.F. *Grazing Management*; Academic Press: San Diego, CA, USA, 2001.
43. Valecchi, D.; Bacci, D.; Gulisano, M.; Sgambati, E.; Sibilio, M.; Lipomas, M.; Macchi, C. Internal jugular vein valves: An assessment of prevalence, morphology and competence by color Doppler echography in 240 healthy subjects. *Ital. J. Anat. Embryol.* **2010**, *115*, 185–189.

bioengineering

MDPI

Article

Effect of Residual and Transformation Choice on Computational Aspects of Biomechanical Parameter Estimation of Soft Tissues

Ankush Aggarwal

Glasgow Computational Engineering Centre, School of Engineering, University of Glasgow, Glasgow G12 8LT, UK; ankush.aggarwal@glasgow.ac.uk

Received: 19 September 2019; Accepted: 28 October 2019; Published: 29 October 2019

Abstract: Several nonlinear and anisotropic constitutive models have been proposed to describe the biomechanical properties of soft tissues, and reliably estimating the unknown parameters in these models using experimental data is an important step towards developing predictive capabilities. However, the effect of parameter estimation technique on the resulting biomechanical parameters remains under-analyzed. Standard off-the-shelf techniques can produce unreliable results where the parameters are not uniquely identified and can vary with the initial guess. In this study, a thorough analysis of parameter estimation techniques on the resulting properties for four multi-parameter invariant-based constitutive models is presented. It was found that linear transformations have no effect on parameter estimation for the presented cases, and nonlinear transforms are necessary for any improvement. A distinct focus is put on the issue of non-convergence, and we propose simple modifications that not only improve the speed of convergence but also avoid convergence to a wrong solution. The proposed modifications are straightforward to implement and can avoid severe problems in the biomechanical analysis. The results also show that including the fiber angle as an unknown in the parameter estimation makes it extremely challenging, where almost all of the formulations and models fail to converge to the true solution. Therefore, until this issue is resolved, a non-mechanical—such as optical—technique for determining the fiber angle is required in conjunction with the planar biaxial test for a robust biomechanical analysis.

Keywords: biomechanics; parameter estimation; nonlinear preconditioning; gradient-based minimization; cirrus

1. Introduction

Characterizing the biomechanical properties of soft tissues remains a crucial starting point for describing and predicting their behavior [1]. Different experimental techniques have been designed, and several decades of research has produced a large amount of stress-strain data for different tissue types, such as aorta [2], myocardium [3], and heart valves [4,5]. Unlike engineered materials, most of these tissues exhibit highly nonlinear and anisotropic responses. In order to describe the wealth of experimental data, different constitutive models have been developed, and nonlinearity and anisotropy remain a hallmark of these models [6].

With the increasing complexity of the constitutive models for soft tissues, the number of associated fitting parameters has also increased. The process of fitting these models to the experimental data, also knowns as parameter estimation, is an important step [7]. It has been observed that reliably estimating the parameters can be a challenge, especially as the number of unknown parameters increase [8]. Moreover, due to the high nonlinearity and anisotropy, the parameters can become correlated, and the experimental data may not be sufficient to uniquely identify them [9,10]. In fact, determining an optimum set of experiments required to uniquely and accurately estimate the model parameters is an active area of research [11].

On the other hand, the effects of parameter estimation techniques on biomechanical characterization remain under-analyzed. Generally, methods originally designed for estimating parameters in linear and isotropic models are used as is, which can suffer from ill-conditioning and slow convergence when applied to the highly nonlinear problems of tissue mechanics. More importantly, in many cases the uniqueness of the estimated parameters cannot be tested, which may result in dubious outcomes. Similarly, extensive research has been carried out in improving the conditioning of linear algorithms [12]; however, these techniques for linear problems may not benefit the nonlinear parameter estimation common in biomechanics. Thus, there is a strong need for advancement in the area of nonlinear parameter estimation to help resolve these issues.

Previously, using simplified analysis and elementary algebra arguments, modifications were proposed to improve the biomechanical parameter estimation [13]. However, that study was focused solely on the case with two unknowns, which has only one local minima. In other words, for two parameters, the iterations always converged to the correct solution, and the improvement was obtained in the speed of convergence. Moreover, one of the proposed modifications was to use a logarithm of the measured stresses, which poses a problem if the stresses are negative. Thus, there is a need to further develop these techniques that are more general and easily applicable.

The aim of this study is to thoroughly test novel techniques in the parameter estimation process for multiple unknowns. To aid easy adaptation and wide applicability, the presented work is restricted to only simple modifications that are straightforward to implement and focus on the issue of non-convergence. In Section 2, details of the methods used are described: how artificial experimental data is generated, which constitutive models are used, details of the algorithm used for parameter estimation, and the transformations tested. In Section 3, results for each model and different formulations are presented. At the end, in Section 4, the significance of the results, limitations and possible future work are discussed before concluding in Section 5.

2. Methods

2.1. Problem Setup

In order to make this study relevant to experimental situations, planar biaxial test protocols—both displacement controlled (DC) and force controlled (FC)—are used. Strain/stress ratios of 0:1, 1:1, and 1:0 in the x– and y–directions are applied. Although, in practice, five or seven stress-strain ratios are used, as using more data is expected to improve the accuracy of the fitted model. That is true when there is noise present in the data and/or the model is not perfect. However, in this study the experimental data is assumed to be noiseless and perfect, which makes the parameter estimation easier and theoretically only two stress-strain ratios are required to estimate all the parameters uniquely. Lastly, shear stresses and deformations are neglected, and plane stress and incompressibility assumptions are used. Thus, only two stress-strain relations remain relevant.

In the DC case, the inputs are stretch ratios λ_x and λ_y, and the outputs are Cauchy stresses σ_{xx} and σ_{yy}. Inversely, in the FC case, the inputs are Cauchy stresses σ_{xx} and σ_{yy}, and the outputs are stretch ratios λ_x and λ_y. The maximum axial stretch applied is 1.1225, while the maximum normal stress applied is 130 kPa. The material parameters are assumed to be homogeneous everywhere, and hence for the DC case, the input-output relation is analytical. Moreover, even for the FC case, an iterative solver based on the Powell hybrid method [14] is used to calculate stretch ratios from the stresses. Thus, no finite element simulations are required, which helps keep the computational expenses reasonable.

2.2. Constitutive Models

Standard notations in large deformation mechanics are adopted [15]: $\mathbf{F}$ is the deformation gradient, $\mathbf{C} = \mathbf{F}^{\top}\mathbf{F}$ is the right Cauchy-Green deformation tensor, and $I_1 = \operatorname{tr}(\mathbf{C})$ is the first invariant of $\mathbf{C}$.

Due to incompressibility, the Jacobian of the deformation $J = \det \mathbf{F}$ is constrained to be unity $J = 1$. If the fiber direction is denoted by a direction vector $\boldsymbol{M}$, the stretch along the fiber is defined by the fourth invariant of $\mathbf{C}$: $I_4 = \boldsymbol{M} \cdot \mathbf{C}\boldsymbol{M}$. For planar tissues with in-plane fibers aligned at an angle θ to the x-axis and normal direction parallel to the z-axis, the fiber direction can be written as $\boldsymbol{M} = [\cos(\theta), \sin(\theta), 0]^\top$.

The Cauchy stress tensor is σ, which is derived from the strain energy density Ψ as

$$\sigma = 2\mathbf{F} \cdot \frac{\partial \Psi}{\partial \mathbf{C}} \cdot \mathbf{F}^\top - p\mathbf{I}, \quad (1)$$

where $\mathbf{I}$ is the identity tensor and p is the hydrostatic pressure acting as a Lagrange multiplier to enforce incompressibility. To define the stress-strain relationship, the following four constitutive models popular in the biomechanics community are used; however, different symbols are used for the material parameters than the literature for consistency across the models in this study.

2.2.1. Gasser-Ogden-Holzapfel (GOH) Model

The GOH model defines the strain energy density function as [16]

$$\Psi(\mathbf{C}) = \frac{c_1}{2c_2}\left(e^Q - 1\right) + \frac{c_0}{2}\left(I_1 - 3\right), \quad (2)$$

where

$$Q = c_2\left[c_3 I_1 + (1 - 3c_3) I_4 - 1\right]^2. \quad (3)$$

The first term is the contribution of fibrous tissue, whereas the second term is assumed to be due to isotropic matrix. c_J ($J = 0, 1, 2$) are material parameters, and the dispersion parameter $c_3 \in [0, 1/3]$ and the fiber angle $c_4 \equiv \theta \in [0, \pi]$ are the structural parameters. Thus, GOH model has a total of $M = 5$ constitutive parameters.

2.2.2. Humphrey Model

Humphrey and Yin proposed the following strain energy density function for soft tissues [17]

$$\Psi(\mathbf{C}) = \frac{c_1}{c_2}\left[e^{c_2(I_1-3)} - 1\right] + \frac{c_3}{c_4}\left[e^{c_4\left(\sqrt{I_4}-1\right)^2} - 1\right]. \quad (4)$$

Here, c_J ($J = 1, 2, 3, 4$) are material parameters, and the fiber angle $c_5 \equiv \theta \in [0, \pi]$ is the structural parameter. Since there is no neo-Hookean term with parameter c_0, the Humphrey model also has $M = 5$ constitutive parameters.

2.2.3. Lee–Sacks Model

Lee et al. proposed the following constitutive model for valve tissue [9]:

$$\Psi(\mathbf{C}) = \frac{c_0}{2}\left(I_1 - 3\right) + \frac{c_1}{2}\left(c_4 e^{c_2(I_1-3)^2} + (1 - c_4)e^{c_3(I_4-1)^2} - 1\right). \quad (5)$$

Here, c_J ($J = 0, 1, 2, 3$) are the material parameters, and the fiber angle $c_5 \equiv \theta \in [0, \pi]$ and $c_4 \in [0, 1]$ are the structural parameters. Thus, Lee–Sacks model has $M = 6$ constitutive parameters.

2.2.4. May-Newman Model

May-Newman and Yin proposed the following strain energy density to define the biomechanical properties of mitral valve tissue [18]

$$\Psi(\mathbf{C}) = c_1\left(e^Q - 1\right) + \frac{c_0}{2}\left(I_1 - 3\right), \quad (6)$$

where

$$Q = c_2 (I_1 - 3)^2 + c_3 \left(\sqrt{I_4} - 1\right)^4. \tag{7}$$

Here, c_J ($J = 0, 1, 2, 3$) are material parameters, and the fiber angle $c_4 \equiv \theta \in [0, \pi]$ is the structural parameter. Thus, May-Newman model has a total of $M = 5$ constitutive parameters.

2.3. Parameter Estimation Algorithm

A general parameter estimation problem for planar biaxial tissues is the following: given the "measured" values of stresses and stretches and a chosen constitutive model, determine the associated constitutive parameters $\boldsymbol{c} = \{c_0, c_1, \ldots, c_M\}$. The experimental input is denoted as x_i and output as $\bar{y}_i$, $i = 1, \ldots, n$ where n is the number of experimental data points. In the DC case, input $x_i = \left[\lambda^i_{xx}, \lambda^i_{yy}\right]$ are the stretches and output $\bar{y}_i = \left[\sigma^i_{xx}, \sigma^i_{yy}\right]$ are the stresses, whereas in the FC case, input $x_i = \left[\sigma^i_{xx}, \sigma^i_{yy}\right]$ are the stresses and output $\bar{y}_i = \left[\lambda^i_{xx}, \lambda^i_{yy}\right]$ are the stretches. The deviation of the chosen model from the measured output is defined as the residual

$$r_i(\boldsymbol{c}) = \langle m(x_i; \boldsymbol{c}), \bar{y}_i \rangle, \tag{8}$$

where m is the input–output function derived using the chosen constitutive model and then evaluated at x_i input and chosen $\boldsymbol{c}$ parameter values. The residual operator $\langle \cdot, \cdot \rangle$ needs to be chosen appropriately. A commonly used option is the uniformly weighted difference

$$r_i^U = m(x_i; c) - \bar{y}_i. \tag{9}$$

In [13], a "log-norm" was proposed which decreased the nonlinearity and improved the convergence. However, logarithm has a drawback that it cannot be applied to negative values. It should be noted that taking a log has the effect of assigning lower weights to higher values. In other words, $\log(y_1) - \log(y_2) = \log(y_1/y_2)$. Inspired by this observation, an alternative residual is tested: a non-uniformly weighted difference

$$r_i^N = \frac{m(x_i; c)}{\bar{y}_i} - 1. \tag{10}$$

While using Equation (10), one must exclude points with measured zero values. In practice, this is easily implemented since exactly zero output is uncommon for anisotropic tissues (except in the load-free reference configuration, which is trivially satisfied by all models and can be simply discarded). Lastly, an objective function is defined (also called the Loss function in literature), as simply the square summation of the residual:

$$\mathcal{F}(\boldsymbol{c}) := \frac{1}{2} \sum_{i=1}^{n} (r_i \cdot r_i). \tag{11}$$

In order to determine the parameters, a minimization problem is formulated to estimate the parameters c_J:

$$\bar{\boldsymbol{c}} = \arg\min_{\boldsymbol{c}} \mathcal{F}(\boldsymbol{c}). \tag{12}$$

The minimization problem is solved using the Gauss-Newton algorithm with a backtracking line search [19]. Details are provided in Algorithm 1.

Since the parameter space size grows exponentially with the number of parameters, it becomes prohibitively expensive to systematically span the space for more than three parameters. Therefore, Latin Hypercube Sampling (LHS) is used to generate 300 samples from the parameter space (see Table 1 for the parameter ranges used). LHS has the advantage of generating uniformly spaced combinations of the parameters while keeping the parameter values random. Using each LHS sample C^k ($k = 1, \ldots, 300$), artificial "measurements" are generated as $\bar{y}_i = m(x_i, C^k)$. Thus, the "true" parameter values are known, and the estimated results can be compared against them. Since nonlinear parameter estimation depends strongly on the initial guess, all other samples $C^l \; \forall l \neq k$ are used as the initial guesses c_0 for the parameter estimation algorithm. Thus, for each test, 300 × 299 = 89,700 minimization simulations are computed. Finally, the histogram of the fraction of cases versus iterations taken to converge are plotted. To test the convergence, the final parameter values are compared with the true parameter values, and the error is calculated. Since the fiber angle θ is cyclic with a period of π, the value of $\cos^2(\theta)$ is compared instead. The non-converged cases are subcategorized into "Unconverged" (U) and "Misconverged" (M): U being the runs that were unable to converge in maximum number of iterations and M being the ones where minimization converged, however, to wrong parameter values.

Algorithm 1: Parameter estimation using Gauss-Newton method with backtracking line search.

Data: Observed data $\overline{y}$ and initial guess c_0, $MAXITER = 30$, $TOL = 10^{-10}$, $\delta = 10^{-5}$
Result: Parameters that fit the model y to observed data $\overline{y}$ by minimizing the functional $\mathcal{F}(c) = \frac{1}{2} r \cdot r$ (11) with the chosen residual (9) or (10)

```
initialization c ← c0;
ITER ← 0;
do
    Calculate the fitting model y(c) and its derivatives J = ∂r(c)/∂c using central finite
      difference ;
    Calculate the search direction and step Δc by solving JᵀJΔc = Jᵀr ;
    Perform line search as follows (backtracking):
    while (ΔF = F(c) − F(c + Δc) < 0) or (calculation of y(c + Δc) failed ) do
        Δc ← Δc/2
    end
    ΔF = F(c) − F(c + Δc);
    c ← c + Δc;
    ITER ← ITER + 1;
while (ΔF > TOL) and (ITER < MAXITER) and (max |Δc| > δ) ;
```

Table 1. Summary of the models used and the ranges of associated parameters. GOH = Gasser-Ogden-Holzapfel model.

Model	c_0	c_1	c_2	c_3	c_4	c_5
GOH Equation (2)	0 to 100 kPa	5 to 100 kPa	20 to 100	0 to 0.3	0 to π	–
Humphrey Equation (4)	–	5 to 100 kPa	1 to 100	5 to 100 kPa	1 to 100	0 to π
Lee–Sacks Equation (5)	0 to 100 kPa	5 to 100 kPa	1 to 100	1 to 100	0 to 1	0 to π
May-Newman Equation (6)	0 to 100 kPa	5 to 100 kPa	1 to 100	1 to 1000	0 to π	–

2.4. Parameter Transformations

The aim is to study the effect of transforming parameter space from c to $\hat{c} = \Gamma(c)$ on the minimization. As the first step, linear parameter transformation is tested, and, in general, a linear transformation can be written as $\hat{c} = \mathbf{P}c$, where $\mathbf{P}$ is a constant matrix. A natural simple linear transformation is rescaling the parameters

$$\hat{c} = [\eta_1 c_1, \ldots, \eta_M c_M], \tag{13}$$

so that the derivative of the residual $\partial \mathcal{F} / \partial \hat{c}_J$ are of the same order. That is,

$$\eta_J = \sum_{i=1}^{n} r_i \frac{\partial r_i}{\partial c_J} \tag{14}$$

calculated from the previous iteration. It should be noted that this rescaling transformation is equivalent to the well-established Jacobi preconditioner [12].

Next, nonlinear transformations are tested. As shown in [13], taking a log of the parameter c_1, i.e., $\hat{c}_1 = \log(c_1)$ was shown to accelerate the convergence for estimating two unknown parameters. To test if this holds true for different models and multiple parameters, the following transformation is used

$$[\hat{c}_1, \hat{c}_2, \hat{c}_3, \ldots] = [\log(c_1), c_2, c_3, \ldots] . \tag{15}$$

Since the model by Humphrey (4) has sum of two exponential terms, an equivalent transformation is tested

$$[\hat{c}_1, \hat{c}_2, \hat{c}_3, \hat{c}_4] = [\log(c_1), c_2, \log(c_3), c_4] . \tag{16}$$

Using each transformation, the number of unknown parameters are gradually increased and their effect on convergence is tested. Lastly, Zhang et al. [20] proposed the following transformation:

$$[\hat{c}_1, \hat{c}_2, \hat{c}_3, \ldots] = \left[c_1 e^{Q_{\max}(c)}, c_2, c_3, \ldots\right] , \tag{17}$$

where $Q_{\max}(\boldsymbol{c})$ is maximum value of the exponent over all applied inputs. Although this is not a simple transformation to implement, for comparison purposes, models with a single exponential term are tested for the DC case.

3. Results

Using linear transformation, there is no effect on the minimization process for any cases (results skipped for brevity). This is not surprising since the number of unknowns in this case is always less than six, which means that the Hessian matrix—even if it has a high condition number—can be inverted with high precision. Thus, the linear transformation has no effect on the presented problem, but it may improve the estimation for heterogeneous problem or while using an algorithm like the steepest-descent where the Hessian is not inverted. Henceforth, only the nonlinear transformations are focused on for different models described in the previous section.

3.1. GOH Model

For the DC case with GOH model (2), as the first check, only two parameters, c_1 and c_2, are estimated while keeping other parameters fixed. Both the log transformation (15) and non-uniformly weighted residual (10) show faster convergence compared to the standard uniform weighted residual and no transformation (Figure 1a). Furthermore, the algorithm is able to find true parameters for all runs and formulations.

If the number of unknowns is increased to include c_0 and c_3 (but keep the fiber angle θ fixed), using the standard formulation (uniform weighted residual and no transformation), there are a small fraction of cases that either do not converge or converge to the wrong solution (Figure 1b). These cases are reduced to almost none when using the log transformation (15) and non-uniformly weighted residual (10). This is an important improvement in addition to faster convergence.

Furthermore, if the fiber angle θ is also treated as an unknown, the number of unconverged and misconverged runs increases dramatically, and only less than a third of the runs converge to the true parameter values (Figure 1c). Although, the log transformation (15) and non-uniformly weighted residual (10) improve the situation slightly, there still remains a large number of non-converged runs.

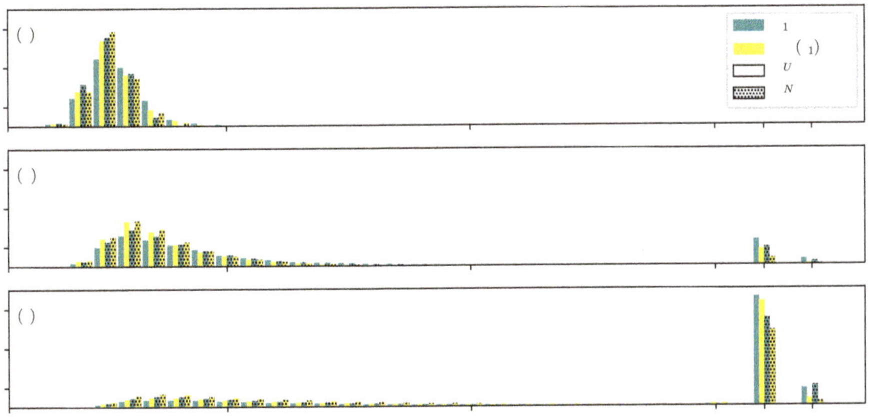

Figure 1. Iterations required using different formulations for GOH model (2) displacement controlled (DC) case when (**a**) only c_1 and c_2 are unknown, (**b**) all parameters except the angle θ are unknown, and (**c**) all parameters are unknown.

The standard formulation (uniformly weighted residual and original parameters) and the best formulation so far (non-uniformly weighted residual and $\log(c_1)$ transformation) are compared with the one proposed by Zhang et al. [20] (Figure 2). Results show that Zhang's transformation helps improve the convergence compared to the standard formulation; however, its performance is sub-par to the one proposed here—both for two and four unknown parameters.

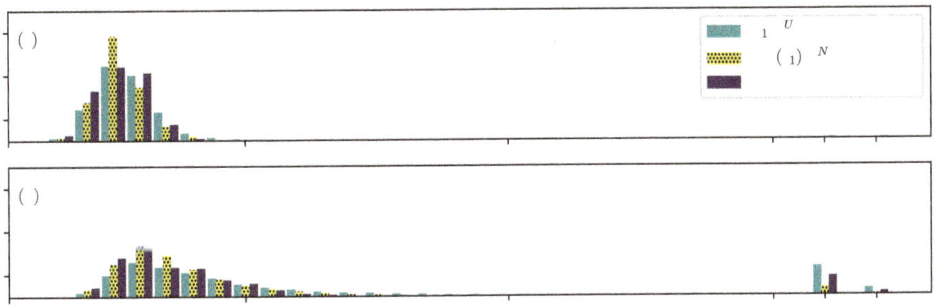

Figure 2. Iterations required using different formulations for the GOH (2) model DC case when (**a**) only c_1 and c_2 are unknown and (**b**) all parameters except the angle θ are unknown.

For the FC case with the GOH model (2), as a start, only two parameters, c_1 and c_2, are estimated and all other parameters are fixed. Results show that all the cases are converged, and that using a log transformation improves the convergence speed (Figure 3a). However, compared to the DC case, no appreciable difference is found by using the non-uniformly weighted residual (10) in this case.

As the number of unknowns is increased to include c_0 and c_3, the behavior remains similar (Figure 3b). The log transformation helps the algorithm by making the convergence faster and decreasing the number of non-converged runs. However, there is no effect of the non-uniformly weighted residuals. If the fiber angle θ is also an unknown, similar to the DC case, the parameter estimation becomes extremely difficult, and only a small fraction of the runs converge (Figure 3c). There is only a small improvement by using the log transformation and non-uniformly weighted residuals. It should be noted that using Zhang's formulation (17) for the FC case is problematic because the strain is unknown and finding the maximum value of the exponent will be expensive. Since no improvement was found in the DC case, the formulation by Zhang is omitted for the FC case.

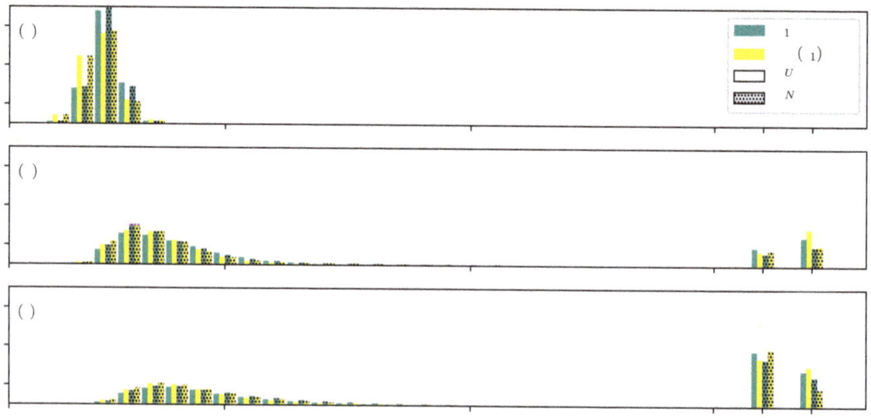

Figure 3. Iterations required using different formulations for the GOH (2) model force controlled (FC) case when (**a**) only c_1 and c_2 are unknown, (**b**) all parameters except the angle θ are unknown, and (**c**) all parameters are unknown.

For the DC case with standard formulation (uniform weighted residual and no transformation) and Humphrey's model (4) with fiber angle θ fixed, there is a significant fraction of runs that are either unconverged or misconverged (Figure 4a). This is caused by the interaction of two exponential terms in the model. By using the log-log transform (16), the situation improves slightly. Furthermore, if the non-uniformly weighted residual (10) is also used, all of the cases converge. Thus, there is an enormous difference in convergence properties by using the transform and non-uniform weighting.

3.2. *Humphrey Model*

If the fiber angle also needs to be determined, the number of converged runs reduces substantially (Figure 4b). The misconverged simulations are reduced to almost none when the log transform is used; however, the number of unconverged cases increase. Thus, overall the total number of non-converged runs remains approximately the same, with only a slight improvement in the convergence using the log transformation (16). Since Humphrey's model has two exponential terms, there is no simple method to find the maximum exponent for each term, and thus the formulation by Zhang et al. (17) is omitted.

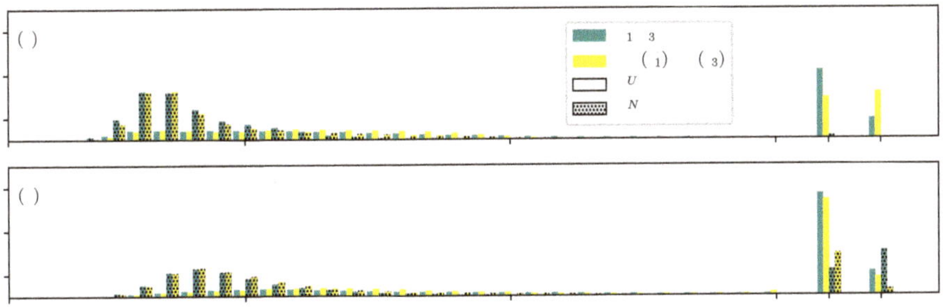

Figure 4. Iterations required using different formulations for the Humphrey model (4) DC case when (**a**) all parameters except the fiber angle θ are unknown and (**b**) all parameters are unknown.

A similar behavior for the FC case is found; when the fiber angle θ is fixed, the parameter estimation is successful for all runs. However, there is no appreciable improvement by using either the log transform or the non-uniformly weighted residual (Figure 5a). If the fiber angle θ is included as an unknown, it becomes challenging to estimate the parameters irrespective of the method used (Figure 5b). Although using non-uniformly weighted residual leads to a decrease in the number of misconverged runs, it also leads to an increase in the unconverged runs with only a slight increase in the number of converged runs.

Figure 5. Iterations required using different formulations for the Humphrey model (4) FC case when (**a**) all parameters except the fiber angle θ are unknown and (**b**) all parameters are unknown.

3.3. Lee–Sacks Model

In Lee–Sacks model (5), when only two parameters, c_1 and c_2, are estimated with the DC case, all of the runs converge irrespective of the formulation (Figure 6a). Using log transformation on c_1 helps speed up the convergence, while using the non-uniformly weighted residual (10) has a limited effect.

If all parameters except the fiber angle θ are estimated, the uniformly weighted residual leads to a large number of unconverged and misconverged runs (Figure 6b). However, changing the residual to non-uniformly weighted one (10) helps improve the situation. When the fiber angle is also estimated, similar to previous two models, a large number of unconverged and misconverged runs are obtained (Figure 6c). However, unlike the previous models, there is only a limited improvement when the residual is changed or the log transformation is used. Similar to the Humphrey's model, Less–Sacks model also has two exponential terms, however, with only one parameter c_1 in front. Thus, it is not clear how to implement the transformation proposed by Zhang et al. (17).

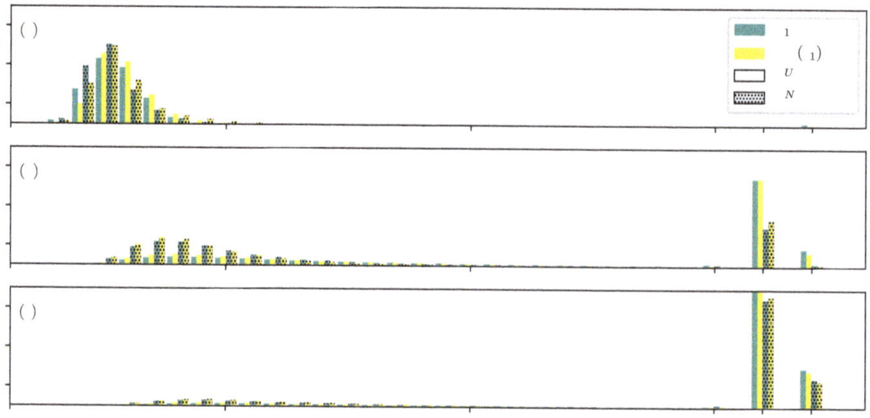

Figure 6. Iterations required using different formulations for the Lee–Sacks model DC case when (**a**) only c_1 and c_2 are unknown, (**b**) all parameters except the fiber angle θ are unknown, and (**c**) all parameters are unknown.

In the FC case with Lee–Sacks model, only two parameters, c_1 and c_2, can be reliably estimated using the standard formulation (Figure 7a). Using the log transformation and non-uniformly weighted residual speeds up the convergence. Furthermore, there is a small number of misconverged runs with only two unknowns using the standard formulation, which disappear when the log transformation is used. However, when the unknown parameters include other parameters, almost none of the FC runs are converged (Figure 7b,c). This happens irrespective of fiber angle θ being fixed or unknown and the choice formulation.

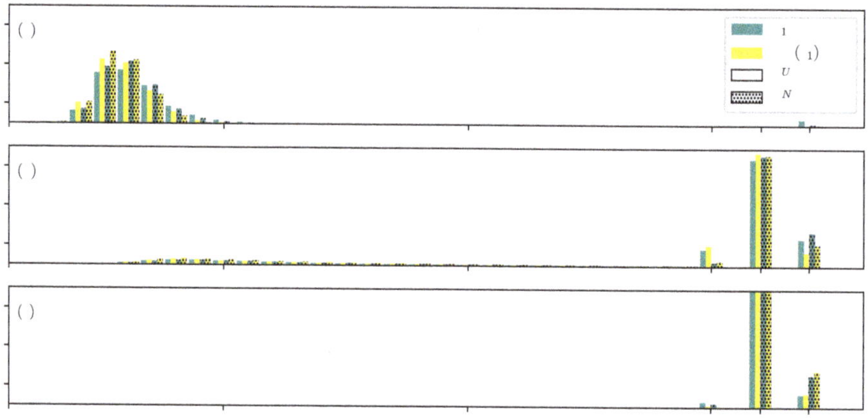

Figure 7. Iterations required using different formulations for the Lee–Sacks model FC case when (**a**) only c_1 and c_2 are unknown, (**b**) all parameters except the fiber angle θ are unknown, and (**c**) all parameters are unknown.

3.4. May-Newman Model

Two parameters c_1 and c_2 in the May-Newman model (6) can be estimated using the DC setup and any of the formulations (Figure 8a). Using the log transformation and non-uniformly weighted residual helps improve the convergence speed. If the number of unknown parameters is expanded to include others, except the fiber angle, most of the runs converge to the correct solution (Figure 8b). The number of non-converged runs becomes lower if the log transformation and/or the non-uniformly weighted residual are used. However, the convergence speed is largely unaffected by the change in formulation.

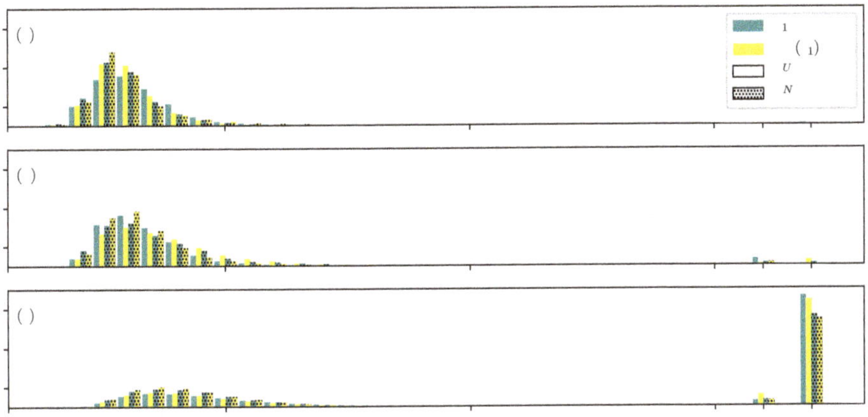

Figure 8. Iterations required using different formulations for the May-Newman model DC case when (**a**) only c_1 and c_2 are unknown, (**b**) all parameters except the fiber angle θ are unknown, and (**c**) all parameters are unknown.

When the fiber angle θ is included as an unknown, the same issue as other models appears where less than a third of the simulations converge to the correct solution (Figure 8c). By using the non-uniformly weighted residual, this problem is mitigated to some extent, although not completely. Furthermore, both the log transformation and non-uniformly weighted residual help reduce the iterations required to converge.

The results using Zhang's transformation [20] are compared with the proposed formulations (Figure 9). Similar to the results for the GOH model, this transformation helps improve the convergence compared to the standard formulation. However, the improvement is less than that using the formulation proposed here.

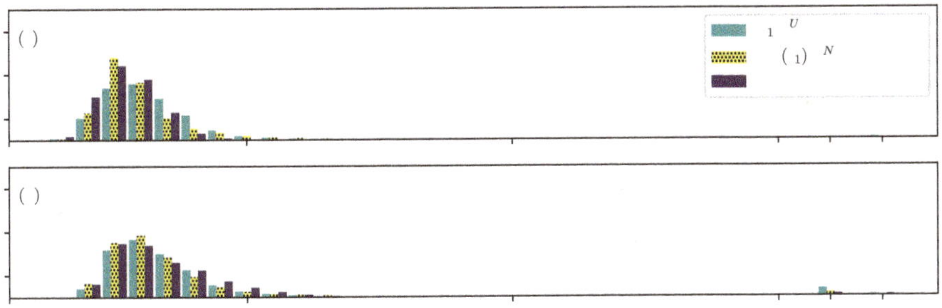

Figure 9. Iterations required using different formulations for the May-Newman (6) model DC case when (**a**) only c_1 and c_2 are unknown and (**b**) all parameters except the angle θ are unknown.

Using the FC setup with May-Newman model, the convergence behavior is similar to other models. When only two parameters c_1 and c_2 are estimated, all runs converge (Figure 10a). In this case, the convergence speed is improved when the log transformation is used, but is unaffected by the residual choice. When all the parameters except the fiber angle θ are estimated, some simulations do not converge (Figure 10b). The number of misconverged runs is reduced slightly by using the log transformation, whereas using the non-uniformly weighted residual has a slight negative effect on the convergence. Lastly, when the fiber angle θ is also an unknowns, all the formulations suffer from poor convergence (Figure10c). Only a small fraction of runs converge, and the choice of formulation has a negligible effect.

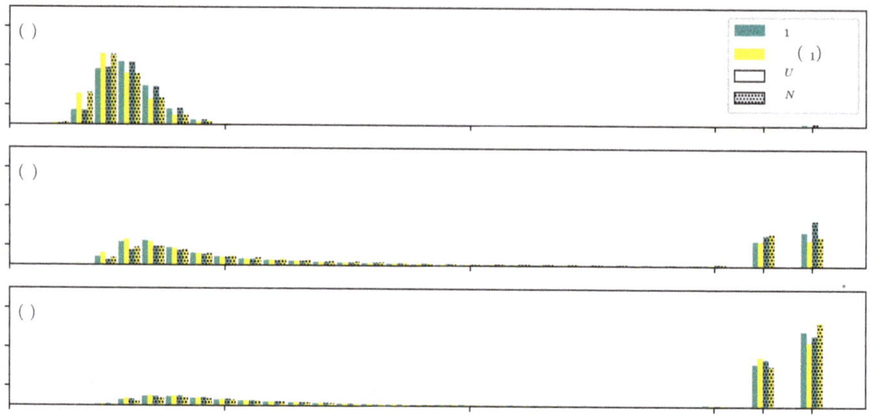

Figure 10. Iterations required using different formulations for the May-Newman model FC case when (**a**) only c_1 and c_2 are unknown, (**b**) all parameters except the fiber angle θ are unknown, and (**c**) all parameters are unknown.

4. Discussion

4.1. Nonlinear Preconditioning

Linear transformation was found to have no effect on the convergence behavior. This is because the number of unknowns is small, and the tissue is assumed to be homogeneous. Thus, the Hessian matrix can be inverted accurately, and therefore linear preconditioners have no advantage for the presented problem. This is an important characteristic of biomechanical problems: the challenges are different from other engineering fields, which necessitates different solutions. The nonlinear transformation proposed here acts as a nonlinear preconditioner, which is a relatively under-explored area [21]. The improvements found in this study will motivate further work along these lines to improve the biomechanical parameter estimation and, therefore, analysis.

4.2. Replacing c_1 with $e^{\hat{c}_1}$

Across almost all models and cases, using a log transform of c_1 led to improvement in the parameter estimation. For many cases, it not only improved the convergence speed, but also helped decrease the fraction on non-converged simulations. It should be noted that taking a log of c_1 while parameter estimation is equivalent to replacing c_1 with $e^{\hat{c}_1}$ in the constitutive models. As noted in [13], this not only helps reduce the nonlinearity but also enforces the constraint $c_1 > 0$. Interestingly, not only this transform helped improve the DC cases, but it also helped improve the FC cases, albeit to a lesser extent. The improvements obtained are all the more impressive considering the minute nature of this change and its implementational simplicity.

4.3. Weighted Residual

Interestingly, the effect of non-uniformly weighted residual was similar to that of the "log"-norm proposed in [13]. It helped improve the parameter estimation for almost all DC cases, but did not have an appreciable—positive or negative—effect on the FC cases. The advantage of this approach over "log"-norm is that it can be used for both positive and negative values. Moreover, the non-uniformly weighted residual is already used in some optimization problems. However, this is the first time it is being compared with the uniformly weighted residual for biomechanical parameter estimation. The approach can be implemented easily and the results show a clear advantage over uniformly weighted residual.

4.4. Adding Fiber Angle as an Unknown

It was not surprising that increasing the number of unknown parameters made their estimation more challenging for all models and cases. However, the most striking differences were observed when fiber angle was added to the unknown list; the fraction of converged results reduced drastically compared to when the fiber angle was considered as a known fixed value. More importantly, the use of log transform or non-uniformly weighted residual had an extremely limited effect when fiber angle was being estimated. Until this issue is resolved, the most practical approach may be to determine the fiber angle using other techniques, such as histology [22] or light scattering[23], and consider it as a fixed unknown during biomechanical parameter estimation. Using optical techniques, it may be possible to estimate the fiber dispersion, as well (c_3 in GOH model and c_4 in Lee–Sacks model). Not surprisingly, this will make the parameter estimation easier (Figure 11); however, the advantages of the log transform and non-uniformly weighted residual remain.

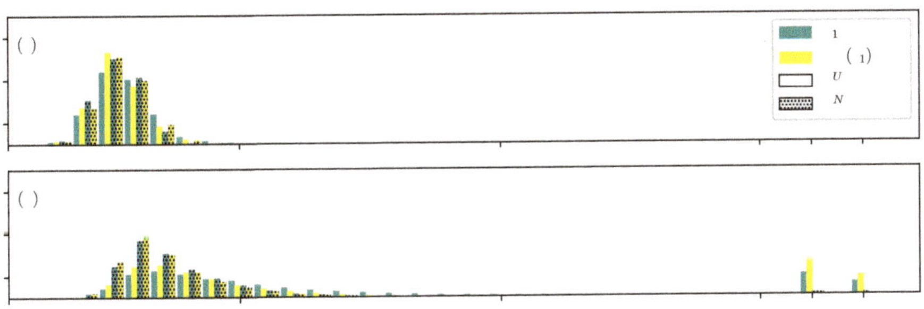

Figure 11. Iterations required using different formulations for (**a**) the GOH model DC case when all parameters except fiber angle θ and dispersion c_3 are unknown, (**b**) the Lee–Sacks model when all parameters except the fiber angle θ and dispersion c_4 are unknown.

Furthermore, the ultimate goal of this study is to have a robust technique that can estimate all the parameters from in-vivo dataset, where inverse models are required [24–26]. Previously, in inverse model developed for the aortic valve, fiber angles had a significant effect on the solution accuracy [10], and therefore needed to be determined by other means. It should be noted that it may not be possible to determine fiber angle separately for in-vivo inverse models. In such situations, determining the population-averaged fiber architectures of native tissues would help make this possible [27,28]. In general, if the parameters can be split into separable subsets, where one subset of the parameters can be estimated before fitting the rest of them and the fitting the second subset does not affect the first one, the estimation process generally becomes easier and faster. However, this requires the two subsets must be theoretically separable. It should be noted that this is not the case for the tow region for any of the models used in this study, as the stiffness of the exponential part is non-zero even at zero deformation ($\mathbf{F} = \mathbf{I}$).

4.5. Displacement Controlled versus Force Controlled

The biaxial testing experimental protocols can be designed to be either displacement controlled or force controlled. In literature, there is some discussion on the implementation difficulty of these two approaches; however, there has been no study comparing them in terms of parameter estimation. The results presented here demonstrate that the two approaches have different convergence properties. For most of the problems, DC cases show better convergence than the FC cases. However, that is not always true, especially when the fiber angles are considered as unknowns. Furthermore, estimating parameters via FC setup requires solving the inverse of stress-strain relationship using an iterative solver, which adds to the computational expense. Therefore, if there is a choice, a DC setup may be easier from the parameter estimation point of view. However, to reliably estimate the fiber angle, FC setup might prove superior.

4.6. Limitations and Future Work

In this study, the focus was limited to simple modifications that are easy to implement. Thus, only a limited number of nonlinear transforms were tested, and those with a clear effect were presented. Although one could develop other nonlinear transforms, there is no clear method to find good candidates. This is especially important for determining the fiber angle, as an improvement is clearly needed in that area to make the parameter estimation robust. Similarly, this analysis could be done for newer models, such as one by Li et al. [29]. However, this model requires integration over a unit sphere for every stress calculation, which makes it computationally unfeasible to solve approximately 90 thousand minimizations. On the other hand, applying Latin Hypercube Sampling approach to Fung's model [30] generates parameter sets with non-convex stress-strain relationship [31], and convexity is difficult to impose in the sampling process.

Due to the high computational cost of this study, the measured data $\bar{y}$ was assumed to be error-free (either modeling or noise). This is a limitation since the presence of error is likely to increase the chances of getting trapped in a local minima and increase the number of non-converged cases. Lastly, it may be possible to apply data-mining techniques on the non-converged cases and obtain insight into the non-convex nature of these parameter estimation problems. However, that is outside the scope of this manuscript, and these directions will be pursued in the future.

5. Conclusions

The aim of this study was to thoroughly analyze the effect of parameter estimation technique on biomechanical characterization of soft tissues using planar biaxial testing. Four invariant-based constitutive for soft tissues were tested, each with their own set of five or six parameters. Because of the large dimension of parameter space, Latin Hypercube Sampling approach was used to randomly generate parameter sets. These parameters were used as "target" parameter values, as well as initial guesses. It was found that small modifications of weighting the residual by experimental data and/or taking a log of the parameter in front of the exponential can significantly improve the parameter estimation process. The advantage was found not only in terms of convergence speed but also that the proposed modifications reduce the possibility of estimating wrong parameter values by getting stuck in a local minima. However, both the standard and modified formulations performed badly when fiber angle was considered as an unknown. Hence, the results suggest that determining the fiber angle using a non-mechanical test, as in, for example, an optical technique, can greatly help the parameter estimation process. Although, this may not be practical for in-vivo situations, for which further research is required to devise reliable parameter estimation techniques.

Funding: This work was supported by the Engineering and Physical Sciences Research Council of the UK (Grant No. EP/P018912/2 to A.A.). This work used the Cirrus UK National Tier-2 HPC Service at EPCC (http://www.cirrus.ac.uk) funded by the University of Edinburgh and EPSRC (EP/P020267/1).

Acknowledgments: I thank Yue Mei for his help in obtaining some preliminary results that led to this study.

Conflicts of Interest: The author declares no conflict of interest.

References

1. Fung, Y.C.; Skalak, R. *Biomechanics: Mechanical Properties of Living Tissues*; Springer: New York, NY, USA, 1981.
2. Bersi, M.R.; Bellini, C.; Di Achille, P.; Humphrey, J.D.; Genovese, K.; Avril, S. Novel methodology for characterizing regional variations in the material properties of murine aortas. *J. Biomech. Eng.* **2016**, *138*, 071005. [CrossRef] [PubMed]
3. Avazmohammadi, R.; Li, D.S.; Leahy, T.; Shih, E.; Soares, J.S.; Gorman, J.H.; Gorman, R.C.; Sacks, M.S. An integrated inverse model-experimental approach to determine soft tissue three-dimensional constitutive parameters: Application to post-infarcted myocardium. *Biomech. Model. Mechanobiol.* **2018**, *17*, 31–53. [CrossRef] [PubMed]
4. Potter, S.; Graves, J.; Drach, B.; Leahy, T.; Hammel, C.; Feng, Y.; Baker, A.; Sacks, M.S. A novel small-specimen planar biaxial testing system with full in-plane deformation control. *J. Biomech. Eng.* **2018**, *140*, 051001. [CrossRef]
5. Ross, C.; Laurence, D.; Wu, Y.; Lee, C.H. Biaxial mechanical characterizations of atrioventricular heart valves. *JoVE (J. Vis. Exp.)* **2019**. [CrossRef]
6. Maurel, W.; Thalmann, D.; Wu, Y.; Thalmann, N.M. Constitutive Modeling. In *Biomechanical Models for Soft Tissue Simulation*; Springer: Berlin/Heidelberg, Germany, 1998; pp. 79–120.
7. Ramião, N.G.; Martins, P.S.; Rynkevic, R.; Fernandes, A.A.; Barroso, M.; Santos, D.C. Biomechanical properties of breast tissue, a state-of-the-art review. *Biomech. Model. Mechanobiol.* **2016**, *15*, 1307–1323. [CrossRef]
8. Aster, R.C.; Borchers, B.; Thurber, C.H. *Parameter Estimation and Inverse Problems*; Elsevier: Amsterdam, The Netherlands, 2018.
9. Lee, C.H.; Amini, R.; Gorman, R.C.; Gorman, J.H.; Sacks, M.S. An inverse modeling approach for stress estimation in mitral valve anterior leaflet valvuloplasty for in-vivo valvular biomaterial assessment. *J. Biomech.* **2014**, *47*, 2055–2063. [CrossRef]
10. Aggarwal, A.; Sacks, M.S. An inverse modeling approach for semilunar heart valve leaflet mechanics: Exploitation of tissue structure. *Biomech. Model. Mechanobiol.* **2016**, *15*, 909–932. [CrossRef]
11. Montgomery, D.C. *Design and Analysis of Experiments*; John Wiley & Sons: Hoboken, NJ, USA, 2017.
12. Saad, Y. *Iterative Methods for Sparse Linear Systems*; SIAM: Philadelphia, PA, USA, 2003; Volume 82.
13. Aggarwal, A. An improved parameter estimation and comparison for soft tissue constitutive models containing an exponential function. *Biomech. Model. Mechanobiol.* **2017**, *16*, 1309–1327. [CrossRef]
14. Moré, J.J.; Garbow, B.S.; Hillstrom, K.E. *User Guide for MINPACK-1*; Technical Report, CM-P00068642; 1980. Avaliable online: http://cds.cern.ch/record/126569/files/?ln=en (accessed on 29 October 2019).
15. Holzapfel, G.A. *Nonlinear Solid Mechanics*; Wiley: Chichester, UK, 2000; Volume 24.
16. Gasser, T.C.; Ogden, R.W.; Holzapfel, G.A. Hyperelastic modelling of arterial layers with distributed collagen fibre orientations. *J. R. Soc. Interface* **2006**, *3*, 15–35. [CrossRef]
17. Humphrey, J.; Yin, F. A new constitutive formulation for characterizing the mechanical behavior of soft tissues. *Biophys. J.* **1987**, *52*, 563–570. [CrossRef]
18. May-Newman, K.; Yin, F.C.P. A Constitutive Law for Mitral Valve Tissue. *J. Biomech. Eng.* **1998**, *120*, 38–47. [CrossRef] [PubMed]
19. Nocedal, J.; Wright, S. *Numerical Optimization*; Springer Science & Business Media: Cham, Switzerland, 2006.
20. Zhang, W.; Zakerzadeh, R.; Zhang, W.; Sacks, M.S. A material modeling approach for the effective response of planar soft tissues for efficient computational simulations. *J. Mech. Behav. Biomed. Mater.* **2019**, *89*, 168–198. [CrossRef] [PubMed]
21. Cai, X.C.; Keyes, D.E. Nonlinearly preconditioned inexact Newton algorithms. *SIAM J. Sci. Comput.* **2002**, *24*, 183–200. [CrossRef]

22. Budde, M.; Annese, J. Quantification of anisotropy and fiber orientation in human brain histological sections. *Front. Integr. Neurosci.* **2013**, *7*, 3. [CrossRef]
23. Sacks, M.S.; Smith, D.B.; Hiester, E.D. A small angle light scattering device for planar connective tissue microstructural analysis. *Ann. Biomed. Eng.* **1997**, *25*, 678–689. [CrossRef]
24. Lei, F.; Szeri, A. Inverse analysis of constitutive models: Biological soft tissues. *J. Biomech.* **2007**, *40*, 936–940. [CrossRef]
25. Martínez-Martínez, F.; Rupérez, M.; Martín-Guerrero, J.; Monserrat, C.; Lago, M.; Pareja, E.; Brugger, S.; López-Andújar, R. Estimation of the elastic parameters of human liver biomechanical models by means of medical images and evolutionary computation. *Comput. Methods Programs Biomed.* **2013**, *111*, 537–549. [CrossRef]
26. Chabiniok, R.; Moireau, P.; Lesault, P.F.; Rahmouni, A.; Deux, J.F.; Chapelle, D. Estimation of tissue contractility from cardiac cine-MRI using a biomechanical heart model. *Biomech. Model. Mechanobiol.* **2012**, *11*, 609–630. [CrossRef]
27. Aggarwal, A.; Ferrari, G.; Joyce, E.; Daniels, M.J.; Sainger, R.; Gorman, J.H., III; Gorman, R.; Sacks, M.S. Architectural trends in the human normal and bicuspid aortic valve leaflet and its relevance to valve disease. *Ann. Biomed. Eng.* **2014**, *42*, 986–998. [CrossRef]
28. Nielsen, P.M.; Le Grice, I.J.; Smaill, B.H.; Hunter, P.J. Mathematical model of geometry and fibrous structure of the heart. *Am. J. Physiol. Heart Circ. Physiol.* **1991**, *260*, H1365–H1378. [CrossRef]
29. Li, K.; Ogden, R.W.; Holzapfel, G.A. A discrete fibre dispersion method for excluding fibres under compression in the modelling of fibrous tissues. *J. R. Soc. Interface* **2018**, *15*, 20170766. [CrossRef] [PubMed]
30. Fung, Y.c. *Biomechanics: Mechanical Properties of Living Tissues*; Springer Science & Business Media: Cham, Switzerland, 2013.
31. Sun, W.; Sacks, M.S. Finite element implementation of a generalized Fung-elastic constitutive model for planar soft tissues. *Biomech. Model. Mechanobiol.* **2005**, *4*, 190–199. [CrossRef] [PubMed]

MDPI
St. Alban-Anlage 66
4052 Basel
Switzerland
Tel. +41 61 683 77 34
Fax +41 61 302 89 18
www.mdpi.com

Bioengineering Editorial Office
E-mail: bioengineering@mdpi.com
www.mdpi.com/journal/bioengineering

www.ingramcontent.com/pod-product-compliance
Lightning Source LLC
Chambersburg PA
CBHW040820250926
51850CB00007BA/1222
* 9 7 8 3 0 3 9 4 3 1 5 0 2 *